Selected Titles in This Series

New Directions in Dirichlet Forms

AMS/IP

Studies in Advanced Mathematics

Volume 8

New Directions in Dirichlet Forms

Jürgen Jost, Wilfrid Kendall,
Umberto Mosco, Michael Röckner,
and Karl-Theodor Sturm

American Mathematical Society • International Press

Shing-Tung Yau, Managing Editor

1991 *Mathematics Subject Classification.* Primary 31C25;
Secondary 35J20, 58E20, 47H20, 60J35.

Library of Congress Cataloging-in-Publication Data

New directions in Dirichlet forms / Jürgen Jost... [et al.].
p. cm. — (AMS/IP studies in advanced mathematics ; v. 8)
Includes bibliographical references.
ISBN 0-8218-1061-8
1. Dirichlet forms. I. Jost, Jürgen, 1956– . II. Series.
QA274.2.N49 1998
519.2—dc21

98-25202
CIP

∞ The paper used in this book is acid-free and falls within the guidelines
established to ensure permanence and durability.
Visit the AMS home page at URL: http://www.ams.org/
Visit the International Press home page at URL: http://www.intlpress.com/

10 9 8 7 6 5 4 3 2 1 03 02 01 00 99 98

Contents

Preface

The present book is based on the lectures delivered by the authors at a Summer School on Dirichlet Forms, sponsored by the European Union, that took place in Anogia in Crete (Greece) during June 22 - 28, 1997. As the title of the book indicates, the purpose of these lectures was to describe recent developments and new perspectives in the theory of Dirichlet forms to an audience of pure and applied mathematicians, from the fields of stochastic analysis, partial differential equations, the calculus of variations, Riemannian geometry, potential theory, and others.

In fact, from the very beginning, the theory of Dirichlet forms combined aspects and insights from those fields of mathematics. It had been known for a long time that there exist deep and close connections between the Dirichlet integral, the Laplace operator, the heat equation, and Brownian motion. It was also clear that these connections extend to more general situations, involving for example the Laplace-Beltrami operator on Riemannian manifolds or the discretized Laplace equation utilized for purposes of numerical analysis.

With the notion of Dirichlet forms, Beurling and Deny provided the proper axiomatic setting for exploring these connections abstractly. In subsequent developments, by Fukushima, Silverstein, and others, Markovian semigroups and probabilistic potential theory based on symmetric Markov processes played a prominent role, and this theory is well documented in the monograph M. Fukushima, Y. Oshima, M. Takeda, Dirichlet Forms and Symmetric Markov Processes, de Gruyter, 1994.

In recent years, research went considerably beyond this "classical" theory, in several directions, and this will be described in the present book. In particular, we shall exhibit deep connections with several other fields of mathematics.

Motivated by questions from particle physics, Albeverio, Ma, and Röckner developed the theory of Dirichlet forms on infinite dimensional spaces. In this sense, the contribution of Michael Röckner describes stochastic analysis on configuration spaces. In particular, this yields a mathematically rigorous approach to the stochastic dynamics of Gibbs measures on configuration spaces, and applications to infinite interacting particle systems.

In the work of Biroli and Mosco, the connections of Dirichlet forms with the theory of subelliptic partial differential equations in the sense of Hörmander were explored and a structural version of the theory of de Giorgi, Nash, and Moser was achieved. The parabolic analogues were investigated by Sturm. The work of Biroli and Mosco also led to applications of Dirichlet forms to homogenization and fractals. Thus, Umberto Mosco's contribution describes how to investigate the dynamic aspects of self-similarity from the point of view of Dirichlet forms. In particular, it is found that the natural dynamic scaling dimension on a self-similar fractal (as occuring for example in the Weyl law for the eigenvalues), which is in general different from (but related to) the Hausdorff dimension, can be interpreted as volume growth dimension of the intrinsic Lagrangian metric.

Questions in Riemannian and algebraic geometry recently naturally led geometers to consider classes of metric spaces that are more general than Riemannian manifolds, for example so-called geodesic length spaces that satisfy generalized curvature bounds in the sense of Alexandrov. On the other hand, Dirichlet forms also naturally define intrinsic metrics, and such geometric aspects of Dirichlet forms are explored in the contribution of Karl-Theodor Sturm. This suggests also a more axiomatic treatment of the function theory of Riemannian manifolds that had been pioneered in the work of S.T. Yau. A stochastic approach to that theory is described in the contribution of Wilfrid Kendall.

Metric spaces of the type alluded to above, in particular those occuring in certain geometric applications often are not locally compact, and thus, as in Michael Röckner's contribution, also in the chapters provided by Jürgen Jost and Karl-Theodor Sturm, the theory is largely developed without the hypothesis of local compactness.

Finally, the contributions of Jürgen Jost and Wilfrid Kendall explore a nonlinear version of Dirichlet forms. While the classical theory of Dirichlet forms is a linear theory based on harmonic functions, here we consider harmonic mappings, i.e. mappings that take their values not in the reals but in some Riemannian manifold, or, more generally, some metric space. For the treatment of such a generalized situation, Jost had introduced $\Gamma-$convergence techniques into the theory and developed a general existence scheme that works without an assumption of local compactness of the target. Jürgen Jost in his contribution describes the analytic aspects, and Wilfrid Kendall the stochastic ones of the geometric theory of harmonic mappings, in particular the connection of harmonic mappings with Brownian motion on Riemannian manifolds as well as a stochastic approach to regularity questions. In both contributions, convexity plays an important role as the appropriate substitute for the linear structure of the classical theory.

Altogether, although the individual contributions treat rather different topics, nevertheless there exist strong connections between those topics, and we interpret this as a sign of the depth, vitality, and fertility of the concept of Dirichlet forms. We hope that our lectures will stimulate further research and important new insights into the directions explored.

We thank all the participants of the Summer School for their interest in our lectures, and we thank Professor Susanna Papadopoulou for making the facilities of the Anogia Conference Center available to us and for administrative help with the organization and her and Sigrid Antoniadis and the staff of the Centre for making our stay in Anogia so pleasant and stimulating. Finally, we thank Birgit Kruschwitz and Harald Wenk for technical help with the editing of the present book.

Jürgen Jost
Max-Planck-Institut für Mathematik
in den Naturwissenschaften
Inselstr. 22-26
D-04103 Leipzig
Germany
email: jost@mis.mpg.de

Wilfrid Stephen Kendall
Department of Statistics
University of Warwick
Coventry CV4 7AL
England
email: w.s.kendall@warwick.ac.uk

Umberto Mosco
Dipartimento di Matematica
Università "La Sapienza" di Roma
I-00185 Roma
Italia
email: mosco@axcasp.caspur.it

Michael Röckner
Fakultät für Mathematik
Universität Bielefeld
Universitätsstr. 25
D-33615 Bielefeld
Germany
email: roeckner@mathematik.uni-bielefeld.de

Karl-Theodor Sturm
Institut für Angewandte Mathematik
Universität Bonn
Wegelerstr. 6
D-53115 Bonn
Germany
email: sturm@wiener.iam.uni-bonn.de

AMS/IP Studies in Advanced Mathematics
Volume 8, 1998

Jürgen Jost: Nonlinear Dirichlet Forms

1.1 Introduction

The theory of Dirichlet forms started with the work of Beurling-Deny [BD1, BD2]. It was conceived as an abstract version of the theory of the Dirichlet integral, i.e. of the variational theory of harmonic functions. It included also the discrete harmonic functions that had received their mathematical treatment already by Courant-Friedrichs-Lewy [CFL]. Courant-Friedrichs-Lewy also pointed out a connection with discretized Brownian motion, and more recently, the stochastic aspects of Dirichlet forms have found an important place in the subject. The theory of Dirichlet forms is well documented in the monographs of Fukushima-Oshima-Takeda [FOT], Ma-Roeckner [MR], and Bouleau-Hirsch [BH]. As explained in other contributions in this volume, Dirichlet forms also constitute the proper axiomatic framework for many aspects of the theory of homogenization, self-similar fractals, or infinite particle systems, to name but a few.

The theory of Dirichlet forms is essentially linear as it is essentially concerned with scalar valued functions which are square integrable. In Riemannian geometry, however, a nonlinear generalization of harmonic functions is quite important. For that purpose, one generalized the classical Dirichlet integral to a so-called energy integral for maps with values in a Riemannian manifold. Minimizers (or other critical points) of this energy integral are called harmonic maps. Harmonic maps have found many applications in Riemannian geometry, but for other applications (as. e.g. by Gromov-Schoen [GS] to p-adic superrigidity) it was found necessary to consider target spaces that are more general than Riemannian manifolds. In that context, existence results were established by Jost [J2] and Korevaar-Schoen [KS]. In order to gain a deeper understanding and at the same time establish the most general existence results, the present author found it desirable to develop a more abstract and axiomatic version of the theory ([J3], [J4], [J5]). In particular, the approach taken in [J5] was to consider a theory of nonlinear or generalized Dirichlet forms as the appropriate setting for these generalized harmonic maps. In a certain sense, this brings together the two quite different generalizations of harmonic functions alluded to above - Dirichlet forms and harmonic maps - and it is hoped that this will lead to new insights in both fields. This theory is the topic of the present lecture notes.

In order to have a good theory of harmonic maps, one needed to impose the condition that the target manifold has nonpositive curvature. This condition can be interpreted as a convexity condition on the energy functional. Convexity conditions will also be fundamental for the abstract theory developed here. In fact, it turns out that convexity is the appropriate replacement for the linear structure in the classical Dirichlet form theory. Convexity will allow to extend many of the basic constructions of that theory to our setting, from resolvents and semigroups to variational aspects and the regularity of minimizers. This will be a basic theme in these notes. An important technical point will be to avoid assumptions of local compactness on the image. We shall concentrate on the analytic and axiomatic aspects. The geometric aspects have been treated already in detail in the author's companion lecture notes [J6]. Therefore, we shall be brief about geometric concepts and do not mention geometric applications. The geometric definitions and results needed for understanding these notes are summarized in an appendix. The reader can consult this appendix whenever he/she encounters an unfamiliar geometric concept in the main body of the text. In particular, the notion of nonpositive curvature in the sense of Alexandrov for a metric space will be important. For further details, we refer to [J6] as already mentioned. For the basic notions of Riemannian geometry and the classical theory of harmonic maps, the reader can also consult the author's textbook [J1].

We shall also not treat the very interesting stochastic aspects of harmonic maps and refer the reader to Kendall's contribution to this volume instead.

Acknowledgement: The research underlying these lecture notes was generously supported by the DFG through the Leibniz program.

1.2 Definition and properties of generalized Dirichlet forms

Let $(X, \mathcal{B}, \mu)$ be a σ-finite measure space. A key notion for us is the space

$$L^2(X,\mu) := \left\{ f : X \to \mathbb{R} \cup \{\pm\infty\}, \mu \text{ - measurable}, \int_X f^2(x)\mu(dx) < \infty \right\}$$

of square integrable functions. (Here and in the sequel, functions that differ only on a set of vanishing μ - measure will be identified.) On $L^2(X, \mu)$, one has the inner product

$$(f,g) := \int_X f(x)g(x)\,\mu(dx)\,,$$

and this product makes $L^2(X, \mu)$ into a Hilbert Space.

The main aim of these lectures is to extend the theory based on this Hilbert space to a more general setting where instead of functions, we consider mappings $f : X \to Y$ with values in some complete metric space (Y, d). To make this generalization meaningful, we need some elementary preparations.

First of all, since L^2 - functions need to be defined only μ - almost everywhere (abbreviated as μ - a.e., or simply a.e. in the sequel), and since L^2 - functions can assume the values $\pm\infty$ at most on a set of μ measure 0, in the definition of $L^2(X, \mu)$ it obviously suffices to consider functions $f : X \to \mathbb{R}$, as we may simply redefine f on the set where it assumes the values $\pm\infty$. Next, we need to explain what a μ - measurable map with values in a complete metric space is. The starting point here is a result that a real - valued function is measurable if it is the pointwise limit of simple functions. In general, we call a function

$$\varphi : X \to Y$$

simple if there exists pairwise disjoint, μ measurable sets $\mathcal{B}_1, \ldots, \mathcal{B}_n \subset X$ whose union is X such that φ is constant on each $\mathcal{B}_j$.

$$f : X \to Y$$

now is called μ - measurable if it is the pointwise limit of simple functions. As in the case of real-valued functions, we then have the following important composition property:

If $f : X \to Y$ is measurable, $\eta : Y \to \mathbb{R}$ continuous, then $\eta \circ f : X \to Y$ is a measurable function in the usual sense. In particular, if $f, g : X \to Y$ are measurable, then

$$d(f(x), g(x))$$

is a measurable function on X.

This observation will now enable us to define L^2-spaces of mappings. Namely, for measurable maps $f, g : X \to Y$, we may define the L^2 - distances as

$$d(f,g) := d_{L^2}(f,g) := \left(\int d^2(f(x), g(x))\, \mu(dx) \right)^{1/2}.$$

Of course, this distance may possibly be infinite for certain maps f, g.

We now make a **digression** that can be omitted at a first reading, in particular by readers who are more interested in the analytic concepts involved than in the geometric applications. For certain applications this notion needs to be redefined slightly. For example, if Y happens to be a strong geodesic length space, one may wish to prescribe a homotopy class for the maps f, g. If f and g are in the same homotopy class, i.e. if there exists a map

$$F : X \times [0,1] \to Y$$

with $F(x,0) = f(x)$, $F(x,1) = g(x)$ and for which $F(x,\cdot)$ is continuous for all $x \in X$, we may consider the homotopy distance between f and g that is defined as follows:
We put

$$d_h(f(x), g(x)) := \inf \{L(\gamma)\colon \gamma\colon [0,1] \to Y \text{ curve with } \gamma(0) = f(x), \gamma(1) = g(x), \gamma(\cdot) \text{ homotopic to } F(x,\cdot) \text{ with fixed endpoints}\}$$

and

$$d_h(f,g) := \left(\int_X d_h^2\left(f(x), g(x)\right) \mu(dx) \right)^{1/2}.$$

Obviously, this homotopy distance does not depend on the particular choice of the homotopy F. If X and Y are manifolds, or more generally if they are amenable to covering space theory, this homotopy distance may be expressed in the following equivalent form. If $\pi_X : \tilde{X} \to X$, $\pi_Y : \tilde{Y} \to Y$ are the universal coverings of X and Y, respectively, we lift $f : X \to Y$ to $\tilde{f} : \tilde{X} \to \tilde{Y}$. Thus we have

$$\pi_Y \circ \tilde{f} = f \circ \pi_X.$$

Given homotopic maps $f, g : X \to Y$, we choose compatible lifts $\tilde{f}, \tilde{g} : X \to Y$, i.e. we choose a lift $\tilde{F}$ of the homotopy F between f and g, (same notations as above) and put $\tilde{f} = \tilde{F}(\cdot, 0)$, $\tilde{g} = \tilde{F}(\cdot, 1)$. Then

$$d_h(f,g) = \left(\int_X d^2(\tilde{f}(x), \tilde{g}(x)) \, \mu(dx) \right)^{1/2}$$

where the integration now is over a fundamental domain for X in $\tilde{X}$, and where we have lifted the measure μ to $\tilde{X}$ in such a manner that it is invariant under all covering transformations. More generally, we may assume that X admits a cover in the following sense: There exists a measure space $(\tilde{X}, \tilde{\mathcal{B}}, \tilde{\mu})$ and a group Γ of measure preserving homomorphisms of $\tilde{X}$ such that $X = \tilde{X}/\Gamma$ and that $\tilde{\mathcal{B}}, \tilde{\mu}$ induce $\mathcal{B}, \mu$ on X under the projection $\pi_X : \tilde{X} \to X = \tilde{X}/\Gamma$. We also require that Y admits a cover $\tilde{Y}$ for which every map $f : X \to Y$ lifts to a map $\tilde{f} : \tilde{X} \to \tilde{Y}$ as above.

We then suppose that we are given a homomorphism

$$\rho : \Gamma \to I(\tilde{Y})$$

from Γ into the isometry group of $\tilde{Y}$ (Note that the metric d of Y lifts to a metric on $\tilde{Y}$ that will again be denoted by d.). $f : \tilde{X} \to \tilde{Y}$ is called ρ-equivariant if for all $x \in \tilde{X}$, $\gamma \in \Gamma$

$$f(\gamma x) = \rho(\gamma) f(x).$$

If $f, g : \tilde{X} \to \tilde{Y}$ are ρ-equivariant, we put

$$d(f,g) := \left(\int_X d^2(f(x), g(x)) \tilde{\mu}(dx) \right)^{1/2}.$$

Again, the integration is over a fundamental domain F_X for X in $\tilde{X}$, i.e. over a maximal set of points in $\tilde{X}$ that are mutually Γ-inequivalent, i.e. given $x, y \in F_X$, there is no $\gamma \in \Gamma$ with $\gamma(x) = y$.

This homotopy distance will be used subsequently without further explicit mentioning.
This **ends** the geometric **digression**.

In order to define an L^2-space for maps from X into a metric space (Y, d), we need to select a μ-measurable base map $f_0 : X \to Y$. In case (Y, d) is $\mathbb{R}$ or $\mathbb{R}^n$ with its standard Euclidean metric, the canonical base map is $f_0 \equiv 0$. In the case of a general Y, one might then take a constant map $f_0 \equiv p$ for some $p \in Y$, but one should take the following two comments into consideration:

1) If X is noncompact, the subsequent construction may depend on the choice of p, as happens already for $X = Y = \mathbb{R}$, and in general there will be no canonical choice for p.

2) In view of the above digression, one may wish to fix a homotopy class of maps from X to Y, and it will then be natural to require that f_0 also belongs to the chosen homotopy class.

In any case, having selected f_0, we put

$$\begin{aligned} L^2(X, Y) &:= L^2(X, \mu; Y, d) := L^2(X, \mu; Y, d; f_0) \\ &:= \{f : X \to Y \ \mu - \text{measurable } : d(f, f_0) < \infty\}. \end{aligned}$$

In the sequel, we shall identify maps that differ only on a set of measure 0, as is customary in the theory of L^p-spaces. Since (Y, d) is a complete metric space, $L^2(X, \mu; Y, d)$ then also becomes a complete metric space. This means that any Cauchy sequence $(f_n)_{n \in \mathbb{N}} \subset L^2(X, Y)$, i.e. any sequence of maps $f_n : X \to Y$ satisfying

$$\begin{gathered} \forall \varepsilon > 0 \; \exists N \in \mathbb{N} \; \forall n, m \geq \varepsilon \\ \int_X d^2(f_n(x), f_m(x)) \, \mu(dx) < \varepsilon \end{gathered}$$

converges to some $f \in L^2(X, Y)$, i.e.

$$\lim_{n \to \infty} \int_X d^2(f_n(x), f(x)) \, \mu(dx) = 0.$$

Let $(X, \mathcal{B}, \mu)$ be given as before. The basic object of study are functionals E defined on

$$\{f \in L^2(X, \mu; Y, d; f_0) \quad \text{for some metric space } (Y, d) \text{ and some base map } f_0 : X \to Y\}$$

and taking values in $\mathbb{R}^+ \cup \{\infty\}$. The letter E here stands for energy, and the original motivation for the theory was to find an axiomatic version for the energy functional maps between Riemannian manifolds that leads to the theory of harmonic mappings.

Given such a functional E, we define $D(E)$ as its domain of definition, i.e. as the set of those f with

$$E(f) < \infty.$$

We point out that while the measure space (X, μ) is fixed, the image or target space (Y, d) is kept variable. For certain purposes, however, one may wish to restrict the class of permitted target spaces (Y, d). Now comes the fundamental

Definition 1.2.1 *A functional E as defined above with domain of definition $D(E)$ is called a generalized Dirichlet form or energy form if the following three conditions hold:*

1) Quadratic contraction property:
Let $f : X \to Y$ be in $D(E)$, $\varphi : (f(X), d) \to (Z, d')$ for some metric space (Z, d') (here $f(X)$ carries the metric induced from the metric d on Y) an L-Lipschitz function, i.e.

$$d'(\varphi(x), \varphi(y)) \leq L\, d(x, y) \text{ for all } x, y \in f(X).$$

Then also $\varphi \circ f \in D(E)$, and

$$E(\varphi \circ f) \leq L^2\, E(f).$$

2) Density:
$D(E) \cap L^2(X, \mu; \mathbb{R})$ is dense in $L^2(X, \mu; \mathbb{R}) = L^2(X, \mu)$. Furthermore, if for some $f \in L^2(X, Y)$,

$$\varphi \circ f \in D(E)$$

for every 1-Lipschitz map $\varphi : (Y, d) \to \mathbb{R}$, then also $f \in D(E)$.

3) Closedness:
E is lower semicontinuous w.r.t. convergence in $L^2(X, \mu; Y, d)$.

Remark: If $f \in L^2(X,\mu;Y,d;f_0), \varphi : (f(X),d) \to (Z,d')$ L-Lipschitz, then $\varphi \circ f \in L^2(X,\mu;Z,d';\varphi \circ f_0)$. Thus, it is useful to allow the base map to vary.

Let E be an energy form. In order to understand how our definition is related to the standard definition of a Dirichlet form as given e.g. in [FOT], we restrict our E to $L^2(X,\mu;\mathbb{R})$, i.e. to $Y = \mathbb{R}$. We start with the following simple observation.

Lemma 1.2.2 *If $f : X \to \mathbb{R}$ is in $D(E)$, $a, \lambda \in \mathbb{R}$, then*

$$E(\lambda f) = \lambda^2 E(f)$$
$$E(f+a) = E(f)$$

Proof: A simple consequence of property 1), considering the Lipschitz functions

$$\varphi_\lambda : \mathbb{R} \to \mathbb{R}, \quad \varphi_\lambda(x) : \lambda x, \quad \varphi_\lambda^{-1}(x) = \varphi_{\frac{1}{\lambda}}(x) = \frac{1}{\lambda} x$$
$$\varphi_a : \mathbb{R} \to \mathbb{R}, \quad \varphi_a(x) : x + a, \quad \varphi_a^{-1}(x) = \varphi_{-a}(x) = x - a.$$

□

By polarization

$$E(f,g) := \frac{1}{4}(E(f+g) - E(f-g))$$

we obtain a Dirichlet form in the usual sense. The closedness and density conditions reduce to the standard ones, but the quadratic contraction property is stronger than the one usually required for Dirichlet forms as can be seen from Lemma 1. Namely, there one requires that

$$E(g,g) \leq E(f,f)$$

whenever $|g(x) - g(y)| \leq |f(x) - f(y)|$ and $|g(x)| \leq |f(x)|$ for all $x, y \in X$. A Dirichlet form $E(\cdot,\cdot)$ is called regular if $D(E) \cap C_0^0(X)$ (where $C_0^0(x)$ is the space of continuous functions with compact support) is dense in $C_0^0(X)$ w.r.t. the C^0-norm as well as dense in $D(E)$ w.r.t. the norm $\|\cdot\|_1$ induced by the product

$$(f,g)_1 := (f,g) + E(f,g)$$

(where $(\cdot,\cdot)$ denotes the L^2-product).

$E(\cdot,\cdot)$ is called strongly local if

$$E(f,g)=0$$

whenever f is constant on some neighborhood of the support of g (or vice versa).

With these definitions, we may state the **representation theorem of Beurling-Deny** (see e.g. [FOT]): *Suppose that X is a locally compact, separable metric space and μ is a positive Radon measure with supp(μ) $= X$. (This means that μ is a non-negative Borel measure on X that is finite on compact sets and strictly positive on non-empty open sets.) Then a regular Dirichlet form E on X admits the following decompositions: For all $f, g \in D(X) \cap C_0^0(X)$*

$$\begin{aligned} E(f,g) &= E^c(f,g) + \int_{X\times X\setminus diagonal} (f(x)-f(y))\,(g(x)-g(y))\,J(dx,dy) \\ &\quad + \int_X f(x)g(x)\,k(dx) \end{aligned}$$

with a strongly local Dirichlet form E^c, a symmetric positive Radon measure J on $X \times X \setminus diagonal$ and a positive Radon measure k on X, the so-called killing measure.

It is clear from the next lemma that our quadratic contraction property implies that for a Dirichlet form that comes from an energy form in our sense, the killing measure has to vanish.

Lemma 1.2.3 *Let E be an energy form.*

1) *If $\varphi : (Y,d) \to (Y,d)$ is an isometry, then for every $f \in D(E)$, $f : X \to Y$*

$$E(\varphi \circ f) = E(f).$$

2) *Every constant map $f_0 : X \to Y$ satisfies*

$$E(f_0) = 0.$$

Proof:

1) If φ is an isometry, so is φ^{-1}, and φ and φ^{-1} both are 1-Lipschitz. Thus, by property 1)

$$E(\varphi\circ f) \le E(f) = E(\varphi^{-1}\circ\varphi\circ f) \le E(\varphi\circ f),$$

and equality has to hold.

2) It is clear from property 1) that $g_0 \equiv 0$ satisfies

$$E(g_0) = 0.$$

Next, by 2) of Lemma 1.1,

$$E(h_0) = 0 \text{ for any constant function } h_0 : X \to \mathbb{R}$$

Finally, using properties 1) and 2) gives the result for constant maps f_0.

Remark: If $\mu(X) = \infty$, constant functions $\neq 0$ are not in $L^2(X, \mu; \mathbb{R})$. Thus, in that case, the reader needs to interprete the preceding statements appropriately.

1.3 Resolvents, semigroups, and variational aspects

In the classical theory of Dirichlet forms, one has a correspondance between such forms, semigroups, resolvents and selfadjoint operators satisfying suitable conditions that we now summarize. For that purpose, one may take any (real) Hilbert space H, not necessarily $L^2(X, \mu)$. The norm and scalar product on H will be denoted by $\|\cdot\|$, $(\cdot, \cdot)$. A reference for the sequel is [FOT].

Definition 1.3.1 *A family $(T_t)_{t>0}$ of symmetric linear operators defined on all of H is called a semigroup if*

(i) $T_t \cdot T_s = T_{t+s}$ *for all* $t, s > 0$

(ii) $\|T_t u\| \leq \|u\| \quad u \in H, t > 0.$

Such a semigroup is called strongly continuous if

(iii) $\lim_{t \searrow 0} T_t u = u$ *for all* $u \in H$

Definition 1.3.2 *A family $(G_\alpha)_{\alpha>0}$ of symmetric linear operators defined on all of H is called a resolvent if*

(i) $G_\alpha - G_\beta + (\alpha - \beta) G_\alpha G_\beta = 0$ *for all* $\alpha, \beta > 0$

(ii) $\|\alpha G_\alpha u\| \leq \|u\|$ *for all* $u \in H, \alpha > 0.$

Such a resolvent is called strongly continuous if in addition (iii) $\lim_{\alpha\to\infty} \alpha G_\alpha u = u$ *for all* $u \in H$.

The following is a basic result in the theory of Dirichlet forms, see [BH],[FOT], or [MR].

Theorem *The following classes of objects correspond to each other:*

(i) *Closed symmetric forms on* H.

(ii) *Non-positive definite selfadjoint operators on* H.

(iii) *Strongly continuous semigroups.*

(iv) *Strongly continuous resolvents.*

This correspondance goes as follows.

A closed symmetric form E corresponds to a non-positive definite self-adjoint operator L via

$$\begin{aligned} D(E) &= D(\sqrt{-L}) \\ E(u,v) &= (\sqrt{-L}u, \sqrt{-L}v) \text{ for all } u, v \in D(E). \end{aligned}$$

If $u \in D(L), v \in H$, this becomes

$$E(u,v) = -(Lu, v).$$

A strongly continuous semigroup $(T_t)_{t>0}$ generates a non-positive selfadjoint operator L as its generator

$$Lu = \lim_{t\searrow 0} \frac{1}{t}(T_t u - u)$$

with $D(L) = \{u \in H \text{ for which this limit exists}\}$.

Conversely, given such an L, a strongly continuous semigroup is generated as

$$T_t = \exp tL = \lim_{n\to\infty} (Id - \frac{t}{n}L)^{-n}.$$

(The limit exists because for all $\lambda > 0, \|(Id - \lambda L)^{-1}\| \leq 1$ as L is non-positive.) A strongly continuous resolvent generates a non-positive selfadjoint operator L via

$$\begin{aligned} \frac{1}{\alpha} Lu &= u - \frac{1}{\alpha} G_\alpha^{-1} u \qquad \text{(this does not depend on } \alpha > 0) \\ D(L) &= G_\alpha(H) \end{aligned}$$

and the converse relation is

$$\alpha G_\alpha = (Id - \frac{1}{\alpha}L)^{-1} \text{ for } \alpha > 0.$$

The resolvent can also be characterized by the fact that

$$J_\lambda u := \frac{1}{\lambda} G_{1/\lambda} u = (Id - \lambda L)^{-1}$$

is the unique minimizer of $\lambda E(v) + \|u - v\|^2$, i.e.

$$\lambda E(J_\lambda u) + \|u - J_\lambda u\|^2 = \inf_{v \in H} (\lambda E(v) + \|u - v\|^2).$$

The semigroup may be obtained from the family $(J_\lambda)_{\lambda > 0}$ via

$$T_t u = \lim_{n \to \infty} J^n_{\frac{t}{n}} x.$$

as follows from the preceding formulae.

The result that a closed densely defined linear operator L on a Banach space B that is non-positive, or, more generally, satisfies

$$\|(Id - \lambda L)^{-1}\| \leq 1 \text{ for all } \lambda > 0$$

defines a semigroup via

$$T_t u = \lim_{n \to \infty} (Id - \frac{t}{n} L)^{-n} u$$

and furthermore that $u(t) := T_t u$ solves the Cauchy problem

$$\frac{du}{dt} = Lu, \; u(0) = v \quad \text{ for } v \in D(L)$$

is the Hille-Yosida theorem. Crandall-Liggett [CL] then showed that this result extends to the case where L is a nonlinear operator on the Banach space B, if the other assumptions continue to hold.

In the case of $H = L^2(X, \mu)$, a Dirichlet form was a symmetric closed form that satisfies a contraction property. In terms of the corresponding semigroup or resolvent, this is characterized by the so-called Markovian property that the T_t or G_α have to satisfy. A linear operator S defined on all of $L^2(X, \mu)$ is called Markovian if

$$0 \leq u \leq 1 \quad \mu - \text{ a.e. } \Rightarrow 0 \leq Su \leq 1 \quad \mu - \text{ a.e for all } u \in L^2(X, \mu).$$

It will turn out that a substantial part of this correspondance extends to the case of energy forms at least if we impose an additional condition on the target space (Y, D), namely to be non-positively curved in the sense of Alexandrov.

Remark: For many applications, it is important to realize that there is a still more general condition than Alexandrov's, namely Busemann's nonpositive curvature condition under which the theory still works. For example, the L^p spaces are nonpositively curved in the sense of Busemann for $1 < p < \infty$, but not in the sense of Alexandrov unless $p = 2$. The resulting theory is developed in [J6]. Since Busemann's condition is not as easy to work with as Alexandrov's, in the present notes we restrict our attention to the latter.

We start with some general definitions. Let Z be simply connected complete geodesic length space. In our applications below, we shall use

$$Z = L^2(X, \mu; Y, d) \tag{1.3.1}$$

equipped with the L^2-metric that we shall also denote by d. Here, we assume that (Y, d) itself is a simply connected complete geodesic length space. Points in Z will be denoted by x, y, z, also in the context of (1.3.1), where these would correspond to L^2 - maps that should rather be called f, g in accordance with the terminology of §1.

Let $D(F) \subset Z$, and let $F : D(F) \to \mathbb{R}$ be a functional. We say that F is densely defined if $D(F)$ is dense in Z. We say that F is convex if whenever $\gamma : [0,1] \to Z$ is geodesic (parametrized proportionally to arclength, as always), and if $\gamma(0), \gamma(1) \in D(F)$, then also $\gamma(t) \in D(F)$ and

$$F(\gamma(t)) \leq t\,F(\gamma(0)) + (1-t)F(\gamma(1)) \tag{1.3.2}$$

for all $0 \leq t \leq 1$.

We extend any convex functional $F : D(F) \to \mathbb{R}$ to a functional

$$F : Z \to \mathbb{R} \cup \{\infty\}$$

by putting

$$F(x) = \infty \quad \text{if } x \in Z \setminus D(F).$$

This extension still satisfies (1.3.2) as the left hand side can only assume the value ∞ if the right hand side does. Consequently, any functional $F : Z \to \mathbb{R} \cup \{\infty\}$ satisfying (1.3.2) is called convex.

Definition 1.3.3 *Let $F : Z \to \mathbb{R} \cup \{\infty\}$. The Moreau-Yosida approximation F^λ of F is*

$$F^\lambda(x) := \inf_{y \in Z} (\lambda F(y) + d^2(x,y)).$$

From now on, we shall make the following general
Assumption: Z is a complete global NPC space.

Lemma 1.3.4 *Let $F : Z \to \mathbb{R}_+ \cup \{\infty\}$ be convex, $\not\equiv \infty$, and lower semicontinuous. Then for every $x \in Z$, and every $\lambda > 0$, there exists a unique $y_\lambda = J_\lambda(x)$ with*

$$F^\lambda(x) = \lambda F(y_\lambda) + d^2(x, y_\lambda).$$

Proof: Let $(y_n)_{n \in \mathbb{N}}$ be a minimizing sequence for F^λ, i.e.

$$\lambda F(y_n) + d^2(x, y_n) \to \inf_{y \in Z} (\lambda F(y) + d^2(x,y)) =: \kappa_\lambda.$$

For $m, n \in \mathbb{N}$, we let $y_{m,n}$ be the midpoint of y_m and y_n. The convexity of F implies

$$F(y_{m,n}) \leq \frac{1}{2} F(y_m) + \frac{1}{2} F(y_n)$$

and the NPC inequality gives

$$d^2(x, y_{m,n}) \leq \frac{1}{2}(d^2(x, y_m) + d^2(x, y_n)) - \frac{1}{4} d^2(y_m, y_n).$$

Together, we get

$$\begin{aligned} \lambda F(y_{m,n}) + d^2(x, y_{m,n}) &\leq \frac{1}{2}(\lambda F(y_m) + d^2(x, y_m)) \\ &+ \frac{1}{2}(\lambda F(y_n) + d^2(x, y_n)) \\ &- \frac{1}{4} d^2(y_m, y_n). \end{aligned}$$

From the definition of κ_λ, the left hand side cannot be smaller then κ_λ, while from the definition of (y_n), the first two terms on the right hand side both tend to $\frac{1}{2}\kappa_\lambda$. It follows that $(y_n)_{n \in \mathbb{N}}$ is a Cauchy sequence. It thus has a uniquely determined limit point y_λ which must realize the above infimum as $d^2(x, \cdot)$ is continuous and F is lower semicontinuous by assumption. □

Lemma 1.3.5 *Assumptions as in the previous lemma. For $x \in Z$, let $y_\lambda = J_\lambda(x)$ as before. If $x \in \overline{D(F)}$, then*

$$x = \lim_{\lambda \to 0} J_\lambda(x).$$

In particular, if F is densely defined, i.e. $\overline{D(F)} = Z$, this holds for all $x \in Z$.

Proof: Since x is assumed to be in the closure of $D(F)$, for every $\rho > 0$, we may find $x_\rho \in B(x, \rho)$ with

$$F(x_\rho) < \infty.$$

Thus

$$\lim_{\lambda \to \infty} (\lambda F(x_\rho) + d^2(x, x_\rho)) \leq \rho^2,$$

and consequently (recall $\kappa_\lambda := \inf_{y \in Z}(\lambda F(y) + d^2(x, y))$)

$$\limsup_{\lambda \to 0} \kappa_\lambda \leq 0. \tag{1.3.3}$$

We shall prove the result by deriving a contradiction from the assumption that for some sequence $(\lambda_n)_{n \in \lambda}$ with $\lambda_n \to 0$ as $n \to \infty$

$$d^2(x_n, y_{\lambda_n}) \geq \beta > 0 \quad \text{for all } n.$$

Then, because of

$$\limsup_{n \to \infty} (\lambda_n F(y_{\lambda_n}) + d^2(x, y_{\lambda_n})) \leq 0 \quad \text{by (1.3.3)},$$

$$F(y_{\lambda_n}) \to -\infty \quad \text{for } n \to \infty$$

contradicting our assumption that F is nonnegative. □

We may now derive the main existence result of [J3].

Theorem 1.3.6 *As always in this §, we assume that Z is a complete NPC space, and that $F : Z \to \mathbb{R}_+ \cup \{\infty\}$ is convex, $\not\equiv \infty$, and lower semicontinuous. For $x \in Z$ and $\lambda > 0$, let $y_\lambda = J_\lambda(x)$ be as in Lemma 2.1. If $(y_{\lambda_n})_{n \in \mathbb{N}}$ is bounded for some sequence $\lambda_n \to \infty$, then $(y_\lambda)_{\lambda > 0}$ converges to a minimizer of F for $\lambda \to \infty$.*

Proof: The proof will be divided into six simple steps.

1) y_{λ_n} minimizes $F(y) + \frac{1}{\lambda_n} d^2(x,y)$. Since (y_{λ_n}) is bounded, it therefore is a minimizing sequence for F.

2) Let $0 < \mu_1 < \mu_2$. The minimizing property of y_{μ_1} implies

$$F(y_{\mu_1}) + \frac{1}{\mu_1} d^2(x, y_{\mu_1}) \leq F(y_{\mu_2}) + \frac{1}{\mu_1} d^2(x, y_{\mu_2}),$$

hence

$$\begin{aligned} F(y_{\mu_1}) &+ \frac{1}{\mu_2} d^2(x, y_{\mu_2}) + \left(\frac{1}{\mu_1} - \frac{1}{\mu_2}\right)(d^2(x, y_{\mu_1}) - d^2(x, y_{\mu_2})) \\ &\leq F(y_{\mu_2}) + \frac{1}{\mu_2} d^2(x, y_{\mu_2}). \end{aligned}$$

The minimizing property of y_{μ_2} therefore implies

$$d^2(x, y_{\mu_1}) \leq d^2(x, y_{\mu_2}).$$

Thus, $d^2(x, y_\lambda)$ is a monotonically increasing function of λ.

3) Since $d^2(x, y_\lambda)$ is monotonically increasing by 2) and bounded on the sequence (y_{λ_n}) by assumption, it follows that $d^2(x, y_\lambda)$ is bounded as $\lambda \to \infty$.

4) From the definition of y_λ,

$$F(y_\lambda) = \inf\{F(y) : d^2(x,y) \leq d^2(x, y_\lambda)\}.$$

Since $d^2(x, y_\lambda)$ is nondecreasing by 2), it follows that $F(y_\lambda)$ is nonincreasing in λ, and

$$\lim_{\lambda \to \infty} F(y_\lambda) = \inf_{y \in Z} F(y) \quad \text{by 1).}$$

5) Let $\varepsilon > 0$. We choose Λ so large that for $\lambda \geq \mu \geq \Lambda$

$$d^2(x, y_\lambda) - d^2(x, y_\mu) < \varepsilon/2 \tag{1.3.4}$$

which is possible by 2), 3).
By 4)

$$F(y_\lambda) \leq F(y_\mu). \tag{1.3.5}$$

We let $y_{\lambda,\mu}$ be the mean value of y_λ and y_μ. Then the convexity of F, (1.3.5), and the NPC condition imply

$$\begin{aligned} & F(y_{\lambda,\mu}) + \frac{1}{\mu} d^2(x, y_{\lambda,\mu}) \\ \leq \; & F(y_\mu) + \frac{1}{\mu}(\frac{1}{2} d^2(x, y_\mu) + \frac{1}{2} d^2(x, y_\lambda) - \frac{1}{4} d^2(y_\lambda, y_\mu)) \\ < \; & F(y_\mu) + \frac{1}{\mu}(d^2(x, y_\mu) + \varepsilon/4 - \frac{1}{4} d^2(y_\lambda, y_\mu)) \quad \text{by (1.3.4)}. \end{aligned}$$

The minimizing property of y_μ then implies

$$d^2(y_\lambda, y_\mu) < \varepsilon.$$

Thus, $(y_\lambda)_{\lambda>0}$ satisfies the Cauchy property for $\lambda \to \infty$.

6) Since Z is complete, (y_λ) therefore converges to some limit y_∞. 4) and the lower semicontinuity of F imply that y_∞ minimizes F. □

In the classical case, where $Z = L^2(x, \mu)$ and F is a Dirichlet form, $J_\lambda(x)$ equals $\lambda G_\lambda x$ where $(G_\lambda)_{\lambda>0}$ is the resolvent assiciated with F. In this light, one should consider the following **resolvent identity**.

Lemma 1.3.7 *Let $F : Z \to \mathbb{R}_+ \cup \{\infty\}$ be a function. Then the resolvent equation*

$$\frac{1}{\mu}(\frac{1}{\lambda} F^\lambda)^\mu = \frac{1}{\lambda + \mu} F^{\lambda+\mu}$$

holds.

Proof:

$$\begin{aligned} \frac{1}{\mu}(\frac{1}{\lambda} F^\lambda)^\mu(x) &= \inf_{y \in Z}(\frac{1}{\lambda} F^\lambda(y) + \frac{1}{\mu} d^2(x, y)) \\ &= \inf_{y \in Z}(\inf_{z \in Z}(F(z) + \frac{1}{\lambda} d^2(y, z) + \frac{1}{\mu} d^2(x, y)). \end{aligned}$$

For each $z \in Z$,

$$\inf_{y \in Z}(\frac{1}{\lambda} d^2(y, z) + \frac{1}{\mu} d^2(x, y))$$

is realized by a unique point y_0, namely the point on the geodesic arc from x to z with

$$d(x, y_0) = \frac{\mu}{\lambda + \mu} d(x, z), \; d(z, y_0) = \frac{\lambda}{\lambda + \mu} d(x, z).$$

y_0 thus satisfies

$$\frac{1}{\lambda} d^2(y_0, z) - \frac{1}{\mu} d^2(x, y_0) = \frac{1}{\lambda + \mu} d^2(x, z)$$

and

$$\frac{1}{\mu}(\frac{1}{\mu} F^\lambda)^\mu(x) = \inf_{z \in Z}(F(z) + \frac{1}{\lambda + \mu} d^2(x, z)) = \frac{1}{\lambda + \mu} F^{\lambda + \mu}(x).$$

□

Another version of the **resolvent identity** is

Corollary 1.3.8 *Let $F : Z \to \mathbb{R}_+ \cup \{\infty\}$ be convex, $\not\equiv \infty$, lower semicontinuous. Let $x \in Z, \lambda > 0, y_\lambda = J_\lambda(x)$ as in Lemma 2.1. Let $\gamma : [0, 1] \to Z$ be the geodesic arc (as always, parametrized proportionally to arclength) with $\gamma(0) = x, \gamma(1) = y_2$, and put*

$$y_{\lambda,s} := \gamma(s).$$

Then

$$J_{(1-s)\lambda}(y_{\lambda,s}) = y_\lambda.$$

Proof: This follows from the proof of Lemma 2.3 if one uses the uniqueness result of Lemma 2.1. □

We continue to derive some further properties of F^λ and J_λ.

Lemma 1.3.9 *Let F be as in Theorem 2.1, $x \in D(F)$, $0 < \mu \leq \lambda$. Then*

$$F(J_\mu x) \leq \left(1 - \frac{d(x, J_\mu x)}{d(x, J_\lambda x)}\right) F(x) + \frac{d(x, J_\mu x)}{d(x, J_\lambda x)} = F(J_\lambda x).$$

Proof: By 2) in the proof of Theorem 2.1,

$$d^2(x, J_\mu x) \leq d^2(x, J_\lambda x).$$

Let x_μ be the point on the geodesic form x to $J_\lambda x$ with $d(x, x_\mu) = d(x, J_\mu x)$. Then by the minimuing property of $J_\mu x$,

$$\begin{aligned} F(J_\lambda x) &\leq F(x_\mu) \\ &\leq (1 - \frac{d(x, x_\mu)}{d(x, J_\lambda x)})F(x) + \frac{d(x, x_\mu)}{d(x, J_\lambda x)}F(J_\lambda x) \end{aligned}$$

by the convexity of F. □

Lemma 1.3.10 *Let F be as in Theorem 2.1. For any x_1, $x_2 \in Z$, $\lambda > 0$,*

$$d(J_\lambda x_1, J_\lambda x_2) \leq d(x_1, x_2).$$

Thus J_λ is a 1-Lipschitz map.

Proof: Put $y_i := J_\lambda x_i, i = 1, 2$, and let $\gamma : [0, 1] \to Z$ be the shortest geodesic with $\gamma(0) = y_1, \gamma(1) = y_2$. Since $F(y_1)$ and $F(y_2)$ are finite and F is convex, F is finite on γ, and we have

$$F(\gamma(t)) + F(\gamma(1-t)) \leq F(y_1) + F(y_2) \text{ for } 0 \leq t \leq 1. \tag{1.3.6}$$

By (1.6.2) of the appendix,

$$\begin{aligned} d^2(\gamma(t), x_1) + d^2(\gamma(1-t), x_2) &\leq d^2(x_1, y_1) + d^2(x_2, y_2) \\ &+ td^2(x_1, x_2) - td^2(y_1, y_2) \\ &+ 2t^2 d^2(y_1, y_2) \\ &- t(d(x_1, x_2) - d(y_1, y_2))^2. \end{aligned}$$

If we had

$$d(y_1, y_2) > d(x_1, x_2), \tag{1.3.7}$$

then for small $t > 0$,

$$d^2(\gamma(t), x_1) + d^2(\gamma(1-t), x_2) < d^2(x_1, y_1) + d^2(x_2, y_2), \tag{1.3.8}$$

and together with (1.3.6) either

$$\begin{gathered} \lambda F(\gamma(t)) + d^2(\gamma(t), x_1) < \lambda F(y_1) + d^2(y_1, x_2) \text{ or} \\ \lambda F(\gamma(1-t)) + d^2(\gamma(1-t), x_2) < \lambda F(y_2) + d^2(y_2, x_2). \end{gathered}$$

This, however, would contradict the minimizing property of y_1 or y_2. Thus, (1.3.7) cannot hold. □

We now wish to construct a strongly continuous semigroup $(T_t)_{t>0}$ with the help of the resolvents. More precisely, we shall extend the Crandall - Liggett theorem [CL] to the present situation.

Let $x \in R(J_\lambda)$ for some $\lambda > 0$, i.e.

$$x = J_\lambda z \text{ for some } z \in Z \tag{1.3.9}$$

Then also $x \in R(J_\mu)$ whenever $0 < \mu \leq \lambda$, namely

$$x = J_\mu(m_{\frac{\mu}{\lambda}}(z, x)) \quad \text{by Cor. 2.1.} \tag{1.3.10}$$

Also

$$\frac{d(x, z)}{\lambda} = \frac{d(x, m_{\frac{\mu}{\lambda}}(z, x))}{\mu}$$

and we denote this value by

$$l_F(x).$$

Note that $l_F(x)$ may depend on the choice of z with $x = J_\lambda z$ above. If $\lambda \geq \mu > 0$, however, we see from (1.3.10) that F having fixed λ and z, $l_F(x)$ does not depend on μ anymore. In order to eliminate the dependence on z, we may take the infimum over all z and $\lambda > 0$ satisfying (1.3.9). For $x \in \overline{D(F)}$, we may then define

$$l_F(x) = \lim_{n \to \infty} l_F(J_{\frac{1}{n}} x)$$

because $J_{\frac{1}{n}} x \in R(J_{\frac{1}{n}})$ and $\lim_{n\to\infty} J_{\frac{1}{n}} x = x$ for $x \in \overline{D(F)}$ by Lemma 2.2. Note that $l_F(x)$ may be infinite for some $x \in \overline{D(F)}$.

Lemma 1.3.11 *Let $x \in R(J_\lambda)$ for some $\lambda > 0$. Then*

$$d(x, J_\lambda x) \leq \lambda l_F(x).$$

Proof: Let $x = J_\lambda z$ as above. Then

$$\begin{aligned} d(x, J_\lambda x) = d(J_\lambda z, J_\lambda(J_\lambda z)) &\leq d(z, J_\lambda z) \text{ by Lemma 2.5} \\ &= d(x, z) = \lambda l_F(x). \end{aligned}$$

□

Lemma 1.3.12 *Let $\lambda \geq \mu > 0, m, n \in \mathbb{Z}$. Then for $x \in D(F) \cap R(J_\lambda)$*

$$(1.3.11) \quad d(J_\mu^n x, J_\lambda^m x) \leq \left(\sqrt{(m\lambda - n\mu)^2 + m\lambda^2} + \sqrt{(m\lambda - n\mu)^2 + \lambda\mu n}\right) l_F(x).$$

Proof: By Cor. 2.1

$$\begin{aligned} a_{n,m} := d(J_\mu^n x, J_\lambda^m x) &= d(J_\mu^n x, J_\mu(m_{\frac{\mu}{\lambda}}(J_\lambda^{m-1}x, J_\lambda^m x)) \\ &\leq d(J_\mu^{n-1}x, m_{\frac{\mu}{\lambda}}(J_\lambda^{m-1}x, J_\lambda^m x)) \text{ by Lemma 2.5} \\ &\leq \frac{\mu}{\lambda} d(J_\mu^{n-1}x, J_\lambda^{m-1}x) + \frac{\lambda - \mu}{\lambda} d(J_\mu^{n-1}x, J_\lambda^m x) \\ &\quad \text{by nonpositive curvature (see (1.6.1)).} \end{aligned}$$

With $\alpha = \frac{\mu}{\lambda}, \beta = \frac{\lambda - \mu}{\lambda}$, we thus have

$$(1.3.12) \qquad a_{n,m} \leq \alpha a_{n-1,m-1} + \beta a_{n-1,m}$$

Using (1.3.12), we now want to prove (1.3.11) by induction. First

$$\begin{aligned} a_{0,m} = d(x, J_\lambda^m x) &\leq \sum_{i=0}^{m-1} d(J_\lambda^i x, J_\lambda^{i+1} x) \text{ by the triangle inequality} \\ &\leq m d(x, J_\lambda x) \text{ by Lemma 2.5} \\ (1.3.13) \qquad &\leq m\lambda l_F(x) \text{ by Lemma 2.6.} \end{aligned}$$

Likewise

$$a_{n,0} \leq n\mu l_F(x).$$

We shall show that if (1.3.12) holds for the pairs (n, m) and $(n, m-1)$ then it also holds for $(n+1, m)$. Induction then yields the claim. From (1.3.12)

$$\begin{aligned}
a_{n+1,m} &\leq \alpha a_{n,m-1} + \beta a_{n,m} \\
&\leq \Big\{\alpha\left(\sqrt{((m-1)\lambda - n\mu)^2 + (m-1)\lambda^2} + \sqrt{((m-1)\lambda - n\mu)^2 + \lambda\mu n}\right) \\
&\quad + \beta\left(\sqrt{(m\lambda - n\mu)^2 + m\lambda^2} + \sqrt{(m\lambda - n\mu)^2 + \lambda\mu m}\right)\Big\} l_F(x) \\
&\quad \text{by our inductive hypothesis} \\
&\leq \Big(\sqrt{\alpha\{((m-1)\lambda - m\mu)^2 + (m-1)\lambda^2\} + \beta\{(m\lambda - n\mu)^2 + m\lambda^2\}} \\
&\quad + \sqrt{\alpha\{((m-1)\lambda - n\mu)^2 + \lambda\mu n\} + \beta\{(m\lambda - n\mu)^2 + \lambda\mu n\}}\Big) l_F(x) \\
&\quad \text{by the Schwarz inequality, noting that } \alpha + \beta = 1 \\
&\leq \left(\sqrt{(m\lambda - (n+1)\mu)^2 + m\lambda^2} + \sqrt{(m\lambda - (n+1)\mu)^2 + (n+1)\lambda\mu}\right) l_F(x) \\
&\quad \text{noting that } \alpha = \frac{\mu}{\lambda} \text{ and } \alpha + \beta = 1.
\end{aligned}$$

This is (1.3.12) for the pair $(n+1, m)$. □

Theorem 1.3.13 *Let $F : Z \to \mathbb{R}_+ \cup \{\infty\}$ be a convex lower semicontinuous functional, $\not\equiv \infty$. Then*

(1.3.14) $$T_t x := \lim_{n\to\infty} J_{\frac{t}{n}}^n x \quad t \geq 0, x \in \overline{D(F)}$$

defines a strongly continuous semigroup of contractions on $\overline{D(F)}$. The convergence here is uniform on bounded subintervals of $[0,\infty)$. If $x \in D(F) \cap R(J_\lambda)$ for some $\lambda > 0$, then for $s, t \geq 0$

(1.3.15) $$d(T_t x, T_s x) \leq 2 l_F(x)|t - s|.$$

Proof: For $x \in D(F) \cap R(J_\lambda)$, we may apply (1.3.11). We choose $\mu = \frac{t}{n}, \lambda = \frac{t}{m}$, $(n \geq m)$ in (1.3.11). Then

(1.3.16) $$d(J_{\frac{t}{n}}^n x, J_{\frac{t}{m}}^m x) \leq \frac{2t}{\sqrt{m}} l_F(x)$$

If $x \in \overline{D(F)}$ and $m, n \to \infty$, then $x_n = J_{\frac{t}{n}} x$ constitutes a sequence in $D(F) \cap R(J_{\frac{t}{n}})$ converging to x (see Lemma 2.2). Then

$$\begin{aligned}
& d(J_{\frac{t}{n}}^n x, J_{\frac{t}{m}}^m x) \\
\leq\ & d(J_{\frac{t}{n}}^n x, J_{\frac{t}{n}}^n x_n) + d(J_{\frac{t}{m}}^m x, J_{\frac{t}{m}}^m x_n) + d(J_{\frac{t}{n}}^m x_n, J_{\frac{t}{m}}^m x_n) \\
\leq\ & 2d(x, x_n) + \frac{2t}{\sqrt{m}} l_F(x_n) \quad \text{by Lemma 2.5, (1.3.11).}
\end{aligned}$$

Therefore, for $T > 0$, we obtain

$$\limsup_{m,n\to\infty} \left(\sup_{0\le t\le T} d(J^n_{\frac{t}{n}}x, J^m_{\frac{t}{m}}x)\right) = 0 \tag{1.3.17}$$

Therefore, as $n \to \infty$, $J^n_{\frac{t}{n}}x$ converges uniformly for $0 \le t \le T$ to some point in X, denoted by $T_t x$.

For $t = 0$, we have

$$T_0 = Id.$$

If we choose $\mu = \frac{s}{n}$, $\lambda - \frac{t}{n}$ $(s \le t)$ and $m = n$ in (1.3.11), we get

$$d(J^n_{\frac{s}{n}}x, J^n_{\frac{t}{n}}x) \le \left(\sqrt{(s-t)^2 + \frac{t^2}{n}} + \sqrt{(s-t)^2 + \frac{st}{n}} \right) l_F(x),$$

hence for $n \to \infty$

$$d(T_s x, T_t x) \le 2(t-s) l_F(x) \tag{1.3.18}$$

which implies the continuity of $T_t x$ w.r.t. t. This also holds for $s = 0$; that latter fact alternatively follows from (2.13), i.e.

$$d(J^n_{\frac{t}{n}}x, x) \le t l_F(x)$$

and passing to $n \to \infty$.

Since all J_λ are contractions by Lemma 2.5, so is T_t. In particular, $T_t x$ is Lipschitz continuous in x. Let us now verify the semigroup property. For $k \in \mathbb{N}$, $t > 0$, we have

$$\begin{aligned} T_{kt} &= \lim_{n\to\infty} J^n_{\frac{kt}{n}} = \lim_{m\to\infty} J^{km}_{\frac{kt}{km}} = \lim_{m\to\infty} (J^m_{\frac{t}{m}})^k \\ &= (\lim_{m\to\infty} J^m_{\frac{t}{m}})^k \text{ since the } J^m_{\frac{t}{m}} \text{ are Lipschitz continuous} \\ &\quad\ \text{with uniform Lipschitz constant 1 (see Lemma 2.5)} \\ &= T^k_t. \end{aligned}$$

For $p, q, r, s \in \mathbb{N}$, one obtains

$$T_{\frac{p}{q}+\frac{r}{s}} = T_{\frac{ps+qr}{qs}} = T^{ps}_{\frac{1}{qs}} \circ T^{rq}_{\frac{1}{qs}} = T_{\frac{p}{q}} \circ T_{\frac{r}{s}}.$$

Consequently, the semigroup property

$$T_{t+s} = T_t \circ T_s$$

holds for rational $t, s \geq 0$, and then also for real $t, s \geq 0$ because of the continuity in t and the Lipschitz continuity in x of $T_t x$. □

Theorem 2.2 is the main result of [Ma]. However, this last result and the proof given here were already known to the present author in September '93. It essentially follows the original proof of the Crandall-Liggett theorem [CL] and also uses the presentation of Miyadera [Mi]. One may ask whether under the assumption of Theorem 2.1, i.e. if we suppose that $T_{t_n}x$ stays bounded for some sequence $t_n \to \infty$, $T_t(x)_{t>0}$ converges to a minimizer of F as $t \to \infty$. This was verified by Mayer [Ma] under various additional assumptions, for example if F is uniformly convex, but the general case is unknown.

Applications of nonlinear resolvents to other geometric variational or evolution problems have been suggested for example by de Giorgi [dG] and Jost [J7].

1.4 Convergence properties

We consider a family of functionals

$$F_n : Z \to \mathbb{R} \cup \{\pm\infty\}$$

defined on some topological space Z. In view of our subsequent applications, we shall assume

$$Z = L^2(X, \mu; Y, d)$$

for some global NPC space (Y, d), but we invite the reader to search for the most general spaces permitted in our subsequent considerations. We also assume that Z satisfies the first axiom of countability; this is just for simplicity, so that we can use sequences instead of filters. We first recall de Giorgi's notion of Γ-convergence, using [dM] as our basic reference.

Definition 1.4.1 $F : Z \to \mathbb{R} \cup \{\pm\infty\}$ *is the* Γ*-limit of the sequence* $(F_n)_{n\in\mathbb{N}}$, $F = \Gamma - \lim_{n\to\infty} F_n$ *if*

(i) whenever $(x_n)_{n\in\mathbb{N}} \subset Z$ *converges to* $x \in Z$, *then*

$$F(x) \leq \liminf_{n\to\infty} F_n(x_n)$$

(ii) for every $x \in Z$, *there exists some sequence* $(y_n)_{n\in\mathbb{N}} \subset Z$ *that converges to* x *and satisfies*

$$F(x) \geq \limsup_{n\to\infty} F_n(y_n)$$

Let us list some properties of Γ-limits:

(1) Γ-limits are lower semicontinuous.

(2) If all the F_n are convex, so is Γ-lim F_n.

(3) If all the F_n satisfy the quadratic contraction property of Def. 2.1, so does Γ-lim F_n.

(4) If x_n is a minimizer for F_n, and if $\lim_{n\to\infty} x_n = x$, then x minimizes F.

(5) The following important compactness result holds in case Z is second countable: Every sequence $(F_n)_{n\in\mathbb{N}}$ contains a Γ-convergent subsequence.

Given a functional F as above, but not assumed to be lower semicontinuous, there exists a greatest lower semicontinuous functional $\underline{F}$ with $\underline{F}(u) \leq F(u)$ for all u. $\underline{F}$ is called the **relaxation** of F. Since $\underline{F}(u)$ is finite whenever $F(u)$ is finite, we have

$$D(F) \subset D(\underline{F}).$$

$\underline{F}$ is given by the formula

$$\underline{F}(u) = \min\{\liminf_{n\to\infty} F(u_n) : u_n \to u \text{ in } Z \text{ for } n \to \infty\}$$

We recall the Moreau-Yosida approximations of a convex functional F

$$F^\lambda(x) = \inf_{y\in Z}\{\lambda F(y) + d^2(x,y)\}$$

and the "resolvents" $J_\lambda x$ characterized by

$$F^\lambda(x) = \lambda F(J_\lambda x) + d^2(x, J_\lambda x).$$

We now extend Mosco's theory [M] to our situation.

Definition 1.4.2 *A family of convex functionals $F_n : Z \to \mathbb{R}_+ \cup \{\infty\}$ converges to a functional F in the sense of Mosco if for every $\lambda > 0$, the Moreau-Yosida approximations $(\underline{F}_n)^\lambda$ of the relaxations of the F_n converge to F^λ pointwise in Z, or equivalently, if the resolvents $J_{n,\lambda}$ associated with the relaxed forms $\underline{F}_n$ converge to the J_λ.*

Lemma 1.4.3 *If $(F_n)_{n\in\mathbb{N}}$ converges to F in the sense of Mosco, then (F_n) also Γ-converges to F.*

Proof:

(i) Suppose that $(u_n)_{n\in\mathbb{N}}$ converges to u in Z. Then

$$\begin{aligned} F_n(u_n) \geq \frac{1}{\lambda} F_n^\lambda(u_n) &\geq \frac{1}{\lambda} \underline{F}_n^\lambda(u_n) \\ &\geq \frac{1}{\lambda} \underline{F}_n^\lambda(u) - 2d(u,u_n)d(u,J_\lambda u). \end{aligned}$$

Since $d(u,u_n) \to 0$ for $n \to \infty$, and $\underline{F}_n^\lambda(u)$ converges to $F^\lambda(u)$ by assumption, we obtain

$$\liminf_{n\to\infty} F_n(u_n) \geq \frac{1}{\lambda} F^\lambda(u) \quad \text{for every } \lambda > 0,$$

hence also by letting $\lambda \to 0$

$$\liminf_{n\to\infty} F_n(u_n) \geq F(u).$$

(ii) Let $u \in Z$. We may find a decreasing sequence $\lambda_n \to 0$ for $n \to \infty$ with

$$F(u) \geq \lim_{\lambda\to 0} \lim_{n\to\infty} \frac{1}{\lambda} F_n^\lambda(u) \geq \lim_{n\to\infty} \frac{1}{\lambda_n} F_n^{\lambda_n}(u).$$

We put $u_n := J_{n,\lambda_n} u$, hence

$$\frac{1}{\lambda_n} F_n^{\lambda_n}(u) = F_n(u_n) + \frac{1}{\lambda_n} d^2(u,u_n).$$

Inserting this into the preceding inequality yields

$$F(u) \geq \limsup_{n\to\infty} F_n(u_n).$$

We may assume $u \in D(F)$ in which case $u_n \to u$ in Z by Lemma 2.2.

Thus, we have verified the two properties required for Γ-convergence. □

Definition 1.4.4 *We say that a functional $F : Z \to \mathbb{R}_+ \cup \{\infty\}$ satisfies the Rellich property if every bounded sequence*

$$(u_n)_{n\in\mathbb{N}} \subset Z$$

with

$$\liminf_{n\to\infty} F(u_n) < \infty$$

contains a subsequence that converges in Z. A sequence $F_n : Z \to \mathbb{R}_+ \cup \{\infty\}$ satisfies an asymptotic Rellich condition if every bounded sequence $(u_n)_{n\in\mathbb{N}} \subset Z$ with

$$\liminf_{n\to\infty} F_n(u_n) < \infty$$

contains a converging subsequence.

Remark: A Rellich condition is typically implied by a Poincaré inequality (as exhibited for example in §4).

Theorem 1.4.5 *A sequence of convex $F_n : Z \to \mathbb{R}_+ \cup \{\infty\}$ satisfying an asymptotic Rellich property Γ-converges to some F if and only if it converges to F in the sense of Mosco.*

Proof: Mosco convergence always implies Γ-convergence by Lemma 3.1. For the reverse direction, we recall that a Γ-limit is automatically lower semicontinuous, and so the F_n Γ-converge to F if and only if the relaxations $\underline{F}_n$ Γ-converge to F. Thus, we may assume w.l.o.g. that the F_n themselves are lower semicontinuous. Let $y_\lambda = J_\lambda u$, $y_{n,\lambda} = J_{n,\lambda} u$ for some $u \in Z$. Then

$$\begin{aligned} F_n(y_{n,\lambda}) + \frac{1}{\lambda} d^2(y_{n,\lambda}, u) &= \frac{1}{\lambda} F_n^\lambda(u) \\ &\leq F_n(u_n) + \frac{1}{\lambda} d^2(u, u_n) \end{aligned}$$

for any u_n. We then choose a sequence $(u_n)_{n\in\mathbb{N}}$ that converges to u and satisfies

$$\limsup_{n\to\infty} F_n(u_n) \leq F(u)$$

according to (ii) of Γ-convergence. Thus, $(y_{n,\lambda})_{n\in\mathbb{N}}$ is bounded in Z, and $F_n(y_{n,\lambda})$ is bounded as well. By the asymptotic Rellich property, after selection of a subsequence, $(y_{n,\lambda})$ converges to some z_λ as $n \to \infty$. We want to show that $z_\lambda = y_\lambda$. For any $v \in Z$, by (ii) of Γ-convergence, we find a sequence $(v_n)_{n\in\mathbb{N}}$ converging to v with

$$\limsup_{n\to\infty} F_n(v_n) \leq F(v).$$

For every n

$$F_n(y_{n,\lambda}) + \frac{1}{\lambda}d^2(u, y_{n,\lambda}) \leq F_n(v_n) + \frac{1}{\lambda}d^2(u, v_n).$$

Using (i) of Γ-convergence for $(y_{n,\lambda})_{n\in\mathbb{N}}$, we then get

$$F(z_\lambda) + \frac{1}{\lambda}d^2(u, z_\lambda) \leq F(v) + \frac{1}{\lambda}d^2(u, v).$$

Since this holds for any v, we must have $z_\lambda = J_\lambda u = y_\lambda$. Therefore, $J_{n,\lambda}u = y_{n,\lambda}$ converges to $J_\lambda u$ as $n \to \infty$. We thus have verified Mosco convergence. □

1.5 Generalized harmonic maps

Although, as mentioned in the introduction, we shall be brief about the geometric aspects in these notes and refer to [J6] for a detailed treatment, it might be useful to introduce the original example that motivated the theory presented here, harmonic maps between compact Riemannian manifolds M and N. For a - sufficiently regular - map $f : M \to N$, we consider the energy functional

$$E(f) = \int_M \|df(x)\|^2 \, d\,vol_m(x) \tag{1.5.1}$$

where $d\,vol_M$ is the measure given by the Riemannian metric of M and $\|df(x)\|$ is the norm of the differential $df(x)$ considered as a linear map from the tangent space T_xM to $T_{f(x)}N$, where the metrics on these spaces are given by the corresponding Riemannian metrics. If $x^\alpha, \alpha = 1, ..., m$, and $f^i, i = 1, ..., n$ are local coordinates in M and N, resp., and if we denote the metric tensors in these coordinates as $(\gamma_{\alpha,\beta})_{\alpha,\beta=1,\dots,m}$ and $(g_{ij})_{i,j=1,\dots,n}$ resp., and if $(\gamma^{\alpha\beta}) := (\gamma_{\alpha\beta})^{-1}$, then in those coordinates

$$\|df(x)\|^2 = \sum_{\substack{\alpha,\,\beta\\ i,\,j}} \gamma^{\alpha\beta}(x) g_{ij}(f(x)) \frac{\partial f^i}{\partial x^\alpha} \frac{\partial f^j}{\partial x^\beta}. \tag{1.5.2}$$

Thus, E is a quadratic variational integral, but since the coefficients g_{ij} depend on the value of the unknown map f, the corresponding Euler-Lagrange equations are nonlinear; they are of the form

$$\Delta_M f + \Gamma(f)(df, df) = 0 \tag{1.5.3}$$

where Δ_M is the Laplace-Beltrami operator of M (a linear second order elliptic differential operator, the Riemannian version of the usual Laplace operator), while the nonlinear term which is quadratic in the first derivatives of f stems from the Riemannian geometry of N. The coeffiecients $\Gamma(f)$ are given by the so-called Christoffel symbols of N which in turn are expressions involving first derivatives of the metric of N. The details are not so important here. Solutions of (1.5.3) are called harmonic maps. (1.5.3) is a semilinear elliptic system, and in general weak solutions need not be regular, or from a more geometric point of view, solutions need not exist in given homotopy classes. It was however discovered in the work of Al'ber [A1, A2] and Eells-Sampson [ES] that these problems disappear if one assumes that N has nonpositive sectional curvature. That condition resulted in a differential inequality

$$\Delta_M \|df(x)\|^2 \geq -\text{const } \|df(x)\|^2 \tag{1.5.4}$$

for a solution f that implied estimates on solutions by standard PDE techniques. We shall not enter into those details here, as our approach will be conceptually different. Namely, we shall translate the curvature condition on N into a convexity condition for the functional E that makes the results of §2 applicable. Also, motivated both by geometric applications and by the intrinsic desire to isolate the essential features of the subject, we shall study a more general situation of L^2-maps between a measure space (X, μ) that will take the role of the Riemannian manifold M (or its universal cover) with the Riemannian volume form and a metric space (Y, d) that will generalize the target N (or its universal cover) with the distance function derived from the Riemannian metric.

This will involve a setting where no differentiable structure is assumed, and so the derivatives occuring in the definition E need to be replaced by expressions involving distances only. The guiding idea for those constructions will be that for a smooth map $f : M \to N$ with M and N having Riemannian distances functions d_M, d_N, resp., we have approximations of the norm of df of the type

$$\|df(x)\|^2 = c(\dim M) \lim_{r \to 0} \int_{B(x,r)} \frac{d_N^2(f(x), f(y))}{d_M^2(x,y)} \, \mu(dy) \tag{1.5.5}$$

with a constant c depending only on the dimension of M, where $B(x,r) := \{y \in M : d_M(x,y) < r\}$, and μ again is the Riemannian volume measure. Alternatively

$$\|df(x)\|^2 = c'(\dim M) \lim_{N \to 0} \frac{\int_{B(x,r)} d_N^2(f(x), f(y))\, \mu(dy)}{\int_{B(x,r)} d_M^2(x,y) \mu(dy)} \tag{1.5.6}$$

which has the advantage of not involving a singular kernel. Of course, there exist many other possibilities for approximating $\|df(x)\|^2$, for example by using Gaussian integral kernels with variance tending to 0.

The point of such approximations is that for example the right hand sides of (1.5.5) and (1.5.6) remain meaningful without assuming differentiable structures and hence can be used to define energy integrals in the more general setting envisioned here.

Let us consider a general situation:
We let $h : X \times X \to \mathbb{R}$ be a nonnegative (but not $\not\equiv 0$) and symmetric (μ-measurable) function, i.e. $h(x,y) \geq 0$ and $h(x,y) = h(y,x)$ for all $x, y \in X$. For geometric applications, we may also wish to impose some invariance condition $h(\gamma x, \gamma y) = h(x,y)$ for all elements γ of some group Γ that operates on X and also leaves the measure μ on X invariant. For a map $f : X \to Y$ into a metric space (Y, d), we then put

$$E_h(f) := \int \int h(x,y) d^2(f(x), f(y))\, \mu(dy)\, \mu(dx).$$

Minimizers of E_h can be characterized by a local mean value property.

Lemma 1.5.1 *$f : X \to Y$ minimizes E_h if and only if for μ-almost all $x \in X$, $f(x)$ is the $h(x,\cdot)$ weighted mean value of f, i.e. if $f(x)$ minimizes*

$$\Phi(p) := \int h(x,y)\, d^2(p, f(y))\, \mu(dy).$$

For the easy proof, see e.g. [J6]. To put this result into a proper perspective, we observe that for a global NPC space (Y, d), the function Φ has a unique minimizer for every x and h as above, because in that case $d^2(\cdot, q)$ is strictly convex for every q.

In a certain sense, harmonic maps between Riemannian manifolds can be considered as satisfying an infinitesimal mean value property, and in this sense, we are trying to approximate such maps by ones that satisfy local mean value properties. (It should be pointed out, however, that the present mean value property is weaker than the one of harmonic functions on domains in spaces of constant curvature which satisfy a mean value property on every distance ball contained in their domain of definition.) We should also point out that Kendall [K] had arrived at the importance of mean value properties in Riemannian geometry from a stochastic point of view.

Lemma 4.1 suggests the following iterative procedure for constructing minimizers of E_h (see [J2]): We start with $f_0 : X \to Y$ in the desired class and define

$$f : X \times \mathbb{N} \to Y$$

via

$$f(x, 0) := f_0(x)$$

and

$$f(x, n+1) := h(x, \cdot)\text{weighted mean value of} f(\cdot, n) \quad \text{for} x \in X, \sim\in \mathbb{N},$$

in order to get the minimizer as

$$\lim_{n\to\infty} f(x, n).$$

As explained above, one may then consider an appropriate family of kernels h in order to also obtain a minimizer of the original functional E. The reader should compare this with the very interesting stochastic mean value constructions in Kendall's contribution.

Remark: As explained in the geometric digression in §1, in order to make the theory applicable in geometry, one should consider ρ-invariant maps and integrate over some fundamental domain for X in a cover $\tilde{X}$, in order to have a simply connected target space. We shall ignore this point here, but we assure the reader that this can easily be taken into account. See e.g. [J6].

Let us consider some

Examples:

1) X is a discrete space, for example a grid as used in numerical analysis, a discrete approximation of a Riemannian manifold, or a discrete group, equipped with a symmetric neighborhood relation, and put

$$h(x, y) = \begin{cases} 1 & \text{if } x \text{ and } y \text{ are neighbors} \\ 0 & \text{otherwise} \end{cases}$$

2) X carries a metric d_X; we put

$$h_\varepsilon(x,y) = \begin{cases} c(\varepsilon) & \text{if } d_X(x,y) < \varepsilon \\ 0 & \text{otherwise} \end{cases}$$

where $c(\varepsilon) > 0$ is so chosen that for example

$$\lim_{\varepsilon \to 0} E_{h_\varepsilon}$$

is nontrivial (i.e. not identically 0 or ∞) (We shall specify the nature of this limit below.). Thus, $c(\varepsilon)$ is just a suitable normalization constant. Taking this limit of the E_{h_ε} for $\varepsilon \to 0$ leads to the energy functional for maps between Riemannian manifolds. In fact, with the appropriate choice of $c(\varepsilon)$, the functionals E_{h_ε} converge to the energy functional E in a monotically increasing manner up to some error term that is controlled by a lower bound on the Ricci curvature of the domain X, see [J1] or [KS]. For other purpose, it might also be useful and interesting to consider

$$\lim_{R \to \infty} E_{h_R}$$

for example if X is a global NPC space, in order to capture the asymptotic geometry of X.

3) We let $h(x,y)$ be some kind of Gaussian kernel, e.g. of the form

$$\frac{1}{t^{n/2}} \exp\left(-\frac{d_X(x,y)}{t^2}\right)$$

if X is a Riemannian manifold of dimension n with distance function d_X, and consider the limit as t goes to 0 or ∞ of the resulting functionals E_h. More generally, one may consider transition density functions $P_t(x,y)$ for the Markov process associated with some Dirichlet form E. In this context, Sturm [St3, St4] investigated the question raised in [J5] under which conditions the resulting E_h monotonically converge for $t \to 0$. Essentially, his result is that one gets monotone convergence provided the target satisfies a suitable lower curvature bound.

Lemma 1.5.2 *Suppose Y is a global NPC space. Then the functionals E_h as defined above are convex on $Z = L^2(X, \mu; Y, d)$.*

Proof: This follows from the fact that the squared distance functions on Y are convex. In the notations of 4) of the Appendix, we have

$$d^2(\gamma(x,t),\gamma(y,t)) \leq (1-t)d^2(f(x),f(y)) + td^2(g(x),g(y))$$

for the geodesic $\gamma(\cdot,t)$ from f to g. Integrating this inequality implies the convexity of E_h. □

As already indicated, we wish to consider energy functionals E obtained as limits of suitable E_{h_ε} for $\varepsilon \to 0$ as in example 2). It was the author's idea in [J2] to use Γ-limits in this context. By property (5) of Γ-limits, we may find suitable sequences $(\varepsilon_n)_{n\in\mathbb{N}}$ that converge to 0 for $n \to \infty$ for which

$$E = \Gamma - \lim_{n\to\infty} E_{h_{\varepsilon n}}$$

exists.

By general properties of Γ-limits as listed in §3, E is lower semicontinuous and satisfies the quadratic contraction property. Furthermore, if Y is a global NPC space, the convexity of the E_h as verified in Lemma 4.2 carries over to the Γ-limit E.

There is still some arbitrariness involved in the choice of the h_ε, and one may wish to exploit this flexibilty to get some additional control over E. For example, it will be convenient to have a representation of E by so-called energy measures $\eta(u,u)$, i.e. (nonnegative) Radon measures satisfying

$$E(u) = \int_X \eta(u,u)\,dx.$$

This has been achieved by Korevaar-Schoen [KS]. In particular, this allows to restrict E to μ-measurable subsets Ω of X by using the characteristic function χ_Ω of Ω and considering

$$E(u;\Omega) := \int_X \chi_\Omega(x)\eta(u,u)\,dx.$$

When we restrict E to scalar valued functions, the existence of the energy measure $\eta(u,u)$ is quite simple, see e.g. [FOT, p.110 f, p. 159 f]. In that case, we have in terms of the polarized form $E(\cdot,\cdot)$ and for $u \in D(E) \cap L^\infty(X;\mathbb{R})$; $\varphi \in D(E) \cap C_0^0(X,\mathbb{R})$

$$
\begin{aligned}
2E(\varphi u, u) - E(u^2,\varphi) &= \int_X \varphi(x)\eta(u,u)\,dx \\
&= \lim_{t\searrow 0} \frac{1}{2t} \int_{X\times X\setminus diagonal} \varphi(x)(u(x)-u(y))^2\, p_t(x,dy)\,\mu(dx), \qquad (1.5.7)
\end{aligned}
$$

where $(p_t)_{t\geq 0}$ is the family of transition functions for the Markov process associated with E, or equivalently for $u, v \in D(E)$

$$
\begin{aligned}
E(u,v) &= \int_X \eta(u,v)\,dx \\
&= \lim_{t\searrow 0} \frac{1}{2t} \int_{X\times X\setminus diagonal} (u(x)-u(y))(v(x)-v(y))p_t(x,dy)\mu(dx). \qquad (1.5.8)
\end{aligned}
$$

(However, as found by Sturm [St3], the corresponding expressions in the general case, namely

$$
\frac{1}{2t} \int_{X\times X\setminus diagonal} d^2(u(x),u(y))p_t(x,dy)\mu(dx)
$$

need not always converge in the same monotonic manner for $t \searrow 0$ as they do in the scalar case.)

One may also wish that E be local in the sense that

$$
E(u;\Omega) = 0
$$

whenever u is constant on the open set Ω. Sturm [St2] has an argument to show the strong locality of ordinary Dirichlet forms (i.e. restrictions to $L^2(X,\mu;\mathbb{R})$ of generalized Dirichlet forms) obtained as Γ-limits of functionals as above for suitable families (h_ε) similar as the ones considered above, and using the quadratic contraction property, one may then deduce locality in our sense. In any case, since by Lemma 4.2 the E_h and hence also the Γ-limit E are convex, we get

Theorem 1.5.3 *The existence result for minimizers of Theorem 2.2 applies to the functionals E_h and E.*

Here we do not want to present the precise existence results because in order to ensure the boundedness of the resolvents J_λ in Theorem 2.1, one needs an additional necessary assumption that can be phrased as a so-called reductivity condition. We refer to [J6] for details. Theorem 4.1, as presented in detail in that reference, includes the previous existence results of Gromov-Schoen [GS], Jost [J2], and Korevaar-Schoen [KS]. As a Corollary, we just quote the original existence result of Al'ber [A2] and Eells-Sampson [ES].

Corollary 1.5.4 *Let M and N be compact Riemannian manifolds, with N of nonpositive sectional curvature. Then every continuous $g : M \to N$ is homotopic to a harmonic map.*

To prove this result, one looks at the universal covers X of M and Y of N. Y then is a global NPC space. The group of covering translations Γ for M and the homotopy class of g determine a homomorphism

$$\rho : \Gamma \to I(Y),$$

and a map $\tilde{f} : X \to Y$ passes to a map $f : M \to N$ that is homotopic to ρ if and only if for all $x \in X$, $\gamma \in \Gamma$

$$\tilde{f}(\gamma x) = \rho(\gamma)\tilde{f}(x).$$

This is the equivariance condition on $\tilde{f}$. Theorem 4.1 then implies that one may minimize E or E_h among all such ρ-equivariant maps in order to produce the desired harmonic map. □

Of course, so far the map has been produced only in the space $L^2(X, Y) \cap D(E)$ and so we need to discuss regularity results for minimizers of energy functionals. If domain and target both are Riemannian manifolds and the latter has nonpositive sectional curvature, regularity results were already known to Al'ber [A1] and Eells-Sampson [ES]. More precise results were obtained by Hamilton [Ha] and Hildebrandt-Kaul-Widman [HKW]. Recently also a very interesting stochastic approach to regularity was put forward e.g. in [K].

If X is a Riemannian manifold and (Y, d) is a locally finite Euclidean Bruhat - Tits building (and thus satisfies an NPC condition), very precise regularity results were obtained in Gromov-Schoen [GS]. If X is still a Riemannian manifold and (Y, d) is an NPC space, Korevaar - Schoen [KS] showed Lipschitz continuity of minimizers. We shall now describe a Hölder continuity result for more general domains.

We need some additional assumptions, in order to put the Harnack inequality of Biroli - Mosco [BM] at our disposal. Let X be a locally compact, separable topological space with a positive Radon measure μ with $\operatorname{supp}\mu = X$. Let E be a generalized Dirichlet form as above with energy measure η and let $E(\cdot,\cdot)$ be the Dirichlet form on $L^2(X,\mu;\mathbb{R})$ obtained by polarization. We assume that $E(\cdot,\cdot)$ is strongly local, i.e. $E(u,v) = 0$ whenever u,v have compact support and v is constant on a neighbarhood of $\operatorname{supp} u$. We assume that $E(\cdot,\cdot)$ admits a μ-separating core C. This is a subspace C of $D(E) \cap C_0^0(X)$ (where $D(E)$ now means the domain of definition of the Dirichlet form on $L^2(X,\mu;\mathbb{R})$ that is dense in $C_0^0(X)$ w.r.t the norm $\|\cdot\|_{L^2}^2 + E(\cdot)$ and which satisfies the condition that for any $x \neq y \in X$, there exists $\psi \in C$ with $\psi(x) \neq \psi(y)$ and

$$\eta(\psi,\psi) \leq \mu$$

on X.

In that case,

$$d_E(x,y) := \sup\{\psi(x) - \psi(y) : \psi \in C, \eta(\psi,\psi) \leq \mu\}$$

yields a metric on X. (If X is a Riemannian manifold and E is the classical Dirichlet integral, then d_E is the original Riemannian metric.) We assume that the topology induced by E is the original topology of X. We also assume the so-called **ball doubling property**.

(A) For $x \in X, r > 0$, we put $B(x,r) := \{y \in X : d_E(x,y) < r\}$, and require that there exist constants $c_1 < \infty, R > 0$ such that

$$\mu(B(x,2r)) \leq c_1 \mu(B(x,r))$$

whenever $x \in X, 0 < r \leq R$.

This ball doubling property should be considered as some kind of generalization of a lower bound for the Ricci curvature in the case of a Riemannian manifold, because such a bound implies a control on the growth of the volume of balls by the Bishop-Gromov volume comparison theorem. It also seems that such a ball doubling property automatically implies that X is locally compact.

Let now a metric space (Y,d) be given. Secondly, we need a **Poincaré inequality**.

(B) For any $X_0 \subset\subset X$, there exist constants $0 < \alpha \leq 1$, $c_2 < \infty$ s.t. for any ball $B(x_0, r) \subset X_0$ and every $f : X \to Y, f \in D(E)$

$$\int_{B(x_0,\alpha r)} d^2(f(x), \bar{f}_{B(x_0,\alpha r)})\mu(dx) \leq c_2 r^2 \int_{B(x_0,r)} \eta(f,f)\, dx.$$

Here $\bar{f}_A$ is the mean value of f on a set A, i.e. the minimum of the function

$$\Phi(q) := \int_A d^2(f(x), q)\, \mu(dx).$$

Another version of the Poincaré inequality that implies the preceding one is

$$\int_{B(x_0,\alpha r)} \int_{B(x_0,\alpha r)} d^2(f(x), f(y))\, \mu(dx)\, \mu(dy) \leq c_3 r^2 \int_{B(x_0,r)} \eta(f,f)\, dx$$

for some constant c_3.

These Poincaré inequalities are generalizations of the corresponding ones for scalar valued functions, and in some (and perhaps all) cases they may be deduced from the latter:

1) If X is a domain in some $\mathbb{R}^d$, Poincaré inequalities for maps $f : X \to Y$ can be derived in the same way as the ones for functions.

2) If (Y, d) is a compact Riemannian manifold, Nash's theorem yields an isometric embedding i of (Y, d) into some Euclidean space, and one applies the Poincaré inequality for $i \circ f$ to get one for f.

Biroli - Mosco [BM] showed that the preceding assumptions, i.e. the ball doubling property and the Poincaré inequality imply a Sobolev inequality which in term implies a Harnack inequality for supersolutions of the nonpositive self-adjoint operator A corresponding to the Dirichlet form $E(\cdot, \cdot)$. This work was extended in [St1]. (Note however that the reverse implications do not always hold, i.e. in the presence of a ball doubling property, a Harnack inequality need not imply a Sobolev inequality, and the latter need not imply a Poincaré inequality.) The Harnack inequality is the following

Lemma 1.5.5 *Let $v \in D(E)$ be a nonnegative, bounded supersolution*

$$Av \leq 0 \text{ in } B(x_0, 4R) \subset\subset X$$

in the weak sense, i.e.

$$\int_{X\cap B(x_0,4R)} \eta(v,\varphi)\,dx \leq 0 \qquad \text{for all } \varphi \geq 0, \varphi \in D(E), \text{ with supp } \varphi \subset\subset X \cap B(x_0, 4R).$$

Then for some $p \geq 1$ and some constant c_4

$$\left(\int_{B(x_0,2R)} |v|^p \mu(dx)\right)^{1/p} \leq c_4 \inf_{B(x_0,R)} v.$$

A consequence of the Harnack inequality is the following Hölder continuity result for minimizers of E with values in a NPC space.

Theorem 1.5.6 *Let the energy form E satisfy the preceding assumptions, let (Y, d) be a complete NPC space, and let $f : X \to Y$ be a local minimizer for E in the sense that*

$$E(f_{|B(x_0,R)}) \leq E(g_{|B(x_0,R)})$$

for all $x_0 \in X, 0 < R < R_0$ for some fixed R_0, and all

$$g \in L^2(B(x_0, R_0), \mu; Y, d) \text{ with } g = f \text{ on } B(x_0, R_0) \setminus B(x_0, R).$$

Then f is locally Hölder continuous on X.

This result was obtained in [J5]. Another proof of Hölder continuity based on a Harnack inequality was given by F. H. Lin [L] (according to the introduction of [L] this was known to him considerably earlier than [L] appeared). The argument of [L] is shorter than the one of [J5], but it assumes the local compactness of the target Y which [J5] does not need.

An interesting question is to find out the appropriate assumptions on X that are needed to improve Hölder continuity to Lipschitz continuity (cf. the result of Korevaar - Schoen [KS] that works if X is a Riemannian manifold).

In order to employ the Harnack inequality for the proof of Theorem 4.2, of course one has to produce supersolutions of A. In fact, by changing signs it is as good to construct subsolutions. This is achieved in

Lemma 1.5.7 *Let $f : X \to Y$ be a local minimizer for E where Y is a global NPC space. Then for any $p \in Y$*

$$Ad^2(f(\cdot), p) \geq 2\eta(f, f) \text{ weakly.}$$

In particular $d^2(f(x), p)$ is a subsolution of A in the weak sense.

Proof: We consider a function $\varphi \in D(E)$ with compact support and $0 \le \varphi \le 1$. For $x \in X$, we consider the geodesic

$$\gamma_x : [0,1] \to Y$$

with $\gamma_x(0) = f(x), \gamma_x(1) = p$. We put

$$f_\varphi(x) := \gamma_x(\varphi(x)).$$

The triangle comparison theorem yields

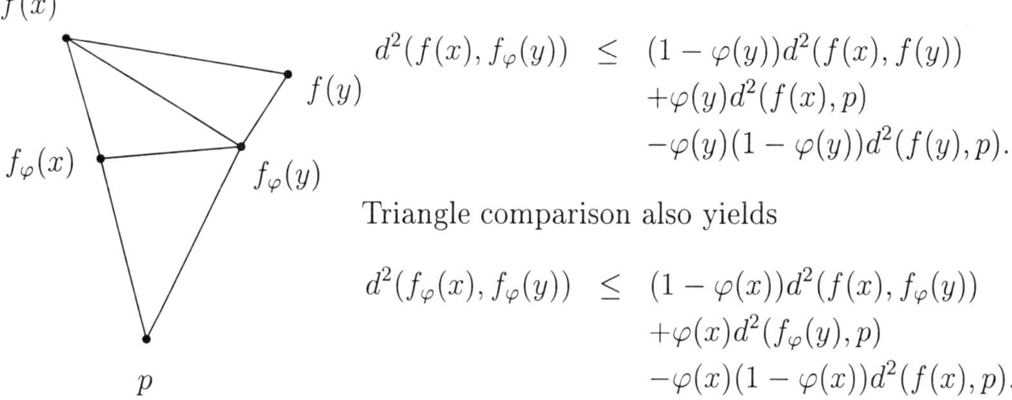

$$\begin{aligned} d^2(f(x), f_\varphi(y)) \;\le\; & (1-\varphi(y))d^2(f(x), f(y)) \\ & +\varphi(y)d^2(f(x),p) \\ & -\varphi(y)(1-\varphi(y))d^2(f(y),p). \end{aligned}$$

Triangle comparison also yields

$$\begin{aligned} d^2(f_\varphi(x), f_\varphi(y)) \;\le\; & (1-\varphi(x))d^2(f(x), f_\varphi(y)) \\ & +\varphi(x)d^2(f_\varphi(y),p) \\ & -\varphi(x)(1-\varphi(x))d^2(f(x),p). \end{aligned}$$

Inserting the first inequality into the second one and using

$$d(f_\varphi(y),p) = (1-\varphi(y))d(f(y),p)$$

yields

$$\begin{aligned} d^2(f_\varphi(x), f_\varphi(y)) \le & (1-\varphi(x))(1-\varphi(y))d^2(f(x), f(y)) \\ & -(\varphi(x)-\varphi(y))(1-\varphi(x))(d^2(f(x),p)-(1-\varphi(y))d^2(f(y),p)). \end{aligned}$$

Since f is minimizing, we have

$$\int \eta(f,f)(dx) \le \int \eta(f_\varphi, f_\varphi)(dx).$$

Combining this inequality with the previous one expressed in terms of $\eta(\cdot,\cdot)$, we get

$$\begin{aligned} 0 \le & -\int 2\varphi(y)\eta(f,f)(dx) \\ & -\int \eta(\varphi(x), d^2(f(x),p))(dx) \\ & + \text{terms that are at least quadratic in } \varphi. \end{aligned}$$

Replacing φ by $t\varphi$ with $t > 0$, dividing the resulting inequality by t and letting t tend to zero eliminates the quadratic and higher terms in φ, and we obtain

$$-\int \eta(\varphi(x), d^2(f(x),p))(dx) \ge 2\int \varphi(x)\eta(f,f)(dx)$$

which is equivalent to the claim. □

1.6 Appendix: Geometric notions

1) Let (Y, d) be a metric space.
A curve in Y is a continuous map

$$\gamma : [0,1] \to Y.$$

The length of a curve γ is defined as

$$L(\gamma) := \sup\{\sum_{i=0}^{n} d(\gamma(t_{i-1}), \gamma(t_i)) \quad 0 = t_0 < t_1 < \ldots < t_n = 1, n \in \mathbb{N}\},$$

γ is called rectifiable if $L(\gamma) < \infty$. (Y, d) is called a length space if for any $x, y \in Y$

$$d(x, y) = \inf\{L(\gamma) : \gamma : [0,1] \to Y \text{ curve with } \gamma(0) = x, \gamma(1) = y\},$$

i.e. if the distance between any two points equals the infimum of the lengths of curves connecting them.
(Y, d) is called a geodesic length space if this infimum is achieved, i.e. if for any $x, y \in Y$, we may find a curve $\gamma : [0,1] \to Y$ with $\gamma(0) = x, \gamma(1) = y$, and

$$L(\gamma) = d(x, y).$$

Any such minimizing curve γ is geodesic. A shortest geodesic is characterized by the property that $d(\gamma(t_1), \gamma(t_2)) = |t_2 - t_1| \, d(\gamma(0), \gamma(1))$ $(= |t_2 - t_1| L(\gamma))$ whenever $0 \leq t_1, t_2 \leq 1$. More generally, a curve $\gamma : [0,1] \to N$ is called geodesic if there exists $\varepsilon > 0$ with

$$L(\gamma_{|[t_1,t_2]}) = d(\gamma(t_1), \gamma(t_2)) \text{ whenever } 0 \leq t_2 - t_1 < \varepsilon (0 \leq t_1, t_2 \leq 1).$$

Obviously, the geodesic property can also be defined for curves defined on an interval different from $[0,1]$, and if a curve is geodesic, its restriction to any subinterval of its domain of definition remains geodesic.
Two curves $\gamma_0, \gamma_1 : [0,1] \to Y$ with the same end points, i.e $\gamma_0(0) = \gamma_1(0) =: x_0$, $\gamma_0(1) = \gamma_1(1) =: x_1$, are called homotopic if there exists a continuous

$$\Gamma : [0,1] \times [0,1] \to Y$$

with

$$\Gamma(t,0) = \gamma_0(t), \Gamma(t,1) = \gamma_1(t) \text{ for all } t \in [0,1]$$
$$\Gamma(0,s) = x_0, \Gamma(1,s) = x_1 \text{ for all } s \in [0,1].$$

We point out that our definition of homotopy between curves with the same end points requires that these end points remain fixed during the homotopy.

We say that (Y, d) is a strong geodesic length space if the infimum of the length of curves in a given homotopy class is always achieved. This means that given

$$\gamma_0 : [0,1] \to Y$$

with $\gamma_0(0) = x, \gamma_0(1) = y$, there exists a curve $\gamma : [0,1] \to Y$ with $\gamma(0) = x, \gamma(1) = y$, homotopic to γ_0, whose length equals the infimum of the length of all such curves homotopic to γ_0.

We point out, however, that we do not require that geodesics are unique in a given homotopy class. However, such uniqueness holds if one makes a suitable assumption about nonpositive curvature as will be done below.

Any rectifiable curve, in particular any geodesic in Y may be parametrized proportionally to arclength. This means that after composition with a suitable homomorphism $\sigma : [0,1] \to [0,1]$ the resulting curve $\gamma : [0,1] \to Y$ satisfies

$$L(\gamma_{|[0,t]}) = tL(\gamma) \text{ for all } 0 \leq t \leq 1.$$

In these notes, we shall automatically assume that all geodesics are parametrized proportionally to arclength.

A function $F : Y \to \mathbb{R}$ is called convex if its restriction to any geodesic $\gamma : [0,1] \to Y$ is a convex function of the arclength parameter t, i.e. $F(\gamma(t)) : [0,1] \to \mathbb{R}$ is a convex function in the usual sense.

2) If (Y, d) is a metric space, we denote by

$$I(Y)$$

the group of isometries of Y, i.e. the group of all homeomorphisms

$$h : Y \to Y$$

with

$$d(h(x), h(y)) = d(x, y) \text{ for any } x, y \in Y.$$

3) A geodesic length space(Y, d) is said to have nonpositive curvature in the sense of Alexandrov or as we shall say for abbreviation, to be an NPC space if for every $p \in Y$, there exists $\rho_p > 0$ such that for any $x, y, z \in B(p, \rho_p)$ and any shortest geodesic $\gamma : [0,1] \to Y$ with $\gamma(0) = x, \gamma(1) = z$ and all $0 \leq t \leq 1$

$$(1.6.1)\; d^2(\gamma(t), y) \leq (1-t)d^2(\gamma(0), y) + td^2(\gamma(1), y) - t(1-t)L(\gamma)^2.$$

(1.6.1) becomes an equality for certain triangles in $\mathbb{R}^2$ with the Euclidean metric. The geometric content of (1.6.1) is that the distance function in (Y, d) is at least as convex as the Euclidean distance function. Thus (1.6.1) is a quantitative convexity statement. In an NPC space, we have the following useful quadrilateral comparison theorem of Reshetnyak [Re]. (Proofs can also be found in [KS] and [J6].)

The statement is that for any geodesics $\gamma_1, \gamma_2 : [0,1] \to Y$ with images continued in some ball $B(p, \rho_p)$ as above and $(0 \leq s, t \leq 1)$

$$(1.6.2)\quad \begin{array}{rl} & d^2(\,\gamma_1(0), \gamma_2(t)\,) + d^2(\,\gamma_1(1), \gamma_2(1-t)\,) \\ \leq & d^2(\,\gamma_1(0), \gamma_2(0)\,) + d^2(\,\gamma_1(1), \gamma_2(1)\,) + 2\,t^2\, d^2(\,\gamma_2(0), \gamma_2(1)\,) \\ & +t\;(\,d^2(\,\gamma_1(0), \gamma_1(1)\,) - d^2(\,\gamma_2(0), \gamma_2(1)\,)\,) \\ & -s\,t\;(\,d(\,\gamma_1(0), \gamma_2(0)\,) - d(\,\gamma_1(1), \gamma_2(1)\,))^2 \\ & -(1-s)\,t\;(\,d(\,\gamma_1(0), \gamma_1(1)\,) - d(\,\gamma_2(0), \gamma_2(1)\,)\,)^2 . \end{array}$$

Again, equality in (1.6.2) holds for certain quadrilaterals in the Euclidean plane.

(Y, d) is called a global NPC space if we may put $\rho_p = \infty$ for some (and hence for all) p in the above condition.

The global NPC condition implies in particular that the geodesic curve between any two points $x, y \in Y$ is unique. If $\gamma_{x,y} : [0,1] \to Y$ is this geodesic, with $\gamma_{x,y}(0) = x$, $\gamma_{x,y}(1) = y$, we put for $t \in [0,1]$

$$m_t(x, y) = \gamma_{x,y}(t).$$

In particular, we shall call

$$m_{\frac{1}{2}}(x, y)$$

the midpoint of x and y. $m_{\frac{1}{2}}(x,y)$ is uniquely determined by the relations

$$d(m_{\frac{1}{2}}(x,y),x) = d(m_{\frac{1}{2}}(x,y),y) = \frac{1}{2}d(x,y). \tag{1.6.3}$$

The uniqeness of geodesics also implies that a global NPC space is simply connected. It is shown in [AB] that the universal cover of an NPC space is a global NPC space. Since in the context of the present notes, one may always lift to universal covers, we shall usually assume that our spaces are global NPC spaces.

4) We now observe that geometric properties of (Y,d) carry over to $Z = L^2(X,\mu;Y,d)$. If (Y,d) is complete, so is Z by definition of L^2-spaces. If (Y,d) is a geodesic length space, so is Z. Namely given $f, g \in Z$, for each $x \in X$, we let $\gamma_x : [0,1] \to Y$ be a shortest geodesic with $\gamma(0) = f(x), \gamma(1) = g(x)$. We put $\gamma(x,y) := \gamma_x(t)$.

$$\Gamma : [0,1] \to Z, \quad \Gamma(t) = \gamma(\cdot,t)$$

is a shortest geodesic in Z between f and g. In fact,

$$\begin{aligned} d(\Gamma(t_1),\Gamma(t_2)) &= \left(\int d^2(\gamma(x,t_1),\gamma(x,t_2))\,\mu(dx)\right)^{1/2} \\ &= |t_1 - t_2| \left(\int d^2(\,f(x),g(x)\,)\,\mu(dx)\right)^{1/2} \end{aligned}$$

by our above characterization, since $\gamma(x,\cdot)$ is a shortest geodesic for every x, and that characterization then conversely shows that $\Gamma(\cdot)$ is geodesic. Thus, geodesics in $Z = L^2(X,\mu;Y,d)$ are just obtained as families of geodesics in Y. One may also perform this construction in fixed homotopy classes. This means that if $F : X \times [0,1] \to Y$ is a homotopy between f and g, one can choose the geodesics from $f(x)$ to $g(x)$ homotopic to $F(x,\cdot)$ to obtain the geodesic family $\Gamma(t) = \gamma(x,t)$ homotopic to $F(x,t)$. Also, curvature properties of Y extend to Z. Namely, if Y is e.g. a global NPC space, the NPC inequality continuous to hold for Z, simply by integrating it over the relevant families of geodesics that constitute the geodesics in Z.

Bibliography

[A1] S. I. Al'ber, On n-dimensional problems in the calculus of variations in the large, Sov. Math. Dokl. **5** (1964), 700-704.

[A2] S. I. Al'ber, Spaces of mappings into a manifold of negative curvature, Soc. Math. Dokl. **9** (1967), 6-9.

[AB] S. Alexander, R. Bishop, The Hadamard-Cartan theorem in locally convex metric spaces, L'Ens. Math. **36** (1990), 309-320.

[BD2] A. Beurling, J. Deny, Dirichlet spaces, Proc. Nat. Acad. Sci. USA **45** (1959), 208-215.

[BD1] A. Beurling, J. Deny, Espaces de Dirichlet I, le cas élémentaire, Acta Math. **99** (1958), 203-224.

[BH] N. Bouleau, F. Hirsch, Dirichlet forms and analysis on Wiener space, de Gruyter, 1991.

[BM] M. Biroli, U. Mosco, A Saint-Venant type principle for Dirichlet forms on discontinuous media, Annal. Mat. Pure Appl. **169** (1995), 125-181.

[CFL] R. Courant, K. O. Friedrichs, H. Lewy, Über die partiellen Differentialgleichungen der mathematischen Physik, Math. Ann. **100** (1928).

[CL] M. Crandall, T. Liggett, Generation of semigroups of nonlinear transformations on general Banach spaces, Amer. J. Math. **93** (1971), 265-298.

[dG] E. de Giorgi, New problems on minimizing movements, in: Boundary value problems for partial differential equations and applications (J. L. Lions, C. Baiochi, eds.), RMA **29**, Masson, Paris, 1993.

[dM] G. dal Maso, An introduction to Γ-convergence, Birkhäuser, 1993.

[ES] J. Eells, J. Sampson, Harmonic mappings of Riemannian manifolds, Amer. J. Math. **85** (1964), 109-160.

[FOT] M. Fukushima, Y. Oshima, M. Takeda, Dirichlet forms and symmetric Markov processes, de Gruyter (1994).

[GS] M. Gromov, R. Schoen, Harmonic maps into singular spaces and p-adic superrigidity for lattices in groups of rank one, Publ. Math. IHES **76** (1992), 165-246.

[Ha] R. Hamilton, Harmonic maps of Riemannian manifolds with boundary, Springer LNM **471**, 1975.

[HKW] S. Hildebrandt, H. Kaul, K. O. Widman, An existence theorem for harmonic mappings of Riemannian manifolds, Acta Math. **138** (1997), 1-16.

[J1] J. Jost, Riemannian geometry and geometric analysis, Springer, 1995.

[J2] J. Jost, Equilibrium maps between metric spaces, Calc. Var. **2** (1994), 173-204.

[J3] J. Jost, Convex functionals and generalized harmonic maps into spaces of nonpositive curvature, Comment. Math. Helv. **70** (1995), 659-673.

[J4] J. Jost, Generalized harmonic maps between metric spaces. in: Geometric analysis and the calculus of variations for Stefan Hildebrandt (J. Jost, ed.), International Press, 1996, pp. 143-174.

[J5] J. Jost, Generalized Dirichlet forms and harmonic maps, Calc. Var. **5** (1997), 1-19.

[J6] J. Jost, Nonpositive curvature: Geometric and analytic aspects, Birkhäuser, 1997.

[J7] J. Jost, A. weak notion of mean curvature and a generalized mean curvature flow for singular sets, in: Advances in Geometric Analysis and Continuum Mechanics (P. Concus, K. Lancester, eds.), International Press, pp. 138-144.

[K] W. Kendall, Brownian motion and partial differential equations. From the heat equation to harmonic maps, Proc. 151, 49th session, Firenze, 1993, pp. 85-101.

[KS] N. Korevaar, R. Schoen, Sobolev spaces and harmonic maps for metric space targets, Comm. Anal. Geom. **1** (1993), 561-659.

[L] F. H. Lin, Analysis on singular spaces, Preprint, 1997.

[M] U. Mosco, Composite media and asymptotic Dirichlet forms, J. Funct. Anal. **123** (1994), 368-421.

[Ma] U. Mayer, Gradient flow on nonpositively curved metric spaces, Thesis, Univ. Utah, Salt Lake City, 1994.

[Mi] Miyadera, Nonlinear Semigroups, Transl. Math. Monographs **109**, AMS, 1991.

[MR] Z. M. Ma, M. Roeckner, An introduction to the theory of (non-symmetric) Dirichlet forms, Springer, 1992.

[Re] Y. G. Reshetnyak, Inextensible mappings in a space of curvature no greater than K, Sib. Math. J. **9** (1968), 683-689.

[St1] K. Sturm, Analysis on local Dirichlet spaces II: Upper Gaussian estimates for the fundamental solutions of parabolic equations, Osaka J. Math. **32** (1995), 275-312.

[St2] K. Sturm, Diffusion processes and heat kernels on length spaces.

[St3] K. Sturm, Monotone approximation of energy functionals for mappings into metric spaces, I, J. reine angew. Math. **486** (1997), 129-151.

[St4] K. Sturm, Monotone approximation of energy functionals for mappings into metric spaces, II, Preprint, 1997.

AMS/IP Studies in Advanced Mathematics
Volume 8, 1998

Wilfrid S. Kendall: From Stochastic Parallel Transport to Harmonic Maps

2.1 Introduction

2.1.1 Aim and objectives.

The aim of this account is to show how to do stochastic differential geometry: how one can work effectively and clearly with random processes which live on geometric objects, particularly by using semimartingale stochastic calculus. We shall do this by:

- explaining how one can carry out semimartingale stochastic calculus effectively in an intrinsic way;
- studying the interplay between particular semimartingales (such as Brownian motion on a manifold) and geometry;
- developing a probabilistic approach to the nonlinear elliptic theory of *harmonic maps*.

2.1.2 History and motivation

We begin with three motivations for stochastic differential geometry.

1. **Transition densities for geometric statistics.** Prompted by the study of datasets of directions of remanent magnetism in rock flows, and also by intellectual curiosity, Fisher [48] introduced the study of probability and

statistical inference for distributions on the sphere. In his introduction he remarked that "Any topological framework requires the development of a theory of errors of characteristic and appropriate form". This comment is an excellent early expression of one of the motivations for work in stochastic differential geometry: it can guide us in the choice of families of probability densities to be used in geometric statistical problems (see [97] for example).

Fisher's work here did not directly involve Brownian motion, but Roberts and Ursell [123] took up some probabilistic aspects of Fisher's paper, particularly concerning limiting behaviour of random walks on the sphere (hence Brownian motion on the sphere, $BM(S^2)$), and connected them with the geometric theory of harmonic spaces. (It is interesting to note in passing that the subsequent development of harmonic spaces has seen at least one substantial contribution using a probabilistic proof [111].)

2. **The stochastic approach to harmonic functions.** It has been known for a long time [21, 68] that there is an intimate and fruitful relationship between, on the one hand, probability (in particular Brownian motion) and, on the other hand, harmonic functions $f : \mathbb{M} \to \mathbb{R}$, stationary values of the Dirichlet form

$$\mathcal{E}(f) \quad = \quad \int_{\mathbb{M}} |\nabla f|^2 \, \mathrm{d}\mu \,. \tag{2.1.1}$$

In particular Brownian motion provides a intuitive and fertile interpretation of the integral representation for bounded harmonic functions using harmonic measure, delivering a stochastic mean-value formula. Many probabilists have used the sample path properties of Brownian motion to derive new and remarkable results for harmonic functions and Euclidean heat kernels. For a simple but effective example see [9, 84].

In the early 1980's Eells and Lemaire [40] asked whether one could apply probabilistic methods to harmonic maps $F : \mathbb{M} \to \mathbb{N}$ (where $\mathbb{N}$ is a curved Riemannian manifold), stationary values of the *nonlinear* Dirichlet form

$$\mathcal{E}(F) \quad = \quad \int_{\mathbb{M}} \| \operatorname{grad} F \|^2 \, \mathrm{d}\mu \,. \tag{2.1.2}$$

The difficulty here is that the relationship between Brownian motion and Eq. (2.1.1) depends crucially on the image $\mathbb{N}$ having the linear structure of the real line, which supplies the tools of the mean value formula and conditional expectation. Curvature intervenes for general $\mathbb{N}$ and destroys this relationship. For a considerable time it was quite unclear how one might establish a relationship in the nonlinear case, and particularly unclear whether such a relationship could ever deliver a way of constructing harmonic maps using stochastic methods.

3. **Intrinsic geometry of diffusions.** In several areas of applied probability one is presented with a diffusion defined by some statistical problem, and it is required to find some succinct expression of the diffusion, whether to deduce a statistical distribution theory of "characteristic errors", as in Fisher's quotation above, or whether simply to present the diffusion in a clear and conceptual way. In particular this issue arises in the *statistical theory of shape*, where one needs to find out about diffusions related to the *shape* (equivalence class under similarities) of configurations of finitely many points diffusing in a Euclidean space. The machinery of stochastic differential geometry does good service in dealing with this kind of task, particularly in combination with implementations of stochastic calculus in computer algebra packages (see for example [72, 83, 86, 97]).

Of course these are not the only reasons for studying random processes on manifolds - for example the idea of stochastic development described below is central to the very exciting recent growth of stochastic analysis for spaces of loops on manifolds. However our treatment will start from the ideas and questions described above, and we will focus in particular on extending to harmonic maps the successful relationship between probability and harmonic functions.

2.1.3 Stochastic calculus formalism

The approach to stochastic differential geometry which we describe here is based on semimartingales and Stratonovich calculus. Good introductions to the basic theory at varying levels of sophistication are to be found in [113, 124, 70].

Recall the *Doob-Meyer decomposition*: a real-valued random process with continuous sample paths is a *semimartingale* if it possesses a "signal-plus-noise" decomposition

$$X = V + M \tag{2.1.3}$$

where V is an adapted continuous process whose sample paths are of locally bounded variation (the "signal") and M is a continuous local martingale (the "noise"). The Doob-Meyer decomposition asserts that the *drift part* V and the *noise part* M of the process X are unique, if X has been identified as possessing such a decomposition. Notice that latent in the above definition is reference to an underlying filtration $\{\mathfrak{F}_t : t \in [0, \infty)\}$ (required to give meaning to the terms "adapted" and "local martingale"): we fix this filtration once and for all and assume that it supports sufficiently many Brownian motions for all our geometric purposes.

In a slight variation on conventional notation, we shall include in the family of semimartingales those random processes which admit a decomposition of the form of Eq. (2.1.3) up to a random stopping time at which they become unbounded.

We also note here that the theory of stochastic calculus extends to include random processes whose sample paths have jumps, so long as the paths remain left-continuous with right-hand limits. This extension has also been pressed into service in stochastic differential geometry [37] and indeed in the probabilistic study of harmonic maps [19, 116], but space constraints do not permit us to give this much attention here.

It is convenient to write equations involving semimartingales using the notation of *stochastic differentials*, for example:

$$\text{(2.1.4)} \qquad \mathrm{d}_I X \quad = \quad \mathrm{d}V + \mathrm{d}_I M\,.$$

The suffix "$_I$" stands for "Itô", to signal that d_I is no ordinary differential. (This is important in the context of the concise notation of differential geometry, where otherwise the symbol "d" would be overloaded to a quite unintelligible extent!) We shall see many such equations below. They are to be interpreted by integration against integrands which are bounded *predictable* functions (limits of continuous random functions which do not depend on the future). Itô [62] gives an algebraic formalism in terms of modules over bounded predictable integrands, which is of importance in understanding *symbolic Itô calculus*, the computer algebra implementation [87, 93] of stochastic calculus mentioned above.

We need to refer to a couple of operations which can be performed on stochastic differentials to yield classical differentials: using the Doob-Meyer decomposition (2.1.3) we define the *Drift* operator

$$\text{(2.1.5)} \qquad \texttt{Drift}\ \mathrm{d}_I X \quad = \quad \mathrm{d}V$$

and the *Bracket* or *quadratic covariation*

$$\text{(2.1.6)} \qquad \mathrm{d}\,[\,X_1, X_2\,] \quad = \quad \mathrm{d}\,[\,M_1, M_2\,]\,.$$

Thus `Drift`$(\mathrm{d}_I X)$ isolates the "trend" V from the Doob-Meyer decomposition $X = M + V$: if the drift vanishes then the semimartingale X is actually a local martingale. The bracket $[M, M]$ of a local martingale M can be defined as the unique continuous increasing process such that

$$M^2 - [M, M]$$

is a continuous local martingale, and $[M_1, M_2]$ can be defined by polarization. However a good intuitive appreciation of the bracket is to think of $\mathrm{d}[X, X]$ as measuring the amount of randomness in the infinitesimal increment $\mathrm{d}_I X$ of the semimartingale X.

Both operations are appropriately module-linear: for example

$$\mathrm{d}[\int f \, \mathrm{d}X_1, X_2] = \mathrm{d}[\int f \, \mathrm{d}M_1, M_2] = \int f \, \mathrm{d}[M_1, M_2] = \int f \, \mathrm{d}[X_1, X_2]$$

whenever f is a bounded predictable function.

Intrinsic formulations of stochastic calculus must take account of the second-order nature of the *Itô formula*, stochastic calculus's fundamental theorem:

$$\mathrm{d}_I f(X) = f'(X) \, \mathrm{d}_I(X) + \frac{1}{2} f''(X) \, \mathrm{d}[X, X] .$$

This holds for twice-differentiable f, though there is an important extension to the case of convex f.

It is immediate from the Itô formula that without further structure one can give no meaningful intrinsic notion of "(local) martingale" on a smooth manifold (though we *can* define a local martingale M on a vector space $\mathbb{V}$ simply by requiring M to satisfy the condition, $\ell(M)$ to be a local martingale whenever $\ell : \mathbb{V} \to \mathbb{R}$ is linear). For general manifolds Schwartz [125, 126] noted the following fundamental point: the Itô formula shows that semimartingales *are* well-defined on smooth manifolds:

Definition 2.1.1 (Semimartingale on a manifold): *A continuous random process X on a manifold $\mathbb{M}$ is a* semimartingale *if and only if its composition $f(X)$ with any C^2 function $f : \mathbb{M} \to \mathbb{R}$ is a real-valued semimartingale.*

As Itô's formula shows, Itô calculus is second-order and so does not fit well to the demands of conventional differential geometry. However *Stratonovich calculus* is first-order: if we define

$$\int Y \, \mathrm{d}_S X = \int Y \, \mathrm{d}_I X + \frac{1}{2} [Y, X]$$

for semimartingales X and Y then an application of the Itô formula shows that

$$\mathrm{d}_S f(X) \quad = \quad f'(X)\,\mathrm{d}_S X$$

whenever f is C^3. Again this can be interpreted in terms of modules, but this time using multiplication by semimartingales rather than by bounded predictable functions. As a consequence of the above formula we *can* define semimartingales intrinsically on manifolds with no particular effort, so long as we use Stratonovich differentials instead of Itô differentials, since the coefficients then transform contravariantly as they should do.

The typical procedure for use of stochastic calculus in stochastic differential geometry is therefore as follows: we define a semimartingale on a manifold using a stochastic differential equation written in terms of Stratonovich differentials. This of course relies on the existence and uniqueness theory for stochastic differential equations (*sde*): typically the coefficients are at least locally Lipschitz, so we may assume the solution is uniquely defined at least up to a so-called explosion time, when the semimartingale first exits the domain of definition of the sde.

It is worth noting that sde solutions are defined up to an event of probability zero. We may therefore choose to interpret an sde for a manifold-valued semimartingale in terms of coordinate patches, in which case we can piece together a countable number of coordinate-patch solutions without danger. However we can alternatively interpret the solution in terms of an extrinsic construction: an embedding of the manifold in high-dimensional Euclidean space, for example. Both approaches are equally valid, and define the same process up to an event of probability zero, so it is a matter of taste and convenience which one might choose in any specific case.

However, as we will see, in practice it is often unnecessary to descend to specific coordinate representations. Given a semimartingale X on a smooth manifold $\mathbb{M}$, and a smooth function $f : \mathbb{M} \to \mathbb{R}$, and some extra geometric structure, we shall in the next section see how to write down an expression for the Itô differential $\mathrm{d}_I f(X)$ in terms of invariant quantities which can be related directly to the geometry of the manifold. (In the event that it *is* necessary to work with a specific coordinate representation, much labour and time can be saved, and the risk of computational errors can be reduced, by using the computer algebra technique of symbolic Itô calculus mentioned above).

The organization of this account is as follows. In section 2.2 we describe the intrinsic approach to semimartingales on manifolds afforded by stochastic development and parallel transport. We use this to introduce a "zoo" of manifold-valued semimartingales, and a geometric version of the fundamental Itô formula. We also introduce the relationship with convexity and harmonic maps which is the recurrent theme of the article, and summarize important geometric inequalities for variations of geodesic distance. This sets the scene for an analysis in section 2.3 of the interaction between distance and semimartingales (particularly Γ-martingales and Brownian motion), which in turn allows us to derive a number of elegant probabilistic proofs of results about harmonic maps. In section 2.4 the distance-based analyses are developed into considerations of *coupling* (when can two random processes meet?) which allow significant development of the previous harmonic map results, and which finally in section 2.5 deliver fully probabilistic treatment of the small-image Dirichlet problem for harmonic maps, resolving the open question alluded to in the discussion surrounding Eq. (2.1.2).

Before moving on to the main part of the exposition we should comment on the literature of expositions and approaches to random processes (particularly semimartingales) on manifolds. The original construction of Brownian motion on a manifold was either by "patching" Itô stochastic differential equations (working from coordinate patch to coordinate patch as in [61, 51, 67]) or by using semigroup methods (it is hard to trace an original reference for this, but [123] is interesting). The use of Dirichlet forms directly to define Brownian motion on a manifold is an immediate consequence of the more general work required to construct a diffusion once one has specified a Dirichlet form, and is covered elsewhere in this volume. An attractive probabilistic development of the semigroup approach is to regard Brownian motion and more general processes as solutions to a *martingale problem*: see [129]. We have mentioned the extrinsic construction, which builds Brownian motion on a manifold by solving a conventional stochastic differential equation in the ambient Euclidean space of an (isometric) embedding of the manifold: [128] describes this for a sphere while [121, 134, 124] develop it for the general case. The approach taken below is to develop the Stratonovich approach using stochastic parallel transport and the frame bundle of the manifold to solve parallelization problems and indeed to supply a highly useful representation: this has its origins in work by Eells and Elworthy and Malliavin (see [41, 109]: see also [37, 38, 64, 119]). Finally, previous expositions and reports by the author on the subject of semimartingale stochastic differential geometry include [78, 80, 81, 92].

Before closing this introductory section, it is a pleasure to acknowledge the thoughtful and careful comments of Dr. Huiling Le of Nottingham University, who read an early draft of this article.

2.2 Stochastic differential geometry

2.2.1 Intrinsic approach to semimartingales

Given the above definition of semimartingale on a manifold, it is immediate that a rather trivial example is given by $\gamma(B)$, where $\gamma : \mathbb{R} \to \mathbb{M}$ is a geodesic on the Riemannian manifold $\mathbb{M}$ (or more generally any smooth curve on a general smooth manifold $\mathbb{M}$) and B is a real-valued Brownian motion (or indeed more generally any continuous semimartingale). This follows by applying the Itô formula to $f(\gamma)$ when $f : \mathbb{M} \to \mathbb{R}$ is a smooth function.

In order to remove non-stochastic complications, we will assume in the following that all our Riemannian manifolds are *complete*, which is to say that their geodesics can be indefinitely extended.

Although the above example is geometrically trivial, it is fundamental to our ability to do stochastic calculus in a geometrically invariant and elegant way.

Less trivial examples are given by Brownian motion $\mathrm{BM}(\mathbb{M})$ on a Riemannian manifold $\mathbb{M}$. If we choose temporarily to define such a Brownian motion as a process X satisfying the *martingale problem*:

$$f(X_t) - \frac{1}{2}\int_0^t \Delta f(X_s)ds \qquad \text{is a martingale},$$

whenever f is a smooth function of compact support and Δ is the Laplace-Beltrami operator on a Riemannian manifold $\mathbb{M}$, then it is immediate that X is a semimartingale on $\mathbb{M}$.

The reader is reminded that the conventional definition of a semimartingale has been extended above to include processes which are semimartingales up to random stopping times. This allows us to include in our class of manifold-valued semimartingales those Brownian motions which can "explode" off the manifold in finite time (Riemannian manifolds with "exploding" Brownian motions are said to be *stochastically incomplete*).

Combining these two ideas in a generalization, note that $F(X)$ is a semimartingale on $\mathbb{N}$ whenever $F : \mathbb{M} \to \mathbb{N}$ is a smooth map and X is $\mathrm{BM}(\mathbb{M})$.

2.2.2 Lifting to the frame bundle

The martingale problem approach referred to above can be generalized to treat a wide range of semimartingales on smooth manifolds in a manner which is more-or-less intrinsic. However it is more appealing to the intuition, and indeed more rewarding in further applications, to work out how to use concepts of differential geometry to "flatten out" a manifold-valued semimartingale X, viewing it as the solution to a system of (Stratonovich) stochastic differential equations driven by some flat semimartingale Z (that is to say, a semimartingale living in a reference Euclidean space $\mathbb{R}^m$ of the same dimension as the manifold).

If one wishes to do this then the first step must be to relate the differential of Z to the differential of X. *Formally* the differential of Z lives in Euclidean space $\mathbb{R}^m$ while the differential of X lives in the tangent space $T_X\mathbb{M}$ to $\mathbb{M}$ at X (though this has no proper mathematical meaning unless in the exceptional case when X is a differentiable function of time). So we need to relate the two formal differentials to each other using a linear isometry

$$\Xi : \mathbb{R}^m \quad \to \quad T_X\mathbb{M} .$$

To make proper sense of this, we should somehow ensure that Ξ is itself a semimartingale, because then we can write down the Stratonovich differential equation

$$\Xi \, \mathrm{d}_S Z \quad = \quad \mathrm{d}_S X .$$

Thus we must view Ξ itself as a manifold-valued semimartingale in the *orthonormal frame bundle*, the smooth manifold $\mathrm{O}(\mathbb{M})$ of *orthonormal frames* for $\mathbb{M}$.

In order for this definition to succeed, we must give a rule for the evolution of the semimartingale Ξ. It is clear what must be done: the only possible way to drive Ξ is to use Z or equivalently X, and therefore what is needed is a way to lift the differential of X up to the frame bundle so as to drive Ξ. Thus we need to supply a linear map

$$H_\xi : T_{\pi\xi}\mathbb{M} \quad \to \quad T_\xi \, \mathrm{O}(\mathbb{M}) ,$$

where $\pi : \mathrm{O}(\mathbb{M}) \to \mathbb{M}$ is the obvious projection map sending $\xi : \mathbb{R}^m \to T_x\mathbb{M}$ to the point $x \in \mathbb{M}$ at which its image space is based.

Clearly not every such map H will be satisfactory. We must be sure to maintain the relationship $\pi(\Xi) = X$, since otherwise the differential $\Xi \, \mathrm{d}_S Z$ will cease to be relevant for X. Differentiating, it follows that we must be sure (whatever the behaviour of the flat semimartingale Z and hence whatever $\mathrm{d}_S X$ might be) that $(\mathrm{d}\pi) H_\xi \, \mathrm{d}_S X = \mathrm{d}_S X$. Applying this to the case when Z is a smooth function of time, we find

$$(\mathrm{d}\pi) H_\xi \quad = \quad \text{identity map on } T_{\pi\xi}\mathbb{M} .$$

A further constraint is imposed by a natural requirement of equivariance under rotations of the initial frame Ξ_0 and the driving semimartingale Z. Together these two constraints exactly comprise the requirement that the map H be the *horizontal lift map* for an *orthonormal connection.* (This will be no surprise to geometers: the above arguments specialize in the smooth case to the Cartan path-based approach to a connection!).

We now impose an additional condition of *zero-torsion*, less pleasant to discuss from our stochastic point of view, (Zero-torsion essentially means that the resulting connection is *symmetric*: if used to differentiate one vector-field U by another vectorfield V then essentially the same result is obtained as if the rôles of U,V were reversed.) A well-known result of geometry is that there is one and only one such orthonormal connection H for each Riemannian manifold, the *Levi-Civita connection.* Since this map H is all that is required to relate our manifold-valued semimartingale X to a flat semimartingale Z, and since H turns out to be a smooth map, we can apply the theory of stochastic differential equations to arrive at the following:

Theorem 2.2.1 (Equations of stochastic development and parallel transport): *To each manifold-valued semimartingale X in a Riemannian manifold $\mathbb{M}$ we can associate a* stochastic anti-development *Z in $\mathbb{R}^m$ and a* stochastic parallel transport *Ξ in $O(\mathbb{M})$ such that*

$$
\begin{aligned}
\mathrm{d}_S X &= \Xi\, \mathrm{d}_S Z\,, \\
\mathrm{d}_S \Xi &= H_\Xi\, \mathrm{d}_S X\,.
\end{aligned}
\tag{2.2.1}
$$

Note furthermore that Z, Ξ are unique when the initial frame Ξ_0 is prescribed (and depend equivariantly on this choice); and on the other hand X can be defined by prescribing the initial condition Ξ_0 (and hence X_0) and the driving flat semimartingale Z.

We make three remarks here.

Remark 2.2.2 *Why* anti-*development? Because the classical notion of development is that the curve Z on the reference space is* developed *onto a curve X on the manifold.*

Remark 2.2.3 *In fact the semimartingale X is defined by the stochastic anti-development only up to its* explosion time, *the random time at which X leaves all compact subsets of $\mathbb{M}$. It is immediate from stochastic differential equation* *theory that Ξ (hence X also) is defined up to the random time at which Ξ leaves all compact subsets on $O(\mathbb{M})$; because of the linearity of the stochastic differential system in the frame directions one can show that Ξ will not explode before X.*

Remark 2.2.4 *The above construction also works for smooth manifolds which are not Riemannian, but for which a smooth connection has been defined. However one must then use the full general linear frame bundle $GL(\mathbb{M})$ rather than the frame bundle of isometries (which of course is not defined for a non-Riemannian manifold!).*

Thus the notion of stochastic development and parallel transport allow us to "flatten out" a manifold-valued semimartingale. This provides an attractive route to the definition of different kinds of semimartingales, which we will now discuss. We will see later that the use of stochastic development goes far beyond this.

2.2.3 The semimartingale zoo

It is now easy to categorize whole families of manifold-valued semimartingales: simply use the familiar flat-space categorizations on their stochastic anti-developments!

1. *Geodesic motion.* Consider the straight-line Euclidean motion $Z_t = v{\cdot}t$, for a fixed velocity vector $v \in \mathbb{R}^m$. The corresponding semimartingale $X = \gamma$ on $\mathbb{M}$ is the geodesic $\gamma : [0, \infty) \to \mathbb{M}$ with initial tangent vector $\Xi_0(v)$: the stochastic differential system for stochastic development converts to the classical Cartan system

$$\begin{aligned} \mathrm{d}\gamma/\,\mathrm{d}t &= \xi(v)\,, \\ \mathrm{d}\xi/\,\mathrm{d}t &= H_\xi\,\mathrm{d}\gamma/dt = H_\xi\xi(v)\,. \end{aligned} \tag{2.2.2}$$

Here we use lower-case symbols γ, ξ to represent *non-random* processes (which is to say, non-random functions of time).

It is convenient here to introduce notation for *left-invariant vectorfields* on $O(\mathbb{M})$:

$$L_v(\xi) = H_\xi \xi v$$

so that

$$(\mathrm{d}\pi)L_v(\xi) \quad = \quad (\mathrm{d}\pi)H_\xi \xi v \quad = \quad \xi v$$

and

$$L_v g(\xi) \cdot \quad = \quad \left[\frac{\mathrm{d}}{\mathrm{d}u} g(\xi(u))\right]_{u=0}$$

so that the Cartan system can be written

$$\begin{aligned} \mathrm{d}\gamma/\,\mathrm{d}t &= \xi(v)\,, \\ \mathrm{d}\xi/\,\mathrm{d}t &= L_v\xi\,. \end{aligned} \tag{2.2.3}$$

2. *Brownian geodesic.* In a mild generalization, we can replace the straight-line motion by a scalar Brownian motion B along a straight line, $Z_t = v \cdot B_t$, to obtain

$$\begin{aligned} \mathrm{d}_S X &= (\Xi(v))\,\mathrm{d}_S B\,, \\ \mathrm{d}_S \Xi &= H_\Xi\,\mathrm{d}_S X\,. \end{aligned} \tag{2.2.4}$$

It is easily checked that the above equations are solved by $\Xi_t = \xi_{B_t}$, $X_t = \gamma(B_t)$ with ξ, γ as given in the first example. So the stochastic system Eq. (2.2.4) for Brownian geodesics can be integrated; this is indeed a rather trivial example! However the following observation will be useful later: if $g : \mathrm{O}(\mathbb{M}) \to \mathbb{R}$ is smooth then the above Stratonovich system can be applied to $g(\Xi)$ to yield the *scalar* stochastic differential equation

$$\mathrm{d}_S g(\Xi) \quad = \quad (L_v g)\,(\Xi)\,\mathrm{d}_S B\,. \tag{2.2.5}$$

We can convert this to Itô form: using the above equation applied in a further iteration to the smooth function $L_v g$ we can expand the drift term $\frac{1}{2}\mathrm{d}\,[\,(L_v g)\,(\Xi), B\,]$ of the Itô stochastic differential equation to find

$$\mathrm{d}_I g(\Xi) \quad = \quad (L_v g)\,(\Xi)\,\mathrm{d}_I B + \frac{1}{2}\,(L_v L_v g)\,(\Xi)\,\mathrm{d}t\,.$$

This specializes when $g = f \circ \pi$ is the lift of a smooth function f on $\mathbb{M}$ to yield

$$\mathrm{d}_I f(X) \quad = \quad \langle \operatorname{grad} f(X), \Xi v\rangle \, \mathrm{d}_I B + \frac{1}{2} \operatorname{Hess} f(X)(\Xi v, \Xi v)\, \mathrm{d}t\,.$$

We note in passing that one can also define a (local) martingale geodesic, by replacing the Brownian motion B with a local martingale M. Then Eq. 2.2.5 converts to

$$\mathrm{d}_S g(\Xi) \quad = \quad (L_v g)\,(\Xi)\, \mathrm{d}_S M\,, \tag{2.2.6}$$

expanding to

$$\mathrm{d}_I g(\Xi) \quad = \quad (L_v g)\,(\Xi)\, \mathrm{d}_I M + \frac{1}{2}\,(L_v L_v g)\,(\Xi)\, \mathrm{d}\,[\,M, M\,]\,. \tag{2.2.7}$$

3. *General semimartingale.* The same idea can be applied to obtain a *geometric Itô formula* for a general semimartingale X with stochastic anti-development Z and stochastic parallel transport Ξ. Working first with a smooth $g : \mathrm{O}(\mathbb{M}) \to \mathbb{R}$, we apply the stochastic differential system for stochastic development to find a scalar Stratonovich stochastic differential equation for $g(\Xi)$:

$$\mathrm{d}_S g(\Xi) \quad = \quad \sum_i \left(L_{v_i} g(\Xi)\right)\, \mathrm{d}_S Z^i$$

where the sum is over an orthonormal basis (v_i) of the reference space $\mathbb{R}^m$, and $Z = \sum_i v_i Z^i$. Following the method described for Brownian and martingale geodesics, we can iterate this once for the smooth function $L_{v_i} g$ and so convert to Itô form

$$\mathrm{d}_I g(\Xi) \; = \; \sum_i \left(L_{v_i} g(\Xi)\right)\, \mathrm{d}_I Z^i + \frac{1}{2} \sum_i \sum_j \left(L_{v_j} L_{v_i} g(\Xi)\right)\, \mathrm{d}\left[\,Z^i, Z^j\,\right]\,.$$

When $g = f \circ \pi$ this reduces to the following concise form.

Theorem 2.2.5 (Geometric Itô formula): *Suppose that $f : \mathbb{M} \to \mathbb{R}$ is a smooth function, and that X is a semimartingale with stochastic anti-development Z and stochastic transport Ξ. Then*

$$\mathrm{d}_I f(X) \quad = \quad \langle \mathit{grad}\, f(X), \Xi\rangle\, \mathrm{d}_I Z + \frac{1}{2}\, \mathit{trace\, Hess}\, f(X)(\Xi, \Xi)\, \mathrm{d}\,[\,Z, Z\,]\,. \tag{2.2.8}$$

Here

$$\begin{aligned} \langle grad\, f(X), \Xi \rangle \, \mathrm{d}_I Z &= \sum_i \langle grad\, f(X), \Xi v_i \rangle \, \mathrm{d}_I Z^i \,, \\ trace\, Hess\, f(X)(\Xi, \Xi) \, \mathrm{d}\,[\,Z, Z\,] &= \sum_i Hess\, f(X)(\Xi v_i, \Xi v_i) \, \mathrm{d}\left[\,Z^i, Z^i\,\right] . \end{aligned}$$

where the v^i form an orthonormal basis.

Thus the apparently trivial notion of a Brownian geodesic leads directly to a fundamental formula for all manifold-valued semimartingales! Notice that there is the possibility of choosing the basis (v_i) adaptively, depending on the past of X, so as to diagonalize the stochastic differential of the matrix-valued bracket process $[\,Z, Z\,]$: we take this up below in the important special case of Γ-martingales.

Remark 2.2.6 *Notice that $\langle grad\, f(X), \Xi \rangle$ defines a random time-varying (indeed, predictable) linear map from $\mathbb{R}^m$ to $\mathbb{R}$, where m is the dimension of the manifold $\mathbb{M}$. Thus the Itô differential $\langle grad\, f(X), \Xi \rangle \, \mathrm{d}_I Z$ makes intrinsic sense.*

4. *Brownian motion.* In the case when the stochastic anti-development is a Euclidean Brownian motion B on $\mathbb{R}^m$ we **define** the resulting X to be Brownian motion BM($\mathbb{M}$) on the manifold $\mathbb{M}$. The geometric Itô formula now tells us that this agrees with our previous provisional definition using the martingale problem:

$$\begin{aligned} \mathrm{d}_I f(X) &= \langle \mathrm{grad}\, f(X), \Xi \rangle \, \mathrm{d}_I B + \frac{1}{2} \, \mathrm{trace\, Hess}\, f(X)(\Xi, \Xi) \, \mathrm{d}\,[\,B, B\,] \\ &= \langle \mathrm{grad}\, f(X), \Xi \rangle \, \mathrm{d}_I B + \frac{1}{2} \sum_i \mathrm{Hess}\, f(X)(\Xi v_i, \Xi v_i) \, \mathrm{d}t \\ &= \langle \mathrm{grad}\, f(X), \Xi \rangle \, \mathrm{d}_I B + \frac{1}{2} \sum_i \left(L_{v_i} L_{v_i} f\right)(X) \, \mathrm{d}t \\ &= \langle \mathrm{grad}\, f(X), \Xi \rangle \, \mathrm{d}_I B + \frac{1}{2} \Delta f(X) \, \mathrm{d}t \,, \end{aligned} \tag{2.2.9}$$

using the fact that different coordinates of a Euclidean Brownian motion have zero bracket with each other (a consequence of these coordinates being independent), and the formula for the Laplace-Beltrami operator as the sum of second derivatives along an orthonormal set of geodesics.

Note that there is an issue here concerning whether or not the equations of stochastic development and parallel transport will define BM($\mathbb{M}$) for all time. It is possible for BM($\mathbb{M}$) to explode in finite (random) time, though simple arguments using comparison results described below show that this cannot happen unless the manifold $\mathbb{M}$ has rapidly growing unbounded negative curvatures. If such an explosion has zero probability of occurring then we say that $\mathbb{M}$ is *stochastically complete*. The corresponding issue for geodesics is of course exactly the question of whether the manifold is *complete* (for emphasis we would say, *geodesically complete*). A criterion for stochastic completeness using this approach is given in [56], which also shows the same criterion excludes *implosion*, which is to say the possibility that the Brownian motion X can be defined over the time interval $(0,T)$ for some positive T but cannot be extended back to time zero because $\lim_{t\downarrow 0} X_t$ diverges to infinity on the manifold. (Implosion is equivalent to the Feller property [95].)

In the case of Brownian motion we can obtain sharper results on explosion and implosion using Dirichlet forms rather than stochastic calculus; however the stochastic calculus approach comes into its own when dealing with more general semimartingales such as the Γ-martingales discussed below.

5. *Γ-martingale.* If the stochastic anti-development is a local martingale $Z = M$ then we *define* the resulting semimartingale to be a *Γ-martingale.* (Note that Brownian motion is the special case when M is a Euclidean Brownian motion.) In that case the geometric Itô formula takes the form

(2.2.10)
$$\mathrm{d}_I f(X) \;=\; \langle \operatorname{grad} f(X), \Xi \rangle \, \mathrm{d}_I M + \frac{1}{2} \operatorname{trace} \operatorname{Hess} f(X)(\Xi,\Xi) \, \mathrm{d}\,[\,M, M\,]\,.$$

The "Γ" in Γ-martingale refers to the traditional notation (using Christoffel symbols) for a connection. Γ-martingales actually depend only on the choice of connection.

Eq. (2.2.10) now corresponds to the Doob-Meyer decomposition, since

$$\langle \operatorname{grad} f(X), \Xi \rangle \, \mathrm{d}_I M$$

is the stochastic differential of a local martingale. Therefore, in the Γ-martingale case, if (v_i) is an orthonormal basis for the reference Euclidean space $\mathbb{R}^m$ and if $M = \sum_i v_i M^i$ then

$$\texttt{Drift}\ \mathrm{d}_I f(X) = \frac{1}{2} \sum_i \sum_j \operatorname{Hess} f(X)(\Xi v_i, \Xi v_j)\, \mathrm{d}\left[M^i, M^j \right] \tag{2.2.11}$$

and as noted above in the discussion on semimartingales we can diagonalize here. Indeed, define the *intrinsic time* τ of the Γ-martingale X in terms of its stochastic anti-development by

$$\mathrm{d}\tau = \sum_i \mathrm{d}\left[M^i, M^i \right] \tag{2.2.12}$$

Then we can choose the orthonormal basis (v_i) adaptively so as to diagonalize (a predictable version of) the density

$$A^{ij} = \frac{\mathrm{d}\left[M^i, M^j \right]}{\mathrm{d}\tau} \tag{2.2.13}$$

for $\mathrm{d}\tau$-almost all time. (This uses the fact that one can define a diagonalizing basis for a symmetric matrix which depends measurably on the matrix.)

Recall that a function $f : \mathbb{M} \to \mathbb{R}$ is said to be *convex* if it is convex when composed with geodesics. From the above diagonalization, and from the fact that $\operatorname{Hess} f(\xi v, \xi v) = \mathrm{d}^2 f(\gamma(t)) / \mathrm{d}t^2$ is nonnegative for smooth convex functions, we deduce that if f is a smooth convex function and X is a Γ-martingale then $\texttt{Drift}\ \mathrm{d}_I f(X) \geq 0$, so a convex function of a Γ-martingale is a submartingale. Note that often a manifold supports *no* nonconstant convex functions, for example if it is compact: however the above relationship can be localized for convex functions defined only within given regions of a manifold, so long as the Γ-martingale is stopped when first it exits the region of definition.

In fact this relationship with convexity characterizes Γ-martingales [26], and extends to non-smooth convex functions [46].

Once again there are issues of explosion (which is to say, *martingale-completeness*) and implosion, for which results can be found in [43, 95].

The interplay between Γ-martingales and convexity is in fact at the heart of this essay, since it combines with the remark (established by considering the Euler-Lagrange equations of extremality for the energy form (2.1.2)) that smooth harmonic maps $F : \mathbb{M} \to \mathbb{N}$ are characterized by the property (again interpreted locally!) that $\phi(F)$ is subharmonic whenever the smooth real-valued function ϕ defined on $\mathbb{N}$ is convex. As a consequence we can verify an observation due to Bismut in about 1981, which is fundamental to the relationship between harmonic maps and probability:

Theorem 2.2.7 (Γ-martingales and harmonic maps): *The smooth map*

$$F : \mathbb{M} \to \mathbb{N}$$

is a harmonic map if and only if

$$F(BM(\mathbb{M})) \text{ is a } \Gamma\text{-martingale}.$$

Note that the natural question, what kind of map carries Brownian motion over into a time-changed *Brownian motion* (not just a Γ-martingale) has also been answered: the map must be a *harmonic morphism* (pull back harmonic functions to harmonic functions). This is equivalent to the requirement that it be a harmonic map which is also horizontally semi-conformal [10], a rather rigid requirement except in dimension 2, where it reduces to the holomorphic or anti-holomorphic property. (But note that nevertheless there has been some interesting work on probabilistic approaches to such maps: see [35, 36].)

We shall have occasion later to use a natural generalization to the case of non-smooth harmonic maps:

Definition 2.2.8 (Finely harmonic maps): *Suppose a map $F : \mathbb{M} \to \mathbb{N}$ is such that $F(BM(\mathbb{M}))$ is a Γ-martingale for quasi-every starting point of the Brownian motion $BM(\mathbb{M})$. Then we say F is a* finely harmonic map.

As we shall see, there is a beautiful three-way interaction between harmonic maps, convexity, and Γ-martingales which will form the theme of this essay.

Aided by the notion of stochastic development, this will allow us to "flatten out" the nonlinear harmonic map theory, and to tie it closely to the linear potential theory of Brownian motion.

6. *Γ-martingales of bounded quasi-conformality and dilatation.* We continue discussion of the Γ-martingale X and its stochastic anti-development M. The matrix $A = \mathrm{d}[M, M] / \mathrm{d}\tau$ has eigenvalues

$$0 \leq \lambda_m \leq \lambda_{m-1} \leq \cdots \leq \lambda_2 \leq \lambda_1$$

and the extent to which these differ measures the extent to which X fails to be a (possibly time-changed) Brownian motion. Two summary measures turn out to be of importance to us:

Definition 2.2.9 (Quasi-conformal Γ-martingales): *We say X is* K^2-quasi-conformal *if*

$$\lambda_1 \quad \leq \quad K^2 \lambda_m .$$

Definition 2.2.10 (Bounded-dilatation Γ-martingales): *We say X is of* K^2-bounded dilatation *if*

$$\lambda_1 \quad \leq \quad K^2 \lambda_2 .$$

Note that a harmonic map F of quasi-conformality or bounded dilatation carries these properties over to the associated Γ-martingale $F(\mathrm{BM}(\mathbb{M}))$.

A Γ-martingale of bounded dilatation has "randomness of at least two substantial degrees of freedom" when viewed in its intrinsic time-scale $\mathrm{d}\tau$, while a quasi-conformal Γ-martingale has "substantial randomness in every direction"when viewed in its intrinsic time-scale. (Reference to this intrinsic time-scale is important: working in the original time-scale dt we may see either kind of Γ-martingale coming to a complete stop.)

It is intuitively clear, and we shall soon substantiate this, that a Γ-martingale of bounded quasi-conformality is very similar to a Brownian motion. However care must be exercised: there *are* differences. For example, if $K^2 > m - 1$ then a K^2-quasi-conformal Γ-martingale can hit prescribed points at positive time, whereas (if $m > 2$) Brownian motion has probability zero of so doing. (This example is due to Krylov [100, §1.3].)

7. *Other possibilities.* Clearly the above possibilities do not exhaust the inventory of the semimartingale zoo. For example, if the manifold $\mathbb{M}$ in question carries a complex structure then we can define *holomorphic* and *anti-holomorphic martingales*: semimartingales on $\mathbb{M}$ which are mapped into time changed complex Brownian motion by holomorphic maps from $\mathbb{M}$ into $\mathbb{C}$. In fact holomorphic martingales are less closely tied to connections than the inhabitants of the semimartingale zoo discussed previously; it is more natural to define them in a sheaf-theoretic way (holomorphic and anti-holomorphic martingales are precisely those semimartingales which are carried into submartingales by the sheaf of plurisubharmonic functions). There has been a considerable amount of work on the resulting relationship between holomorphic martingales and several-complex-variable theory [50, 69, 125], also [60, 2nd edition]; however this lies largely outside the scope of our essay (though we shall note some important points of overlap).

The issue of sheaf-theoretic definitions propagates back to the consideration of Γ-martingales: as we noted above, Darling has characterized Γ-martingales as those semimartingales which are carried into submartingales by the sheaf of convex functions. This thought can be further developed by asking the question, suppose we reverse the stochastic development idea and ask what connections are compatible with a specific class of semimartingales forming Γ-martingales? Emery [44] has addressed this question: two connections have the same Γ-martingales if they give the same connection when symmetrized. It follows from this that much of the theory of Γ-martingales can be very naturally developed using convexity of functions as a primitive notion, as indeed is done in [44]: however the stochastic development approach comes into its own when one wishes to study and to exploit relationships between different Γ-martingales defined on the same probability space and manifold, as we will do below.

2.2.4 Intrinsic geometry

As noted for example in [60], a smooth elliptic semimartingale diffusion X leads to a natural Riemannian geometry on its state space. This is easily seen from the stochastic development construction: starting with an arbitrarily chosen connection we can develop X to deliver a semimartingale $\tilde{Z}$ on the reference space $\mathbb{R}^m$ and the corresponding stochastic parallel transport $\tilde{\Xi}$. This leads (as in our discussion of Γ-martingales) to a random symmetric matrix process (defined as a matrix-valued Radon-Nikodym derivative)

$$\tilde{A} \quad = \quad \frac{\mathrm{d}\left[\tilde{Z},\tilde{Z}\right]}{\operatorname{trace}\mathrm{d}\left[\tilde{Z},\tilde{Z}\right]}$$

and by the Markov property of the diffusion and its regularity we can realize $\tilde{A}$ in terms of a smooth quadratic form Q on the tangent bundle $T(\mathbb{M})$ *via*

$$\tilde{A}(v,v) \quad = \quad \langle \tilde{\Xi} v, Q(X)\tilde{\Xi} v\rangle\,.$$

But now this quadratic form Q yields a new Riemannian metric on $\mathbb{M}$ which is intrinsic to the diffusion. Moreover if we rework the stochastic development procedure using the new intrinsic Riemannian metric and its associated Levi-Civita connection then the resulting stochastic anti-development Z will be Euclidean Brownian motion B plus a superimposed drift $U\,\mathrm{d}t$:

$$\mathrm{d}_I Z \quad = \quad \mathrm{d}_I B + U\,\mathrm{d}t$$

and this drift can be related (again using the Markov property of the diffusion) to an *intrinsic drift vectorfield* χ on $\mathbb{M}$ by

$$\chi(X) \quad = \quad \Xi U$$

where Ξ is the parallel transport of X with respect to the new metric.

Thus each semimartingale which is a smooth elliptic diffusion can be characterized in distribution by an intrinsic Riemannian metric and an intrinsic drift vectorfield.

Note that a much deeper generalization of this phenomenon (intrinsic geometry of diffusion) can be obtained for processes defined by rather general Dirichlet forms: see Sturm's contribution to this volume.

2.2.5 Diffusions and statistical shape

As mentioned in the introduction, a rather non-trivial example of the intrinsic geometry of diffusions arises when one considers the *statistical shape* of a configuration of m points in d-dimensional Euclidean space, where the points are moving with independent Euclidean Brownian motions. The shape here is the equivalence class of a configuration under changes of scale, orientation, and location; and it forms a Riemannian manifold away from a singularity set. The shape diffuses on the resulting *shape-manifold* as a semimartingale. In case $m = 3$, $d = 2$ the semimartingale can be viewed as a random time-change of $\mathrm{BM}(S^2)$, Brownian motion on the sphere [72]; more generally the semimartingale is *not* a time-change of Brownian motion [83, 86].

An alternative shape geometry, at least for three points in the plane, arises if we require the three points to move not independently but as if they are being transformed according to Brownian motion on the special linear group. This is more appropriate in biological applications of the statistics of shape. The resulting geometry is that of the hyperbolic plane of curvature -2 [97]: this will be of no surprise to people who know symmetric space theory! As previously noted, most of these shape theory results were obtained using a mixture of stochastic differential geometry and computer algebra, more specifically symbolic Itô calculus.

2.2.6 Geodesic variational formulae

The formula (2.2.11) for the drift of a function $f(X)$ of a Γ-martingale X makes it plain that control of $f(X)$ depends on control of the second derivative of f along geodesics, since

$$\operatorname{Hess} f(X)(\Xi v, \Xi v) \quad = \quad L_v L_v f \circ \pi(\Xi) \quad = \quad \left[\frac{\mathrm{d}^2}{\mathrm{d}u^2} f(\gamma(u))\right]_{u=0}$$

for a unit-speed geodesic γ started at X with initial vector Ξv (note that $|v| = 1$ as the geodesic γ is unit-speed). In stochastic differential geometry we typically require to control such second derivatives when f is the distance from something: a point, a geodesic, or another Γ-martingale. We do this using a number of important and classical curvature-related bounds which we summarize for future use below.

Before doing this we note a technical difficulty. Distance functions on manifolds are not everywhere smooth! For example, the distance from a fixed point

$\mathbf{o}$ is certainly not smooth at $\mathbf{o}$ itself, though this usually does not matter (either the Γ-martingale is Brownian motion, which will never hit $\mathbf{o}$, or in any case the required control need be exercised only away from $\mathbf{o}$). But also the distance function is not smooth at the *cut-locus* $C(\mathbf{o})$, the locus where geodesics emanating from $\mathbf{o}$ fail to be length-minimizing. If the underlying manifold $\mathbb{M}$ is simply-connected and of non-positive curvature (a *Cartan-Hadamard manifold*) then the cut-locus is empty, but in general it cannot be ignored.

2.2.7 Distance from a fixed point

Suppose then that we are considering $\ell(u) = \operatorname{dist}(\gamma(u), \mathbf{o})$, the distance from $\mathbf{o}$ of a unit-speed geodesic $\gamma(u)$. Suppose further that $\gamma(u)$ does not meet $C(\mathbf{o})$, so the technical difficulty mentioned above does not arise. Set $\ell(0) = \ell$. If the sectional curvatures of the underlying manifold $\mathbb{M}$ are pinched between k_- and k_+,

$$k_- \quad \leq \quad \text{sectional curvatures} \quad \leq \quad k_+ \,,$$

and if $\ell\sqrt{k_+} < \pi$ in the case when $k_+ > 0$, then we have the following bounds on the second derivative of ℓ:

$$\left[\frac{\mathrm{d}^2}{\mathrm{d}u^2}\ell(u)\right]_{u=0} \quad \begin{matrix}\geq\\ \leq\end{matrix} \quad |(\gamma')^\perp|^2 \,\frac{C_{k_\pm}(\ell)}{S_{k_\pm}(\ell)} \tag{2.2.14}$$

where we take $\geq$ for k_+, $\leq$ for k_-, set $(\gamma')^\perp$ to be the component of γ' normal to the radial geodesic from $\mathbf{o}$, and set

$$\begin{aligned}
&\left.\begin{aligned} S_k(\ell) &= \frac{\sin(\sqrt{k}\ell)}{\sqrt{k}} \\ C_k(\ell) &= \cos(\sqrt{k}\ell) \end{aligned}\right\} \quad \text{if } k > 0\,;\\
&\left.\begin{aligned} S_k(\ell) &= \ell \\ C_k(\ell) &= 1 \end{aligned}\right\} \quad \text{if } k = 0\,;\\
&\left.\begin{aligned} S_k(\ell) &= \frac{\sinh(\sqrt{-k}\ell)}{\sqrt{-k}} \\ C_k(\ell) &= \cosh(\sqrt{-k}\ell) \end{aligned}\right\} \quad \text{if } k < 0\,.
\end{aligned} \tag{2.2.15}$$

The proof is basically a variational argument using the Index lemma for Jacobi vectorfields, combined with a simple comparison argument for a second-order differential equation. For a modern treatment of these ideas in textbook form see [66]. Also note that the comparison result given above can be obtained as a simple consequence of curvature bounded by k in the sense of Alexandrov, as described in Jost's contribution to this volume.

There is a variation on the above result which will be important: if in addition $k_+ > 0$ then

$$\left[\frac{\mathrm{d}^2}{\mathrm{d}u^2}\frac{1}{k_+}\left(1 - C_{k_+}(\ell(u))\right)\right]_{u=0} \quad \geq \quad C_{k_+}(\ell)\,. \tag{2.2.16}$$

(In case $k_+ = 0$ then a valid variant replaces $\frac{1}{k_+}\left(1 - C_{k_+}(\ell(u))\right)$ by $\ell(u)^2/2$.)

Remark 2.2.11 *The inequality fails for $k_+ < 0$ simply because then we have $|C_{k_+}| > 1$.*

2.2.8 Distance between two moving points (I)

With curvature bounds as above, and in particular supposing that if $k_+ > 0$ then $\ell\sqrt{k_+} < \pi$, let $\ell(u)$ now be the distance between two different geodesics $\gamma_0(u)$ and $\gamma_1(u)$. Suppose that $\gamma_0(u)$ avoids the cut-locus of $\gamma_1(u)$ (this implies that $\gamma_1(u)$ avoids the cut-locus of $\gamma_0(u)$ as well). We find

$$\begin{aligned}\left[\frac{\mathrm{d}^2}{\mathrm{d}u^2}\ell(u)\right]_{u=0} \quad &\begin{matrix}\geq\\ \leq\end{matrix} \\ &\left[|(\gamma_0')^\perp|^2 - 2\frac{\langle\xi_0^{-1}(\gamma_0')^\perp, \xi_1^{-1}(\gamma_1')^\perp\rangle}{C_{k_\pm}(\ell)} + |(\gamma_1')^\perp|^2\right]\frac{C_{k_\pm}(\ell)}{S_{k_\pm}(\ell)}\,.\end{aligned} \tag{2.2.17}$$

Here ξ is the parallel transport for the geodesic γ, so that the inner product $\langle(\xi_0^{-1}\gamma_0')^\perp, \xi_1^{-1}(\gamma_1')^\perp\rangle$ makes sense.

Once again, we take $\geq$ for k_+, $\leq$ for k_-.

Convenient bounds for applications follow by completing the square ....

The proof is essentially as for the previous bound.

2.2.9 Distance between two moving points (II)

Suppose now we are interesting in bounding the distance between two *Brownian motions*. It turns out that we need to bound a quantity similar to that given in the previous subsection, but *averaged* over two sets of orthonormal geodesics,

$\gamma_0^{(i)} : i = 1, \ldots, m-1$ intersecting at a shared point $b = \gamma_0^{(i)}(0)$, and $\gamma_1^{(i)} : i = 1, \ldots, m-1$ intersecting at a shared point $a = \gamma_1^{(i)}(0)$, such that all geodesics are perpendicular to the distance minimizing geodesic between a and b and such that parallel transport along the intervening geodesic carries $\gamma_0^{(i)}$ to $\gamma_1^{(i)}$. Suppose the curvature condition is replaced by

$$(m-1)k_- \quad \leq \quad \text{Ricci curvatures of } \mathbb{M}$$

(recall the Ricci curvature is the sum of curvatures for $m-1$ planes sharing a common vector and otherwise defined by a perpendicular set of $m-1$ vectors). For $k_- > 0$, suppose additionally that $\ell\sqrt{k_-} < \pi$. Let $\ell^{(i)}(u)$ be the distance between $\gamma_0^{(i)}(u)$ and $\gamma_1^{(i)}(u)$. Then

$$\sum_{i=1}^{m-1} \left[\frac{\mathrm{d}^2}{\mathrm{d}u^2} \ell^{(i)}(u) \right]_{u=0} \quad \leq \quad -2(m-1)k_- \frac{S_{k_-}(\ell/2)}{C_{k_-}(\ell/2)} . \tag{2.2.18}$$

The proof follows the same lines as before, except that a further argument is required exploiting the fact that in the case of constant curvature Jacobi vector-fields which are perpendicular at both end-points of a geodesic are perpendicular across the whole intervening range.

A similar result holds for the sum of second geodesic derivatives of radial distance from a single point, and results in upper bounds on $\operatorname{dist}(\mathrm{BM}, \mathbf{o})$ using lower Ricci curvature bounds.

There are further variations on these comparison results, for example distance from a geodesic [75, 57] or more generally a totally geodesic submanifold; also distance in a manifold with varying radial curvature bounds depending on distance from a fixed reference point $\mathbf{o}$ [54]. However the above bounds suffice for most purposes of stochastic differential geometry (and indeed arguments both for the more general bounds and for their applications are largely technical variations on what is given here.)

2.3 Distance and semimartingales

Distance is a primitive notion in Riemannian geometry. Therefore a natural starting point for stochastic differential geometry is to conduct an Itô analysis of the distance process $\operatorname{dist}(X, \mathbf{o})$, for X a semimartingale (typically a Γ-martingale or Brownian motion), and $\mathbf{o}$ a fixed reference point. Variations include: distance from a totally geodesic hyperplane, distance from another (independent or coupled) Brownian motion or Γ-martingale.

2.3.1 Distance of a Γ-martingale from a point

The simplest case, which also carries all the main ideas, is that of the Itô analysis of the distance of a Γ-martingale from a reference point $\mathbf{o}$. Consider a Γ-martingale X on a manifold $\mathbb{M}$ all of whose sectional curvatures are bounded above by $-H^2 < 0$ (but which is not necessarily simply connected). Let M be the stochastic anti-development, and let Ξ be the parallel transport, for X. Set

$$\begin{aligned} r(x) &= \operatorname{dist}(x, \mathbf{o}) \\ R &= r(X) \end{aligned} \tag{2.3.1}$$

and use the equations of stochastic development and the bound Eq. (2.2.14) on the second derivative of geodesic arc-length to compute the following comparison, valid up to the first time that X hits the cut-locus of the reference point $\mathbf{o}$:

$$\begin{aligned} \mathrm{d}_I R &= \langle \operatorname{grad} r(X), \Xi \rangle \, \mathrm{d}_I M + \frac{1}{2} \sum_i \operatorname{Hess} r(X)(\Xi v_i, \Xi v_i) \lambda_i \, \mathrm{d}\tau \\ &\geq \langle \operatorname{grad} r(X), \Xi \rangle \, \mathrm{d}_I M + \frac{H}{2} \coth(HR) \left(\sum_i |v_i^{\perp}|^2 \lambda_i \right) \mathrm{d}\tau \end{aligned} \tag{2.3.2}$$

where (v_i) is an adapted basis which diagonalizes the Hessian quadratic form $\operatorname{Hess} r(X)(\Xi, \Xi)$, the (λ_i) are the associated eigenvalues, and $\Xi v^{\perp}$ is the component of Ξv which is normal to $\operatorname{grad} r(X)$, and the inequality on stochastic differentials is interpreted as meaning that the difference of left-hand side minus right-hand side is a *classical* differential of a non-negative random measure. The above stochastic differential inequality is valid for general semimartingales but is not then so useful, as in the general case the Itô differential term will contribute a further drift term.

In the Γ-martingale case we now change time-scale according to the quadratic variation of R:

$$\mathrm{d}\sigma = \sum_i |\langle \operatorname{grad} r(X), \Xi v_i \rangle|^2 \lambda_i \, \mathrm{d}\tau = \sum_i |v_i^{\parallel}|^2 \lambda_i \, \mathrm{d}\tau$$

where $v^{\parallel}$ is defined by requiring $\Xi v^{\parallel}$ to be the component of Ξv which is *parallel* to $\operatorname{grad} r$. Now a continuous martingale in the time-scale of its quadratic variation differential $d\sigma$ is a standard Brownian motion (stopped at a random time if $d\sigma$ integrates up to a finite limit). Therefore the stochastic differential inequality can be reformulated in the $d\sigma$ time-scale as

$$d_I R \quad \geq \quad d_I W + \frac{H}{2}\coth(HR)\left(\frac{\sum_i |v_i^\perp|^2\lambda_i}{\sum_i |v_i^{\|}|^2\lambda_i}\right) d\sigma \tag{2.3.3}$$

where W is a standard Brownian motion in the $d\sigma$ time-scale.

Thus a negative curvature upper bound has a tendency, bounded below in drift by a term with coefficient $\frac{H}{2}\coth(HR)$, to push the Γ-martingale away from the reference point. If in addition we can bound from below the expression

$$\left(\frac{\sum_i |v_i^\perp|^2\lambda_i}{\sum_i |v_i^{\|}|^2\lambda_i}\right) \tag{2.3.4}$$

(note that this depends implicitly on $\operatorname{grad} r$ and X through the adapted basis (v_i)) then we can use the theory of standard Brownian motion with constant drift to deliver lower bounds on the behaviour of $R = r(X)$. Such bounds on the expression (2.3.4) are automatic in the case of Brownian motion (for then any basis (v_i) will serve and the expression equals $m-1$): for general Γ-martingales we need a condition such as quasiconformality or bounded dilatation of the Γ-martingale.

It is straightforward to modify this for cases when the upper curvature bound is non-negative, or when there is a lower curvature bound (giving rise to an upper bound on the radial process). We leave such considerations as an exercise for the reader, though we shall shortly have occasion to use the instance of a lower curvature bound.

Notice that the above analysis holds only till the first time that X hits the cut-locus of $\mathbf{o}$. In the case of a Cartan-Hadamard manifold (simply-connected, non-positive curvature) there is no cut-locus and so this is not an issue. In other cases it presents a technical obstacle for the application of stochastic calculus to differential geometry: we shall discuss its solution later.

2.3.2 Probabilistic approaches to Picard theorems

The main question to be faced now is, how can we use such bounds as are given above to establish links between Brownian motion and harmonic maps? To make a start on this we review Burgess Davies' delightful Brownian motion proof of the Picard little theorem from complex analysis [32].

Suppose $f : \mathbb{C} \to \mathbb{C} \setminus \{\pm 1\}$ is a holomorphic map. The Picard little theorem tells us f must be constant. On the other hand Lévy's celebrated theorem tells us that $f(\mathrm{BM}(\mathbb{C}))$ is a time-changed complex Brownian motion $\mathrm{BM}(\mathbb{C})$. Burgess Davies showed that if we consider $\mathrm{BM}(\mathbb{C}\setminus\{\pm 1\})$, and connect end-point to start-point with a straight-line segment, then with probability one for all sufficiently large times the resulting loop will be topologically entangled in $\mathbb{C} \setminus \{\pm 1\}$ (of course one has to fix-up the construction of the loop when the segment crosses one of the excluded points $\{\pm 1\}$).

Now suppose that f is not constant. Burgess Davies points out that if we consider $f(Z)$ for Z a $\mathrm{BM}(\mathbb{C})$ then by Lévy's theorem we know that $f(Z)$ must be a $\mathrm{BM}(\mathbb{C})$ up to a random time-change. Omitting some technical details, this means we can apply the entangling result and the recurrence of the original $\mathrm{BM}(\mathbb{C})$ to create a null-homotopic loop in $\mathbb{C}$ which is mapped by f into a homotopically non-trivial loop. This is a contradiction, and we are forced to conclude that f is constant.

How can this be generalized to the case of harmonic maps? We should not expect to exploit Brownian entanglement results, as we will want to work in higher dimensional situations. This generalization has already been addressed by differential geometers working in harmonic map theory: for example [52] observes that by lifting to universal covers we can resolve the Picard little theorem into a Liouville theorem for holomorphic maps into the complex unit disk and thus one can formulate a generalization. Since the complex unit disk carries a natural metric of constant negative curvature, the generalization of Picard theorem for harmonic maps should run like this: maps F in the following are always constant!

F	:	$\mathbb{M}$	$\to$	$\mathbb{N}$
harmonic map and further conditions		*suitable conditions (almost Euclidean)*		*simply connected, enough negative curvature*

In [52] and elsewhere geometric methods (volume and distance comparison) have been used to supply Picard little theorems (for example, with a bounded dilatation condition of F, nonnegative Ricci curvature for $\mathbb{M}$, and curvature bounded above away from zero for $\mathbb{N}$). The question of how a probabilistic approach might be made now becomes clearer: one should try to prove results for suitable Γ-martingales on $\mathbb{N}$ which are incompatible with properties valid for Brownian motion on $\mathbb{M}$ unless there are no non-constant F of the required form.

2.3.3 Γ-martingale convergence theorem

The first step on this road (though not the first step chronologically!) is to try to prove a Liouville theorem for harmonic maps using Brownian motion. But to do this we need to know more about Γ-martingales. Now the fundamental result for ordinary (that is, real-valued!) martingales is the martingale convergence theorem: in a simple form, bounded martingales always converge at time infinity. It was natural to attempt to generalize this result to Γ-martingales and this was done at the very earliest stages of the theory: see Darling [27, 28] and Zheng [138]. They established the following result:

Theorem 2.3.1 (Darling-Zheng Γ-martingale convergence theorem): *If X is a Γ-martingale on a manifold $\mathbb{M}$ with intrinsic time τ then*

1. *(Darling): if $\tau(\infty) < \infty$ then X converges in $\mathbb{M} \cup \{\infty\}$, the one-point compactification of $\mathbb{M}$;*
2. *(Zheng): if X converges in $\mathbb{M}$ then $\tau(\infty) < \infty$.*

We shall prove only the first part, which is all we will need below.

Proof: (part (i) only.) A *localization* argument (stopping X at the exit times from progressively larger geodesic balls) shows we may suppose $\tau(\infty) \leq T < \infty$ for some fixed T.

We can cover $\mathbb{M}$ by geodesic balls in a locally finite cover, such that a ball centre x radius ε_x supports a uniformly strictly convex function

$$\phi_x(y) = \frac{1}{k_x}\left(1 - \cos(\sqrt{k_x}\,\mathrm{dist}(x, y))\right)$$

We choose k_x to be a nonnegative upper bound on curvature in the ball: this together with Eq. (2.2.16) means that ϕ_x is strictly convex as long as the ball is sufficiently small.

The strict convexity of ϕ_x in the corresponding ball means that $\phi_x(X)$ behaves as a submartingale in the ball. Together with the finiteness of the intrinsic time $\tau(\infty)$, this allows us to show there can only be a finite number of downcrossings of X from $\phi_x^{-1}(\varepsilon_x)$ to $\phi_x^{-1}(\varepsilon_x/2)$, using analysis of $\mathrm{d}_I\phi_x(X)$ (the drift being controlled by the quantification of the strict convexity of ϕ_x given by Eq. (2.2.16)) and the finiteness of T.

Now a *topological* downcrossings argument completes the proof: if X fails to converge in $\mathbb{M}$ then it must eventually leave any compact subset (since it cannot oscillate between ϕ_x levels too much by the above), and this exactly says it must then converge to ∞ in the one-point compactification. □

2.3.4 An "Ultimate" Liouville Theorem

We get an immediate pay-off from the Γ-martingale convergence theorem: a probabilistically proved Liouville theorem which, when it was proved, superseded most of the then-known Liouville theorems, and which has a very pleasant "categorical" appearance. To state this we need the notion of a geodesic ball which is not too big:

Definition 2.3.2 (Regular geodesic ball): *A* regular geodesic ball $\mathcal{B}$ *is a geodesic ball (centre* **o**, *radius* r*) such that*

(1) $\mathcal{B}$ *does not intersect the cut-locus of the centre* **o**;

(2) the sectional curvatures in $\mathcal{B}$ *are all smaller than* $\left(\frac{\pi}{2r}\right)^2$.

A good way to visualize this is to think of regular geodesic balls as "small hemispheres" (compact subsets of open hemispheres). Essentially all the geometric difficulties are carried by this specific example.

Theorem 2.3.3 ("Ultimate" Liouville Theorem *[59, 81], see also [85]***):** *Consider the functions* f *and* F *in the following diagram:*

$$\begin{array}{ccccc} & & F & \\ & \mathbb{M} & \longrightarrow & \mathcal{B} \\ \text{(harmonic) } f & \downarrow & \text{harmonic} & \text{regular} \\ & [0,1] & \text{map} & \text{geodesic ball} \end{array} \tag{2.3.5}$$

If all such harmonic functions f *are constant, then all such harmonic maps* F *are also constant.*

Proof: Consider X, a Brownian motion $\mathrm{BM}(\mathbb{M})$ started at $x \in \mathbb{M}$ say, and let $Y = F(X)$. If k_+ is the non-negative strict upper bound for the curvatures of the regular geodesic ball $\mathcal{B}$ (as given in Definition 2.3.2 (2)), then we can use Eq. (2.2.16) to show that

$$\phi(y) \quad = \quad \frac{1}{k_+}\left(1 - \cos(\sqrt{k_+}\,\mathrm{dist}(\mathbf{o}, y))\right)$$

(for $y \in \mathcal{B}$, and taking $\mathbf{o}$ to be the centre of the regular geodesic ball $\mathcal{B}$) supplies a strictly convex function which delivers a finite upper bound on the expected total intrinsic time $\mathbb{E}[\tau(\infty)]$. This allows us to apply the first part of the Darling-Zheng Γ-martingale convergence theorem, since now we know that $\tau(\infty) < \infty$ with probability one. So the limit $Y(\infty)$ exists in $\mathcal{B}$ (no need to add the compactification $\{\infty\}$ since $\mathcal{B}$ is itself compact).

The hypothesis on $\mathbb{M}$ is that bounded harmonic functions are all constant, and of course this means that $Y(\infty)$ must be non-random: we set

$$y \quad = \quad Y(\infty) \quad = \quad \lim_{t\to\infty} F(X(t)) \,.$$

(Otherwise we could use $\mathbb{P}[Y(\infty) \in A | X(0) = x]$ to generate non-constant bounded harmonic functions on $\mathbb{M}$.)

Now we use some non-trivial geometry of regular geodesic balls. Again we can use the geodesic comparison arguments for distance from a specified point as given in the previous subsection and so show the following function is bounded non-negative convex on $\mathcal{B}$:

$$\psi_y(z) \quad = \quad \frac{1}{k} \frac{1 - \cos(\sqrt{k}\,\mathrm{dist}(z,y))}{\cos(\sqrt{k}\,\mathrm{dist}(y,\mathbf{o}))} \,.$$

(Here $\mathbf{o}$ is the centre of the regular geodesic ball $\mathcal{B}$ of interest, and $\psi_y(z)$ is defined for $y, z \in \mathcal{B}$.)

It follows that $\psi_y(Y)$ is a bounded negative submartingale (convexity properties of ψ_y) which converges to $\psi_y(Y(\infty)) = \psi_y(y) = 0$. It then follows from the property defining a submartingale that $\psi_y(Y) = 0$ always, and so $Y = y$ for all time. But this implies F must be constant, which establishes the result. □

It should be noted that the choice of the form for the function ψ came from the geometry paper [63], which used partial differential equations to establish a special case of the above Liouville theorem. A translation of the above probabilistic proof into non-stochastic (potential-theoretic) terms may be found in [85]. Note that there are Liouville theorems which are *not* special cases of the above result, for example Cheng's Liouville theorem for harmonic maps of linear growth [18] (but Stafford [127] has produced a probabilistic proof of this).

2.3.5 Picard little theorem for harmonic maps

Goldberg and coworkers have established a Picard little theorem for harmonic maps: any such F in the following has to be constant!

(2.3.6)

F	:	$\mathbb{M}$	$\rightarrow$	$\mathbb{N}$
harmonic map of bounded dilatation		*nonnegative Ricci*		*Sectional curvatures bounded above and away from 0*

A standard covering argument shows that we may suppose both $\mathbb{M}$ and $\mathbb{N}$ to be simply-connected.

Bearing in mind Burgess Davies' proof for the holomorphic Picard little theorem, and the Ultimate Liouville Theorem, one might try for a probabilistic proof along the following lines: show that the Γ-martingale of K^2-bounded dilatation given by $F(\mathrm{BM}(\mathbb{M}))$ has a random *limiting direction* if F is nonconstant, and so derive a contradiction in the manner of the proof of the Ultimate Liouville theorem, using the fact that manifolds with non-negative Ricci curvatures have no non-constant harmonic functions [137]. This program has been carried out with a large degree of success in [73, 74] and generalized in [24, 53]: [4] provides an alternative stochastic approach using gradient estimates. In fact most of the angle-based work *preceded* the conceptually simpler Ultimate Liouville theorem: it was informed and motivated by work by Dynkin and others [39, 98, 120] concerning limiting directions of Brownian motion on Cartan-Hadamard manifolds.

2.3.6 Limiting directions for Brownian motion

A Cartan-Hadamard manifold $\mathbb{N}$ has no cut-locus and so we may extend geodesic polar coordinates (r, θ) over all $\mathbb{N}$. Hence it makes sense to discuss whether there might exist a limit at time infinity for the angle process $\Theta = \theta(\mathrm{BM}(\mathbb{N}))$ of Brownian motion.

In fact it is easy to show (by skew-product representation arguments) that such a limit cannot exist for the angle process for the simplest Cartan-Hadamard manifold, namely Euclidean space of dimension 2 or more. This argument has been extended to cover manifolds of rotational symmetry [110]: skew-product arguments use the rotational symmetry to show that the angle process has a genuinely random limit if the curvature is less than a negative constant (and indeed even if the curvature is negative and does not decay too fast).

If there is no rotational symmetry then the argument becomes much harder. In order to control the angle of Brownian motion one needs an extra condition:

either that the manifold is two-dimensional [57, 75], or a non-radial gradient condition on the curvature [98], or a lower bound on the curvature [103, 120, 130]. The lower curvature bound is the most attractive condition: we state it as a theorem and indicate its proof.

Theorem 2.3.4 (Limiting angle in pinched Cartan-Hadamard manifolds): *Suppose $\mathbb{N}$ is a Cartan-Hadamard manifold with negative curvature pinched between two negative constants:*

$$-L^2 \quad \leq \quad \textit{Sectional curvatures} \quad \leq \quad -H^2$$

Then $BM(\mathbb{N})$ has a random limiting direction.

Proof: (Sketch.) We sample the BM($\mathbb{N}$) process X at the times when it moves unit distance from the most recent sampling point:

$$T_0 = 0, \quad T_{k+1} \quad = \quad \inf\{t > T_k : \operatorname{dist}(X_t, X_{T_k}) = 1\}$$

Conditioning on $\mathfrak{F}_{T_k}$, the past at time T_k, an Itô analysis modifying the analysis of Eq. (2.3.3) to apply to *lower curvature bounds* shows that $u \mapsto \operatorname{dist}(X_{T_k+u}, X_{T_k})$ can be bounded *above* by Brownian motion plus constant positive drift.

Hence the strong law of large numbers can be applied to show that $T_k \to \infty$ and moreover there is $\beta > 0$ such that as $K \to \infty$ so

$$\mathbb{P}\left[\, T_{K+k} \geq \beta k \text{ for all } k > 0 \mid X_0 \,\right] \quad \to \quad 1$$

uniformly in the starting point X_0.

Now on the other hand the discussion following Eq. (2.3.3) shows that the *upper curvature bound* implies that R is bounded *below* by Brownian motion plus a constant positive drift, and so elementary Brownian motion theory (related to the strong law of large numbers) shows that for some positive α as $R_0 \to \infty$ so

$$\mathbb{P}\left[R_t \geq \frac{1}{2} R_0 + \alpha t \right] \to 1$$

Together with a comparison of angles with those defined by comparable triangles on the hyperbolic plane,

$$\angle(\Theta(T_{K+k}), \Theta(T_{K+k+1})) \quad \leq \quad c_0 \exp(-c_1 R(T_{K+k}))$$

for c_0, $c_1 > 0$. This allows us to deduce convergence to a limiting angle which is random (since as the starting distance tends to infinity so it is uniformly increasingly probable that the limiting angle is close to the initial angle). □

The obvious question here is, whether one can drop the lower bound on the curvature. This is an open problem! As noted above, the lower bound is not required in the two-dimensional case. Treatments of variable lower curvature bounds are to be found in [58, 103]: it is noteworthy that the best lower curvature bound corresponds approximately to the lower curvature bound required to prevent the possibility of explosion/stochastic incompleteness (Brownian motion leaving all compact sets within a finite random time). Ancona [2] describes an informative simply-connected three-dimensional example using a warped metric of negative curvature bounded above away from zero, in which the limiting angle exists but is constant. It is conceivable that a variation on this example would lead to a case in which the limiting angle does not exist. However the interesting question here attaches to the *randomness* of the limiting angle. Although the limiting angle in Ancona's example is non-random, nevertheless one can identify limiting pseudo-angles which are random, so the invariant σ-algebra of the Brownian motion in question does not satisfy a zero-one law. It is still an open question whether it is possible to construct a Cartan-Hadamard manifold with curvature bounded above away from zero for which the invariant σ-algebra satisfies a zero-one law, equivalently, for which all bounded harmonic functions are constant.

We should also note the use of Gromov hyperbolicity, visibility, and a condition on the Laplacian of the distance function, all to replace the negative upper bound on the curvature of the Cartan-Hadamard manifold: see for example [1, 99].

2.3.7 Probabilistic proofs for Picard little theorems

The way is now clear for a probabilistic proof of a generalized Picard little theorem. We simply generalize the proof of limiting angle from Brownian motion to Γ-martingales. It is easily seen that we need at least a bounded dilatation condition (since for example Brownian geodesics cannot possess a limiting angle!)

Theorem 2.3.5 (Limiting angle for Γ-martingales): *Let $\mathbb{N}$ be a Cartan-Hadamard manifold with sectional curvatures pinched between two negative constants. If X is a Γ-martingale of bounded dilatation on $\mathbb{N}$ then it possesses a limiting angle at time infinity: moreover this angle is genuinely random.*

Proof: The techniques carry over largely unchanged from the Brownian motion case above. There are two points to watch. First, the intrinsic time τ of X may be finite, so that we cannot exclude the possibility that X converges within $\mathbb{N}$. However this leads immediately to the conclusion that a limiting angle exists: randomness of this angle can then be derived from an Itô analysis of the distance of X from any prescribed geodesic line.

Second, a simple geometric argument is required to verify that the bounded dilatation condition deploys to enable a lower bound of $\operatorname{dist}(X, \mathbf{o})$ (in intrinsic time-scale) by a standard Brownian motion with constant drift. □

A Picard little theorem for harmonic maps is an immediate corollary of the above.

Theorem 2.3.6 (Picard little theorem for harmonic maps I): *Suppose that $\mathbb{N}$ is a manifold with sectional curvatures pinched between two negative constants, and $\mathbb{M}$ is a manifold whose universal cover supports no non-constant bounded harmonic functions. If the map*

$$F : \mathbb{M} \quad \to \quad \mathbb{N}$$

is harmonic and of bounded dilatation then it must be constant.

Proof: By a covering argument we can suppose that $\mathbb{N}$ is simply connected. But now $F(\mathrm{BM}(\mathbb{M}))$ is a Γ-martingale of bounded dilatation, and therefore has a random limiting angle unless F is constant. Since a random limiting angle would supply non-constant bounded harmonic functions, we conclude that F must be constant. □

By [137] the Ricci-nonnegative condition on $\mathbb{M}$ implies no non-constant harmonic functions on its universal cover, hence the geometric Picard little theorem for harmonic maps, as stated above, is nearly a special case of the probabilistically proved version. However the probability proof requires the lower curvature bound, and that this restriction is non-trivial is a consequence of the open question for Brownian motion angle.

This contrast shows that the probability proof is genuinely *different* from the geometric proof: a price has to be paid in terms of a lower curvature bound, but in return one gains a tight link to (linear) potential theory on the domain manifold. This link is characteristic of probabilistic approaches to harmonic map theory.

Some gain on the lower curvature bound can be achieved by a more careful analysis along the same lines as above but using varying radial curvature bounds [54] in the manner of the approach to the Brownian motion problem described in [58]: see [78] for an indication of the Γ-martingale approach. (It should also be noted that [53] describes how to weaken the *upper* curvature bound: see also [57].) However such an analysis does not add anything substantial to our understanding of the lower curvature bound problem here.

Before leaving the topic of limiting angle, note that in [24] it is shown how to replace the upper curvature bound in the above by a combination of Gromov hyperbolicity, a visibility condition, and requirement that the Laplacian of the distance function is sufficiently negative in sufficiently many balls in $\mathbb{N}$.

2.3.8 Life on the cut-locus

The previous analysis depends on the Itô analysis, for X a Γ-martingale with stochastic anti-development M and parallel transport Ξ,

$$\mathrm{d}_I r(X) = \langle \operatorname{grad} r(X), \Xi\rangle \, \mathrm{d}_I M + \frac{1}{2} \operatorname{trace}\left(\operatorname{Hess} r(X)(\Xi, \Xi)\, \mathrm{d}\,[\, M, M\,]\right) \tag{2.3.7}$$

which, as we have noted, holds true only up to the first time that X first hits the cut-locus of the reference point $\mathbf{o}$. In the case of Cartan-Hadamard manifolds this is not a hindrance, since the cut-locus is always empty. However other manifolds typically *do* have cut-locus (for example, the Cheeger-Gromov theorem says that manifolds with non-negative Ricci curvature and no cut-locus must be Euclidean: see [82] for a probabilistic perspective on this result). So a more detailed analysis is required once we move away from Cartan-Hadamard manifolds, as is required even in generalizations of the above (see [5]).

2.3.9 Historical perspective

In fact it is well-known in the analytic and geometric context what has to be done: the Laplacian of the distance function can be extended over the cut-locus so long as it is there interpreted in a distributional sense as a non-positive measure [13, 136, 137]. The question for probabilists is, how to understand this result in terms of the trajectory of Brownian motion.

2.3.10 Cut-locus correction by upper bound

The first point to note is that by one means or another the analytic interpretation of the Laplacian of the distance function translates directly into a simple correction term to be added to Eq. (2.3.7). In [79] the geometric treatment given in [13] is adapted to the Brownian path in order to show the following.

Theorem 2.3.7 (The radial part of Brownian motion on a manifold): *Suppose that X is BM($\mathbb{M}$) with stochastic anti-development B and parallel transport Ξ. If R is its radial part ($R = \operatorname{dist}(X, \mathbf{o})$) then*

$$\text{(2.3.8)} \qquad d_I r(X) \quad = \quad \langle \mathit{grad}\, r(X), \Xi \rangle \, d_I B + \frac{1}{2} \Delta\, r(X)\, dt - dL$$

where L is an increasing random process which changes only when X visits the cut-locus of the reference point $\mathbf{o}$.

There is nothing surprising in this: L is the potential corresponding to the measure part of $-\Delta\, r$. Informally speaking, this result tells us that the effect of the cut-locus is simply to pull the Brownian motion closer in to the reference point $\mathbf{o}$.

An alternative more analytic derivation of the probabilistic representation is to be found in [56], which applies it to non-explosion and non-implosion (Feller property) results for Brownian motion. The non-implosion results are extended to Γ-martingales (to answer conjectures by Emery) in [95], where the method of proof is also streamlined to link it tightly to convexity and Topogonov's comparison theorem after the analytic approach of Wu [136].

It is reasonable to ask for a more detailed representation of the dL term, in particular for an explicit relationship in terms of local times on submanifolds and the local geometry of the cut-locus. For Brownian motion this is to be found in [25]: in fact

$$\text{(2.3.9)} \qquad -dL \quad = \quad \frac{1}{2} \left[D_+ r(\nu) + D_- r(-\nu) \right] dL^{C(\mathbf{o})}$$

where

- ν is a measurable unit normal vectorfield to the cut-locus $C(\mathbf{o})$;
- $D_+ r(\pm\nu)$ is the one-sided directional derivative of r in the direction $\pm\nu$;
- $L^{C(\mathbf{o})}$ is "geometric local time on the cut-locus".

These terms make sense because old work on the cut-locus establishes that it is a $(m-1)$-*dimensional rectifiable set*: up to Hausdorff $(m-2)$-measure zero it is the countable union of C^1 hypersurfaces in $\mathbb{M}$. Consequently ν makes sense. Moreover $-r$ can be related to convex functions using Topogonov's theorem, so that the required one-sided derivatives exist. Finally the $(m-1)$-regularity of $C(\mathbf{o})$ means that $L^{C(\mathbf{o})}$ can be interpreted as a σ-finite local time measure.

This representation has been extended to suitable Γ-martingales (those which do not hit the conjugate part of the cut-locus) in [8, 104].

It is noteworthy that currently applications of Eq. (2.3.8) do not use the detailed nature of $\mathrm{d}L$ as expressed in Eq. (2.3.9): applications which *did* do this would be of considerable interest. For a typical application of Eq. (2.3.8) see Le [8, 101], who extends Γ-martingale inequalities of Darling [29] to the case of non-trivial cut-locus.

It should also be noted that the full radial representation Eq. (2.3.8) fills a gap in the probabilistic proof in [33] of the constancy of bounded harmonic functions on manifolds of non-negative Ricci curvature (special case of a result by Yau [137]).

2.4 Coupling and parallel transport

Stochastic calculus and the probabilistic approach come into their own when considering and applying the Itô analysis of $\operatorname{dist}(X, Y)$, for X, Y two co-adapted semimartingales (typically Γ-martingales or Brownian motions). This analysis is fundamental for applications of the *coupling* idea to stochastic differential geometry.

The probabilistic notion of *coupling* is as follows. One constructs two random processes (for example two Brownian motions started at different initial points) on the same probability space, in such a way that one can draw immediate conclusions about the random process from the nature of the construction. Thus for example a partial domain monotonicity result for Neumann heat kernels in convex regions can be derived from a coupling construction of reflected Brownian motions X_1, X_2 started from two different initial points, in which the coupling is constructed so as to ensure that if X_1 is initially further from a reference point $\mathbf{o}$ than X_2 then this persists for all time [84]. A more involved coupling construction can be used to show that this convex domain monotonicity does not hold in general [9].

The original idea of coupling is due to Doeblin [34], who used it to prove convergence to equilibrium for appropriate Markov chains. It has been used intensively by workers in interacting particle systems (see for example [105]) and epidemics [7, 112]. A systematic exposition has been given by Lindvall [107] who was also the first to introduce the method in the context of Brownian motion [106]. Lindvall's *Brownian motion reflection coupling* is seminal for the theory described below: we can couple two m-dimensional Euclidean Brownian motions so that they almost surely meet at some random time! (In such a case we say that the coupling *succeeds.*) If B_1 is one of the Brownian motions, begun at $B_1(0) = b_1$, and we require to construct the other B_2 to be begun at $B_2(0) = b_2$, then this may be achieved with great simplicity just by constructing B_2 as the reflection of B_1 in the hyperplane which is the perpendicular bisector of the line segment between b_1 and b_2.

As we shall see, coupling constructions for Brownian motion can also be used to derive gradient inequalities for harmonic functions and harmonic maps, and bounds on the second eigenvalue. (Their importance has recently grown still further, with the advent of coupling-based algorithms for "perfect (or exact) simulation" introduced by Propp and Wilson [47, 96, 122], which has motivated recent work on coalescing Brownian flows [11]: however we shall say no more of that here.)

It is fundamental to the above example that B_1 and B_2 are not independent. Notice also that this coupling is *co-adapted*, in the sense that the Brownian motions can be adapted to the same filtration (so that with respect to this filtration both B_1 and B_2 have the correct first- and second-order stochastic calculus characteristics for Brownian motions). There is a whole taxonomy of different kinds of couplings which progressively weaken this latter requirement, for example *raw couplings* which would involve conditional constructions of B_2 arranged to couple with B_1 at specific times as much as possible under the constraint that B_2 *viewed on its own* should be a Brownian motion (this last kind of coupling has been used to great practical effect in [108]). We shall not consider this taxonomy here, but simply make two remarks:

1. In general the behaviour of raw couplings is more closely connected to potential theory, since existence conditions for raw couplings are typically equivalent to potential theoretic conditions such as constancy of all bounded harmonic functions;

2. On the other hand existence and nonexistence results and constructions for co-adapted couplings typically are more directly related to essentially local

conditions based on curvature bounds, and adapt easily to the nonlinear theory of harmonic maps.

The second of these remarks gives the reason why we shall here concentrate on co-adapted couplings: using the machinery of stochastic development we can construct such couplings and use them to deduce geometric results arising from curvature conditions.

It is natural to follow on from Lindvall's reflection coupling for Euclidean Brownian motion to ask when the same can be done for Brownian motion on a curved manifold. Similar ideas work for Brownian motion on spheres; viewing the sphere as embedded in Euclidean space one can use reflection in the perpendicular bisector hyperplane for the initial points b_1, b_2 viewed in the ambient Euclidean space (this is followed up in [11]). However one runs into difficulties considering hyperbolic space. If $\mathbb{H}^2$ is the hyperbolic plane then an Itô analysis of the hyperbolic distance between B_1 and B_2 shows that reflection coupling cannot succeed. In fact a symmetry argument shows that half this distance is the distance $\operatorname{dist}(B_1, \gamma)$ from B_1 to the geodesic γ acting as perpendicular bisector between b_1, b_2, and variations on the methods of the previous subsection show that $\operatorname{dist}(B_1, \gamma)$ can be bounded below by Brownian motion with constant positive drift, which has positive probability of drifting off to plus infinity.

Actually a potential-theoretic argument (based on the existence of non-constant bounded harmonic functions) shows that even raw couplings for $\mathrm{BM}(\mathbb{H}^2)$ must have positive probability of not succeeding. However we want to work with curvature conditions rather than potential theory, and to derive results for Γ-martingales not just Brownian motion;we therefore address the following question.

If $\mathbb{N}$ is a Cartan-Hadamard manifold with sectional curvatures bounded above by a negative constant $\kappa^2 < 0$, then when can we construct co-adapted couplings for Γ-martingales which succeed almost surely?

2.4.1 Distance between two Γ-martingales.

Suppose that Γ-martingales X_1, $X_2 \in \mathbb{N}$ have respective stochastic anti-developments M_1, M_2 and respective parallel transports Ξ_1, Ξ_2. Consider the Itô analysis of $S = \operatorname{dist}(X_1, X_2)$. By viewing (X_1, X_2) as a Γ-martingale on the Riemannian product manifold $\mathbb{N} \times \mathbb{N}$, and applying the geometric Itô formula

Eq. (2.2.8) we can deduce

(2.4.1)
$$\begin{aligned} \mathrm{d}_I S &= \langle \operatorname{grad} \operatorname{dist}(X_1, X_2), (\Xi_1, \Xi_2) \rangle (\, \mathrm{d}_I M_1,\, \mathrm{d}_I M_2) + \\ &+ \tfrac{1}{2} \operatorname{trace} \left(\operatorname{Hess} \operatorname{dist}(X_1, X_2)((\Xi_1, \Xi_2), (\Xi_1, \Xi_2))\, \mathrm{d}\left[\, (M_1, M_2), (M_1, M_2)\, \right]\right) \end{aligned}$$

until the first time that the geodesic between X_1 and X_2 is conjugate or non-unique.

Given suitable curvature bounds, the drift term trace(Hess...) can be bounded using Eq. (2.2.17). We also need to control the bracket process for the martingale part $\langle \ldots \rangle (\, \mathrm{d}_I M_1,\, \mathrm{d}_I M_2)$, and for this we require some kind of bounded dilatation or quasi-conformality condition on the two martingales M_1, M_2.

To see that some sort of control on the martingale term is necessary, consider the case where X_1, X_2 are Brownian geodesics running along the same shared geodesic. Then S is bounded above by the distance between X_1, X_2 measured along the geodesic, which is a nonnegative martingale up to the time of first coupling. Basic martingale theory then shows either S converges to a positive constant or to zero. The second is easily arranged, for example by constructing X_1, X_2 to be independent, and in this case X_1, X_2 will couple with probability one *whatever* curvature conditions are imposed!

Theorem 2.4.1 (Γ-martingales of K^2-bounded dilatation and coupling): *Suppose that X_1 and X_2 are Γ-martingales of K^2-bounded dilatation, begun at different initial points $X_1(0) = x_1$, $X_2(0) = x_2$ of a Cartan-Hadamard manifold with negative sectional curvatures bounded above by a negative constant $-L^2$. Let the Γ-martingales have parallel transports Ξ_1, Ξ_2 and stochastic anti-developments M_1, M_2. Then X_1 and X_2 have positive probability of never meeting.*

Proof: (Sketch: see [77] for full details.) We start with the Itô analysis of S described above in Eq. (2.4.1). There are no conjugate or non-unique geodesics on a Cartan-Hadamard manifold, so Eq. (2.4.1) holds for all time. Using the bound (2.2.17) and completing the square, the drift term $\mathrm{d}\Lambda$ is bounded below using

$$\begin{aligned} &\operatorname{trace} \left(\operatorname{Hess} \operatorname{dist}(X_1, X_2)((\Xi_1, \Xi_2), (\Xi_1, \Xi_2))\, \mathrm{d}\left[\, (M_1, M_2), (M_1, M_2)\, \right]\right) \\ &\geq L \tfrac{\cosh(LS) - 1}{\sinh(LS)} \left[(\, \mathrm{d}\tau_1 - \mathrm{d}\left[\, \langle \Xi_1 V_1, M_1 \rangle \,\right]) + (\, \mathrm{d}\tau_2 - \mathrm{d}\left[\, \langle \Xi_2 V_2, M_2 \rangle \,\right]) \right] . \end{aligned}$$

Here $\mathrm{d}\tau_1$, $\mathrm{d}\tau_2$ are the intrinsic times for the Γ-martingales X_1, X_2; the vectors V_1, V_2 are unit vectors in $\mathbb{R}^m$ such that $\Xi_1 V_1$, $\Xi_2 V_2$ are parallel to the geodesic connecting X_1 to X_2.

The bounded dilatation condition preserves us from the possibility that $\mathrm{d}\tau_1 - \mathrm{d}[\langle \Xi_1 V_1, M_1\rangle]$ and $\mathrm{d}\tau_2 - \mathrm{d}[\langle \Xi_2 V_2, M_2\rangle]$ might both vanish in the $(\mathrm{d}\tau_1 + \mathrm{d}\tau_2)$-timescale. But to prove our result we must consider S in its natural timescale, and therefore must obtain a lower bound on $\mathrm{d}\Lambda/\mathrm{d}[N,N]$ where N is the martingale part

$$\mathrm{d}_I N \quad = \quad \langle \operatorname{grad}\operatorname{dist}(X_1,X_2), (\Xi_1,\Xi_2)\rangle(\mathrm{d}_I M_1,\, \mathrm{d}_I M_2)\,.$$

The bounded dilatation condition and elementary but tedious arguments from Euclidean vector geometry (detailed in [77]) allow us to obtain an upper bound on $\mathrm{d}[N,N]/(\mathrm{d}\tau_1 + \mathrm{d}\tau_2)$, and this delivers the required result. □

It is natural to ask whether the constant negative upper bound on curvature may be relaxed. So long as uniqueness of geodesics is preserved it should be straightforward to do this using the variable curvature comparisons described for example in [54]. However *some* negative curvature is required if coupling is not to be almost surely successful: see the results of the next subsection.

2.4.2 Coupling Brownian motions in the case of nonnegative Ricci curvature

We have seen that the reflection coupling for Brownian motions is almost surely successful in the case of Euclidean space of whatever dimension. When trying to generalize this result to Riemannian manifolds, it is reasonable to impose a condition of nonnegative *Ricci* curvature, since the rotational symmetry of the Brownian stochastic differential lends itself to Ricci curvature bounds rather than sectional curvature bounds; see the bound Eq. (2.2.18).

In general, excepting the constant curvature cases where there is much symmetry, we cannot make sense of reflection couplings using single fixed perpendicular bisector hyperplanes. However notice that the purpose of the reflection is simply to transform one Brownian differential $\mathrm{d}B_1$ into another $\mathrm{d}B_2$. Instead of doing this using a fixed hyperplane, we can use parallel transport along a minimal geodesic connecting the two Brownian motions, together with reflection in the hyperplane normal to this geodesic. This is the essence of the proof of the following result.

Theorem 2.4.2 (Brownian coupling and nonnegative Ricci curvature): *Let $\mathbb{M}$ be a Riemannian manifold of nonnegative Ricci curvature. Given any two initial points x_1, x_2, it is possible to construct two co-adapted Brownian motions X_1, X_2 begun at x_1, x_2 respectively, such that they couple successfully with probability one.*

Proof: (Sketch: for full details consult any one of [22, 76, 81].) Let $S = \text{dist}(X_1, X_2)$, and let Ξ_1, Ξ_2 be the stochastic parallel transports and B_1, B_2 be the stochastic anti-developments of X_1, X_2. We can use the Itô analysis of Eq. (2.4.1), but only till the first time that the minimal geodesic connecting X_1, X_2 is conjugate or non-unique.

Setting aside this difficulty for a moment, we have to consider how to express the transformation of $\mathrm{d}B_1$ into $\mathrm{d}B_2$. Let the *mirror map* $\mathfrak{M}_{(x_1,x_2)}$ map the tangent space $T_{x_1}\mathbb{M}$ to the tangent space $T_{x_2}\mathbb{M}$ using parallel transport along the minimal geodesic from x_1 to x_2 followed by reflection in the hyperplane normal to the geodesic. Then we define $\mathrm{d}_I B_2 = [(\Xi_2)^{-1}\mathfrak{M}_{(X_1,X_2)}\Xi_1]\,\mathrm{d}_I B_1$. Notice that $[(\Xi_2)^{-1}\mathfrak{M}_{(X_1,X_2)}\Xi_1]$ is a (random but predictable) linear mapping of $\mathbb{R}^m$ into itself, despite involving maps between tangent spaces on manifolds (compare the use of $\langle \text{grad}\, f(X), \Xi\rangle\,\mathrm{d}_I Z$ in Theorem 2.2.5). Thus it is proper to use it as the integrand for an Itô differential.

We can now define the coupling between X_1 and X_2 up to the first time that the geodesic between the two Brownian motions fails to be unique, using a stochastic differential system which mixes Itô and Stratonovich terms:

$$\begin{aligned} \mathrm{d}_S X_1 &= \Xi_1\,\mathrm{d}_S B_1 \\ \mathrm{d}_S \Xi_1 &= H_{\Xi_1}\,\mathrm{d}_S X_1 \\ \mathrm{d}_S X_2 &= \Xi_2\,\mathrm{d}_S B_2 \\ \mathrm{d}_S \Xi_2 &= H_{\Xi_2}\,\mathrm{d}_S X_2 \\ \mathrm{d}_I B_2 &= \left[(\Xi_2)^{-1}\mathfrak{M}_{(X_1,X_2)}\Xi_1\right]\,\mathrm{d}_I B_1 \end{aligned} \tag{2.4.2}$$

Substituting in to the Itô analysis Eq. (2.4.1) of S, we find the effect of the mirror map is to make the martingale part become

$$\langle \text{grad}\,\text{dist}(X_1, X_2), (\Xi_1, \Xi_2)\rangle(\,\mathrm{d}_I B_1,\, \mathrm{d}_I B_2) \quad = \quad 2\langle V_1,\, \mathrm{d}_I B_1\rangle$$

where $\Xi_1 V_1$ is the unit tangent vector at X_1 defining the geodesic from X_1 to X_2. So the martingale part of S is simply a scalar Brownian motion of rate $2^2 = 4$.

Using the mirror map, the drift part of S simplifies to half a sum of $m-1$ second geodesic derivatives of $\operatorname{dist}(y_1, y_2)$, evaluated at $y_1 = X_1$, $y_2 = X_2$, where a typical geodesic variation causes y_1 to move by a geodesic defined by $\Xi_1 V$ and causes y_2 to move by a geodesic defined by $\mathfrak{M}_{(X_1,X_2)}\, \Xi_1 V$, and V ranges through an orthonormal set of unit vectors such that $\Xi_1 V$ is normal to the geodesic from X_1 to X_2. This is exactly the context in which we can apply the bound (2.2.18), and results in the conclusion that the quantity

$$\frac{1}{2}\sum_i \Big(\operatorname{Hess}\operatorname{dist}(X_1, X_2)((\Xi_1, \Xi_2), (\Xi_1, \Xi_2)) \times \\ \times \mathrm{d}\,[\,(M_1, M_2), (M_1, M_2)\,]\Big) \tag{2.4.3}$$

is negative.

It follows that, up till the first time that uniqueness or non-conjugacy fails for the minimal geodesic from X_1 to X_2, we may bound S above by a scalar Brownian motion of rate 4. If this bound could persist for all time then we would have almost sure successful coupling, since scalar Brownian motion will eventually hit zero. The question is, how to deal with breakdown of the stochastic system Eq. (2.4.2)?

The original approach of [76, 81] involves a complicated accounting for the different ways in which breakdown can happen. Failure of uniqueness of minimal geodesic can be disposed of with relative ease. Failure of non-conjugacy is fixed by allowing the two Brownian motions X_1, X_2 to evolve independently for a short time near the locus of conjugacy; analysis of Jacobi vectorfields shows that for independent evolution near conjugacy the drift of S is nearly negatively infinite. A simpler and more attractive treatment in [22] finesses matters: again one allows the two Brownian motions to evolve independently for brief intervals when the system Eq. (2.4.2) breaks down, but now one argues using the ideas of the cut-locus correction described above in §. This shows that in the limit, as the brevity of the brief intervals shrinks to zero, so one achieves a construction which gives S the characteristics of a scalar Brownian motion of rate 4 with a nonpositive drift. This then establishes that the coupling is almost surely successful. □

Again one may ask whether the curvature condition may be relaxed, now to allow regions of negative Ricci curvature. This has been achieved recently by Wang [135].

2.4.3 Applications

We comment briefly on applications of the idea of coupling Brownian motions on manifolds.

We begin by remarking that some caution is appropriate when making simplifying assumptions on the geometry of manifolds. It is tempting to wonder what might be lost in Theorem 2.4.2 simply by assuming that, in addition to nonnegative Ricci curvature, any two points are connected by a unique geodesic. However the celebrated splitting theorem of Cheeger and Gromoll [15] implies that the only such manifolds are Euclidean spaces! (A proof using Brownian motion and the cut-locus correction may be found in [82].)

Given the work above, we can immediately establish a toy version of Yau's [137] result on constancy of nonnegative harmonic functions on manifolds of $\mathbb{M}$ of nonnegative Ricci curvature. For suppose that $f : \mathbb{M} \to [0, 1]$ is a *bounded* harmonic function. For x_1, $x_2 \in \mathbb{M}$ consider Brownian motions X_1, X_2 constructed as above so as to begin at x_1, x_2 and to couple almost surely. Then $f(X_1) - f(X_2)$ is a bounded martingale which vanishes at time infinity. Basic martingale results tell us that $\mathbb{E}\left[f(X_1(t)) - f(X_2(t)) \right] = f(x_1) - f(x_2)$, and also

$$\mathbb{E}\left[f(X_1(t)) - f(X_2(t)) \right] \quad = \quad \lim_{s \to \infty} \mathbb{E}\left[f(X_1(s)) - f(X_2(s)) \right] \quad = \quad 0\,.$$

Hence $f(x_1) = f(x_2)$.

Of course establishing successful coupling involves more work than the alternative probabilistic proof referred to earlier at the end of §2.3.10. However the coupling approach generalizes to allow one to establish probabilistic proofs of gradient inequalities [22]: here a rather careful analysis is required not only of the distance $\operatorname{dist}(X_1, X_2)$ but also, simultaneously, of the two distances $\operatorname{dist}(X_1, \mathbf{o})$ and $\operatorname{dist}(X_2, \mathbf{o})$. Careful estimation bounds the probability that X_1 and X_2 couple before leaving a small geodesic ball centred on $\mathbf{o}$, bounding in terms of the initial distance $\operatorname{dist}(X_1(0), X_2(0))$. The gradient inequality then follows by a simple martingale argument somewhat on the lines above. (See [133, 132] for an alternative probabilistic approach to gradient estimates using differentiation formulae.)

The coupling approach also permits a very satisfactory probabilistic proof of the original generalized Picard little theorem for harmonic maps (see Eq. 2.3.6). For if $\mathbb{M}$ has nonnegative Ricci curvatures then for x_1, $x_2 \in \mathbb{M}$ we can construct co-adapted Brownian motions X_1, X_2 which begin at x_1, x_2 and which couple almost surely. Suppose that $F : \mathbb{M} \to \mathbb{N}$ is a harmonic map of K^2-bounded dilatation from $\mathbb{M}$ into a Cartan-Hadamard manifold $\mathbb{N}$ of sectional curvatures bounded above by a negative constant. Then $F(X_1)$ and $F(X_2)$ are co-adapted Γ-martingales which couple almost surely. By Theorem 2.4.1 above it follows $F(x_1) = F(x_2)$, and hence we may deduce that F is constant.

Notice that the above proof does not require the irritating lower bound on curvatures of $\mathbb{N}$ which was a feature of the angle-based methods described in the previous subsection. The price paid in return is a more restrictive condition on $\mathbb{M}$: the angle-based methods required only that the universal cover of $\mathbb{M}$ support no non-constant bounded harmonic functions. Clearly the coupling proof generalizes beyond nonnegative Ricci curvature (Wang's result [135] demonstrates this): we only require that $\mathbb{M}$ have the *Brownian coupling property* set out below.

Definition 2.4.3 (Brownian coupling property): *A Riemannian manifold* $\mathbb{M}$ *has the* Brownian coupling property *(BCP) if for all* x_1, $x_2 \in \mathbb{M}$ *we can construct co-adapted Brownian motions* X_1, X_2 *begun at* x_1, x_2 *which almost surely couple successfully.*

We have just seen, in the proof of the toy version of Yau's result described above, that the BCP implies that all bounded harmonic functions are constant. It would be most interesting to find an example of a manifold for which all bounded harmonic functions are constant but the BCP fails. (Apart from intrinsic interest this might help guide current investigations into the achievable limits of efficient perfect simulation [11, 12].)

We note in passing that the possibilities for co-adapted coupling of processes on manifolds are more extensive than one might at first glance believe: see [5] which demonstrates successful coupling for some basic examples of hypoelliptic diffusions.

Further geometric applications of coupling include analysis of the ends of a manifold of nonnegative Ricci curvature [23], and a factorization theorem for harmonic maps arising from a factorization result for Brownian motion on a vector bundle [42]. In the next subsection we describe how coupling ideas combined with convexity considerations take one much further in a probabilistic treatment of harmonic maps. Before turning to this we should also note the important bounds on spectral gaps obtained by Chen and others using these coupling ideas [16, 17].

2.5 Harmonic map Dirichlet problem

Probability has much to contribute to the question of how to solve the Dirichlet problem for harmonic functions. If we are required to extend $f : \partial\Omega \to [0,1]$ from the boundary $\partial\Omega$ to the whole of a domain Ω, then the probabilistic approach delivers a stochastic mean-value representation

$$f(x) \quad = \quad \mathbb{E}\left[\, f(X(T)) \mid X(0) = x \,\right]$$

(here X is a Brownian motion run until the random time T when it first leaves the domain Ω) which is valid for "sensible" domains Ω, and also gives a stochastic interpretation of what "sensible" means (domains for which the boundary $\partial\Omega$ corresponds to the different ways in which X can leave Ω) and a way to fix things up when the domain is not "sensible" (redefine the boundary $\partial\Omega$ using the Martin boundary for the diffusion X).

But what about harmonic maps? The above approach is road-blocked: harmonic maps are defined using a *nonlinear* Dirichlet form as in Eq. 2.1.2. If we attempt to mimic the probabilistic construction for the linear case we come up against problems:

- How should one define expectation (a linear operation) on a nonlinear manifold?
- Even if one can make such a definition, the "law of iterated expectations" fails.

In this section it is shown how nevertheless the Γ-martingale approach can be developed into methods for solving Dirichlet problems for harmonic maps.

2.5.1 The rôle of convexity and coupling

The rôle of convexity in stochastic differential geometry has been strongly elucidated by Emery and Zheng [44, 46]. It is fundamental for harmonic map Dirichlet problems.

First observe that the convexity-based ideas of the Ultimate Liouville Theorem (Theorem 2.3.3), together with coupling ideas, suggest a strategy for determining when one has *uniqueness* for a harmonic map Dirichlet problem. Consider a fixed codomain $\mathcal{B} \subset \mathbb{N}$, a domain $\mathbb{M}$ which is a compact Riemannian manifold with ("sensible"!) boundary $\partial\mathbb{M}$, and a Dirichlet problem defined by data of a boundary map $\partial F : \partial\mathbb{M} \to \mathcal{B}$. Two different extensions F_1, F_2 of ∂F

lead to two different Γ-martingales $Y_1 = F_1(\mathrm{BM}(\mathbb{M}))$, $Y_2 = F_2(\mathrm{BM}(\mathbb{M}))$ which necessarily agree once the $\mathrm{BM}(\mathbb{M})$ in question has hit the boundary (and part of the definition of a "sensible" boundary is that $\mathrm{BM}(\mathbb{M})$ *will* eventually hit the boundary!). So non-uniqueness for the Dirichlet problem implies that one can construct two Γ-martingales in $\mathcal{B}$ with the same end-value.

Now of course in Theorem 2.3.3 the end-value was a non-random end-point on $\mathcal{B}$, allowing us to compose with a bounded nonnegative convex function vanishing only at the end-point. In general this will not be the case here. However we *will* have a product Γ-martingale (Y_1, Y_2) on $\mathcal{B} \times \mathcal{B}$, which will surely hit the diagonal $\mathcal{D} = \{(x, x) : x \in \mathcal{B}\}$ when the individual Γ-martingales couple, as we know they must do. So if we can exhibit a bounded nonnegative convex function vanishing only on the diagonal $\mathcal{D}$ then the argument of Theorem 2.3.3 allows us to deduce that (Y_1, Y_2) must lie on the diagonal for all time. This means $Y_1 = Y_2$ for all time, and hence (bearing in mind we can choose the starting point of the $\mathrm{BM}(\mathbb{M})$ arbitrarily) we deduce $F_1 = F_2$ hence uniqueness for the Dirichlet problem.

This simple conceptual argument motivates the next definition and justifies the two results following.

Definition 2.5.1 (Convex geometry): *A subset $\mathcal{B} \subset \mathbb{N}$ of a Riemannian manifold $\mathbb{N}$ is said to possess* convex geometry *if the following holds:*

(1) $\mathcal{B}$ is convex;

(2) there is a nonnegative convex function

$$\Phi : \mathcal{B} \times \mathcal{B} \to [0, 1]$$

such that Φ vanishes only on the diagonal: $\Phi(x_1, x_2) = 0$ if and only if $x_1 = x_2$.

Theorem 2.5.2 (Γ-martingale uniqueness): *Suppose that $\mathcal{B} \subset \mathbb{N}$ possesses convex geometry. If Y_1, Y_2 are two Γ-martingales on $\mathcal{B}$ which almost surely couple successfully then they must be almost surely equal for all time.*

Theorem 2.5.3 (Uniqueness for small-image harmonic map Dirichlet problem): *Suppose that $\mathcal{B} \subset \mathbb{N}$ possesses convex geometry. Then there can be at most one harmonic map extension to any boundary data $\partial F : \partial\mathbb{M} \to \mathcal{B}$.*

These results deliver a probabilistic proof of the existence result for the Dirichlet problem for harmonic maps as expounded in [6, 55].

It should be noted that in Theorem 2.5.3 the boundary $\partial\mathbb{M}$ must be "sensible" in the undefined sense mentioned above: however it is easy to see in this probabilistic approach that the linear potential theory boundary is exactly the correct boundary for the harmonic map problem.

There is of course the important question of whether the definition above is largely vacuous. Convex geometry is possessed by all sufficiently small geodesic balls, as a consequence of an argument in [44], though this is not quantified. However in [85, 88] an explicit construction is given of a Φ which works for all regular geodesic balls $\mathcal{B}$: exactly the context in which uniqueness has been obtained by analytic means using partial differential equation theory [63]:

Theorem 2.5.4 (Convex geometry for regular geodesic balls): *Suppose that $\mathcal{B} \subset \mathbb{N}$ is a regular geodesic ball, with nonnegative upper curvature bound k, centre $\mathbf{o}$ and radius $\arccos(h) < \pi/(2\sqrt{k})$. Then convex geometry for $\mathcal{B}$ is provided by*

$$\Phi(x,y) = \left(\frac{1-\cos(\sqrt{k}\,\mathrm{dist}(x,y))}{\cos(\sqrt{k}\,\mathrm{dist}(x,\mathbf{o}))\cos(\sqrt{k}\,\mathrm{dist}(y,\mathbf{o})) - \tilde{h}^2}\right)^{\nu+1} \tag{2.5.1}$$

if $\nu \geq 1$ and $2\nu\tilde{h}^2(h^2 - \tilde{h}^2) \geq 1$.

Further work in [85] establishes a continuous dependence of the initial point $Y(0)$ of a Γ-martingale Y in $\mathcal{B}$ on its terminal value $Y(\infty)$ (viewed in the L^2 sense): this is a consequence of local Lipschitz regularity of the convex function endowing $\mathcal{B}$ with convex geometry.

2.5.2 Construction of Γ-martingales

Given the above, when the co-domain $\mathcal{B}$ has convex geometry we can reduce the existence aspect of the harmonic map Dirichlet problem to a purely probabilistic question, a *Γ-martingale reachability problem*:

- Given a random variable Z taking values in a region $\mathcal{B}$ of convex geometry, can one construct a Γ-martingale Y for which Z is a terminal value?

For if we can do this then for each $x \in \mathbb{M}$ we can construct the unique Γ-martingale $Y^{(x)}$ in $\mathcal{B}$ attaining $\partial F(X^{(x)}(t))$, where $X^{(x)}$ is BM($\mathbb{M}$) begun at x.

Hence we obtain a map $F : x \mapsto Y^{(x)}(0)$. Arguments involving the Markov property of $\mathrm{BM}(\mathbb{M})$, and aided by the continuous dependence mentioned above, then show that F is a *finely harmonic map* (see Definition 2.2.8) and indeed continuous. This solves the harmonic map Dirichlet problem at least up to the question of whether we can translate "finely harmonic" to "harmonic": by the Itô formula this resolves to the regularity question of whether F is actually twice differentiable. Actually the same issue applies for the probabilistic resolution of the ordinary harmonic function Dirichlet problem: we will say more about this below.

We have to be careful here about what is prescribed in the Γ-martingale reachability problem: in the context of harmonic maps we are prescribing the proposed terminal value Z as a function $\partial F(X(T))$ of a Riemannian Brownian motion, consequently Z *must* be viewed as a random variable defined on a probability space $(\Omega, \mathfrak{F}, \{\mathfrak{F}_t : t \geq 0\}, \mathbb{P})$. In fact this filtered probability space may be viewed as generated by the Riemannian Brownian motion, and the stochastic differential system Eq. (2.2.1) of stochastic development and parallel transport permits a "flattening" simplification of supposing the generating process to be ordinary Euclidean Brownian motion. (The case of construction of the Γ-martingale when the filtration is *not* specified is covered in [45].)

One way of proceeding from this point, as described in detail in [85], is to exploit the "flattening" simplification together with continuous dependence as follows. Viewing Z as defined on a probability space $(\Omega, \mathfrak{F}, \{\mathfrak{F}_t : t \geq 0\}, \mathbb{P})$ generated by a Euclidean Brownian motion X, we may consider it as a function of the Brownian path: $Z = G(X)$. Because of the continuous dependence property it is enough to find a solution when Z depends on only finitely many values of X: $Z = G(X(t_1), \ldots, X(t_k))$; moreover we may suppose G to be a smooth map, constant off a compact set. The Markov property can now be employed to reduce this to the case $k = 1$ and so (by Brownian scaling) to the case $t_1 = 1$: $Z = G(X(1))$.

2.5.3 Existence using barycentres

Finally we must construct a sequence of approximate Γ-martingale solutions to the reachability problem when $Z = G(X(1))$. We do this using *Riemannian centres of mass* or *barycentres*, a concept due originally to Cartan [14] (see also [49, 71]).

Definition 2.5.5 (Riemannian barycentres): *Let $\mathcal{B} \subset \mathbb{N}$ be a convex neighborhood in a Riemannian manifold, and let μ be a probability measure on $\mathcal{B}$. We say that $x \in \mathcal{B}$ is a* (Riemannian) barycentre *of μ if it is a local minimum of*

$$Q(x) \quad = \quad \frac{1}{2} \int_{\mathcal{B}} \operatorname{dist}(x,u)^2 \, \mu(\, du) \, .$$

If the minimum is unique we write

$$x \quad = \quad \mathcal{E}(\mu) \, .$$

By extension to random variables we write $x = \mathcal{E}(Z)$ if x is the unique barycentre of the distribution of $Z \in \mathcal{B}$, and let $\mathcal{E}(Z|\mathfrak{F}_s)$ be the $\mathfrak{F}_s$-adapted random variable which is the unique barycentre of the conditional distribution of Z given $\mathfrak{F}_s$.

We note in passing that the notion of barycentre can be made to depend only on the connection if we consider extremals, satisfying

$$\int_{\mathcal{B}} \operatorname{dist}(x,u) \operatorname{grad}_1 \operatorname{dist}(x,u) \, \mu(\, du) \quad = \quad 0 \, .$$

This is discussed in [45]. Furthermore the notion of barycentre has recently found application in statistical geometry [114] and the statistics of shape [102].

Summarizing the argument of [85], view the barycentre as an approximate expectation (albeit *not* obeying the crucial iterated expectation property: $\mathcal{E}(\mathcal{E}(Z|\mathfrak{F}_s)) \neq \mathcal{E}(Z)$ in the presence of curvature [45]) it is now clear how to construct approximate Γ-martingales which reach the required target $Z = G(X(1))$: simply chain together conditional barycentres: set $Y^{[n]}(1) = Z$ and

$$Y^{[n]}(k/n) \quad = \quad \mathcal{E}(Y^{[n]}((k+1)/n)|\mathfrak{F}_{k/n}) \, .$$

These constructions can now be used to define "space-time maps" $F^n : \mathbb{R}^m \times [0,1] \to \mathcal{B}$ by piecewise geodesic interpolation from

$$Y^{[n]}(k/n) \quad = \quad F^n(X(k/n), k/n) \, .$$

We can now do simple calculations with Euclidean Brownian motion exploiting the convex function Φ endowing $\mathcal{B}$ with convex geometry; these show that the family of maps F^n is equicontinuous and hence one can apply the Arzela-Ascoli theorem to deduce a converging subsequence with a limit $F : \mathbb{R}^m \times [0,1] \to \mathcal{B}$. But the resulting random process $Y(t) = F(X(t),t)$ yields a submartingale whenever composed with a convex function; hence [26] it is a Γ-martingale, which clearly has the required reachability property!

The attractions of this approach are as follows:

- It offers easy extensions to the case when the domain is furnished not with Riemannian geometry, but with a suitably regular diffusion. Essentially what one requires is that the diffusion can be co-adaptively coupled: this includes smooth elliptic diffusions but also at least some hypoelliptic diffusions, as demonstrated in [5].
- It connects very directly with linear potential theory. For example note that [85] provides a theorem which systematically relates the question of existence of harmonic map Fatou radial limits to the linear case, thus generalizing analytically based work of [6].
- It offers an intriguing set of equivalences, implications, and counterexamples relating harmonic maps to convexity, Γ-martingales, and geodesics: see the last subsection in this account for more details.

2.5.4 Other approaches

We remark briefly on a variety of other probabilistic approaches uncovered in the last five years (indeed this variety contrasts remarkably with the apparent inability of probability to contribute to existence questions for harmonic maps before 1990!).

Picard [117] uses techniques based on Malliavin calculus to derive existence results. In a second approach Picard formalizes the notion of "Γ-martingale with jumps" (resembling the Γ-martingale approximations described above, but more sophisticated) and conscripts this into the service of harmonic maps [116, 118]. See also Cohen [19] and Arnaudon [3].

It is possible to rewrite the equations of stochastic development and parallel transport for a Γ-martingale $Y = F(X)$ and its generating Brownian motion X, when F is a harmonic map, as a *backwards stochastic differential equation.* This idea, introduced by Pardoux and Peng [115] in the context of stochastic control, is of strikingly wide application. Unfortunately the standard method of solution for backwards sde requires the use of the contraction mapping principle, and this breaks down in the harmonic map / Γ-martingale case because of a quadratic dependence arising from the Levi-Civita connection. Notwithstanding, in a *tour de force* involving a clever localization Darling [30] shows how to use backwards sde theory to construct Γ-martingales and hence harmonic maps. In [31] he extends the methods to deal with Γ-martingales on non-compact manifolds.

Finally Thalmaier [131] describes an approach which is different yet again, this time addressing the question of probabilistic approaches to harmonic map

Dirichlet problems where uniqueness fails. Clearly Γ-martingale reachability problems also fail to have unique solutions in such cases: Thalmaier analyses these by considering the martingale solving the reachability problem with the least mean intrinsic time, and relates this to harmonic map singularities.

Although not strictly a probabilistic approach, the work of Jost [65], described elsewhere in this volume, has strong links to probability. This approach constructs harmonic maps *directly* using barycentres, in a manner reminiscent of the method [85] discussed above. Notice however that Jost's approach works for metric spaces which are more general than manifolds: the question of extending Γ-martingale theory to encompass Jost's approach raises intriguing questions concerning the "correct" definitions of Γ-martingale and convexity.

2.5.5 Smoothness *via* gradient estimates

The probabilistic methods described above all deliver (continuous) finely harmonic maps rather than harmonic maps, and as mentioned above it is an interesting question how one can produce a probabilistic approach which delivers regularity. This has been answered in [94], which uses coupling ideas to deliver gradient estimates for finely harmonic maps (from which full regularity is easily deduced). The idea is based on the work of Cranston [22] with two interesting variations:

1. The arguments of Cranston are simplified by using co-adapted couplings which involve *time-changed* Brownian motions. In two dimensions, a Brownian motion subjected to a Markov time-change can be recognized as a Brownian motion in a conformally equivalent metric. This is no longer the case in higher dimensions: one must add a non-vanishing drift. However in the case of gradient estimates it can be shown that this extra drift assists coupling to happen, and therefore does not affect the resulting gradient estimate. This allows a simplification of the Cranston argument, as one needs to deal only with the distance between the two coupled processes and not their distances from some reference point.

2. In the case of harmonic maps whose image may possess positive curvature, these arguments only deliver Hölder($\frac{1}{2}$) estimates, not sufficient for regularity. However a careful modification of reflection coupling, which actually introduces independence as the two Γ-martingales come closer together, can be deployed to bootstrap the Hölder($\frac{1}{2}$) estimate up to a full gradient estimate.

2.5.6 Equivalences

To close this account we mention briefly some work which ties together convexity, coupling of Γ-martingales, harmonic maps, and stochastic control.

The very term "stochastic reachability" suggest control theory ideas, and indeed *any* coupling construction can be viewed as the solution of a (typically non-standard) stochastic control problem, aiming to maximize the probability of a successful coupling. The *value* of such a control problem turns out to be exactly the best candidate for a function to produce convex geometry. In [89, 91] it is shown that a convex domain $\mathcal{B}$ has convex geometry if and only if the Γ-martingale reachability problem has unique solutions depending continuously on the data.

Given the above work it is natural to wonder whether this equivalence extends to include the harmonic map Dirichlet problem with values in $\mathcal{B}$, and though no results have been obtained on this nevertheless one feels it should be the case. Emery [44] made the bold conjecture that the equivalence is implied by a much simpler phenomenon: that any two points in $\mathcal{B}$ are connected by one and only one geodesic (one could view this as a geodesic Dirichlet problem!). In fact this is not the case: [90] provides a two-dimensional counterexample, the "Propeller", using Riemannian barycentres. (A further related counterexample has been derived which is of significance in statistical geometry: [20].)

Bibliography

[1] A. Ancona. Théorie du potentiel sur les graphes et les variétés. In *Ecole d'Été de Probabilités de Saint-Flour XVIII-1988*, volume 1427 of *Lecture Notes in Mathematics*, pages 5–109, New York–Heidelberg–Berlin, 1990. Springer-Verlag.

[2] A. Ancona. Convexity at infinity and Brownian motion on manifolds with unbounded negative curvature. *Rev. Mat. Iberoamericana*, 10:189–220, 1994.

[3] M. Arnaudon. Sur l'existence dans une variété de martingales de valeur terminale donné. Research report, IRMA and University Louis Pasteur, Strasbourg, 1993.

[4] M. Arnaudon and A. Thalmaier. Complete lifts of connections and stochastic Jacobi fields. Research report, IRMA and University Louis Pasteur, Strasbourg, 1997.

[5] G. Ben Arous, M. Cranston, and W.S. Kendall. Coupling constructions for hypoelliptic diffusions: Two examples. In M. Cranston and M. Pinsky, editors, *Stochastic Analysis: Summer Research Institute July 11-30, 1993.*, volume 57, pages 193–212, Providence, RI Providence, 1995. American Mathematical Society.

[6] P. Aviles, H. Choi, and M. Micallef. Boundary behavior of harmonic maps on non-smooth domains and complete negatively curved manifolds. *Journal of Functional Analysis*, 99:293–331, 1991.

[7] F.G. Ball, D. Mollison, and G. Scalia-Tomba. Epidemics with two levels of mixing. *Annals of Applied Probability*, 7:46–89, 1997.

[8] D. Barden and H.L. Le. Some consequences of the nature of the distance function on the cut locus in a Riemannian manifold. Research report, Mathematics Department, University of Nottingham, Nottingham, 1995.

[9] R. Bass and K. Burdzy. On domain monotonicity of the Neumann heat kernel. *Journal of Functional Analysis*, 116:215–224, 1993.

[10] A. Bernard, E.A. Campbell, and A.M. Davie. Brownian motion and generalized analytic and inner functions. *Ann. Institute Fourier*, 29:207–228, 1979.

[11] K. Burdzy and W.S. Kendall. Brownian coalescence on high-dimensional spheres. In preparation, 1997.

[12] K. Burdzy and W.S. Kendall. Efficient Markovian couplings. In preparation, 1997.

[13] E. Calabi. An extension of E. Hopf's maximum principle with an application to geometry. *Duke Mathematical Journal*, 25:45–56, 1958.

[14] H. Cartan. *Leçons sur la géometrie des espaces de Riemann.* Gauthiers-Villars, Paris, 1928.

[15] J. Cheeger and D. Gromoll. The splitting theorem for manifolds of non-negative Ricci curvature. *J. Differential Geometry*, 6:119–128, 1971.

[16] M.-F. Chen. Optimal couplings and application to Riemannian geometry. In B. Grigelionis *et al.*, editor, *Probability theory and Mathematical Statistics*, volume 1, page 15 pp. VPS / TEV, 1994.

[17] M.-F. Chen and F.-Y. Wang. Application of coupling method to the first eigenvalue on manifold. *Science in China (Ser. A)*, 37:1–14, 1994.

[18] S.-Y. Cheng. Liouville theorem for harmonic maps. *Proc. Symp. Pure Math.*, 36:147–151, 1980.

[19] S. Cohen. Géometrie différentielle avec sauts. *Comptes Rendus des Séances de l'Académie des Sciences. Série I. Mathématique*, 314:767–770, 1992.

[20] J.M. Corcuera and W.S. Kendall. Riemannian barycentres and geodesic convexity. Research report 325, Department of Statistics, University of Warwick, 1998. Submitted for publication.

[21] R. Courant, K. Freidrichs, and H. Lewy. Über die partiellen differenzengleichungen der mathematischen physik. *Mathematische Annalen*, 100:32–74, 1928.

[22] M. Cranston. Gradient estimates on manifolds using coupling. *Journal of Functional Analysis*, 99:110–124, 1991.

[23] M. Cranston. A probabilistic approach to Martin boundaries for manifolds with ends. *Probability Theory and Related Fields*, 96:319–333, 1993.

[24] M. Cranston, W.S. Kendall, and Yu. Kifer. Gromov's hyperbolicity and Picard's little theorem for harmonic maps. In I.M. Davies, A. Truman, and K.D. Elworthy, editors, *Stochastic Analysis and Applications, Gregynog, (9th-14th July 1995)*, pages 139–164, Singapore, 1996. World Scientific Press.

[25] M. Cranston, W.S. Kendall, and P. March. The radial part of Brownian motion II: Its life and times on the cut locus. *Probability Theory and Related Fields*, 96:353–368, 1993.

[26] R.W.R. Darling. Martingales in manifolds - definitions, examples, and behaviour under maps. In *Séminaires des Probabilités XVI supplément, Springer Lecture Notes in Mathematics*, volume 921, pages 217–236. Springer-Verlag, New York-Heidelberg-Berlin, 1982.

[27] R.W.R. Darling. Convergence of martingales on a Riemannian manifold. *Kyoto University. Research Institute for Mathematical Sciences. Publications*, 19:753–763, 1983.

[28] R.W.R. Darling. Convergence of martingales on manifolds of negative curvature. *Annales de l'Institut Henri Poincaré. Probabilités et Statistique*, 21:157–175, 1985.

[29] R.W.R. Darling. Exit probability estimates for martingales in geodesic balls, using curvature. *Probability Theory and Related Fields*, 93:137–152, 1992.

[30] R.W.R. Darling. Constructing gamma-martingales with prescribed limit, using backwards sde. *The Annals of Probability*, 23:1234–1261, 1995.

[31] R.W.R. Darling. Martingales on non-compact manifolds: maximal inequalities and prescribed limits. *Annales de l'Institut Henri Poincaré. Probabilités et Statistique*, 32:431–454, 1996.

[32] B. Davies. Picard's theorem and Brownian motion. *Trans. Amer. Math. Soc.*, 213:353–362, 1975.

[33] A. Debiard, B. Gaveau, and E. Mazet. Théorèmes de comparaison en geometrié Riemannienne. *Kyoto University. Research Institute for Mathematical Sciences. Publications*, 12:391–425, 1976.

[34] W. Doeblin. Exposé de la theorie des chaines simples constantes de Markov à un nombre fini d'etats. *Rev. Math. Union Interbalkanique*, 2:77–105, 1938.

[35] F. Duheille. On the range of $\mathbb{R}^2$ or $\mathbb{R}^3$-valued harmonic morphisms. *Comptes rendus de l'academie des sciences serie I - mathematique*, 320:1495–1500, 1995.

[36] F. Duheille. A probabilistic elementary proof of a result of P. Baird and J.C. Wood. *Annales de l'institut Henri Poincaré - probabilités et statistiques*, 33:283–291, 1997.

[37] T.E. Duncan. Estimation for jump processes in the tangent bundle of a Riemannian manifold. *Applied Mathematics and Optimization*, 4:265–274, 1978.

[38] T.E. Duncan. Stochastic systems in Riemannian manifolds. *Journal of Optimization Theory and Applications*, 27:399–426, 1979.

[39] E.B. Dynkin. Non-negative eigenfunctions of the Laplace-Beltrami operator and Brownian motion in certain symmetric spaces. *Doklady Akademii Nauk SSSR*, 141:288–291, 1961.

[40] J. Eells and L. Lemaire. *Selected Topics in Harmonic Maps.* American Mathematical Society, Providence, RI, 1983.

[41] K.D. Elworthy. *Stochastic differential equations on manifolds.* Cambridge University Press, Cambridge, 1982.

[42] K.D. Elworthy and W.S. Kendall. Factorizations of Brownian motions and harmonic maps. In K.D. Elworthy, editor, *From local times to global geometry, control, and physics*, volume 150 of *Pitman Research Notes in Mathematics*, pages 75–83. Longmans, 1986.

[43] M. Emery. Convergence des martingales dans les variétés. In *Colloque en l'Honneur de Laurent Schwartz (vol 2)*, volume 132, pages 47–62. Société Mathématique de France, 1985.

[44] M. Emery and P.A. Meyer. *Stochastic Calculus in Manifolds, with an Appendix by P.A. Meyer*. Springer-Verlag, New York-Heidelberg-Berlin, 1989.

[45] M. Emery and G. Mokobodzki. Sur le barycentre d'une probabilité dans une variéte. In *Séminaire de Probabilités XXV, Lecture Notes in Mathematics*, volume 1485, pages 220–233. Springer-Verlag, New York-Heidelberg-Berlin, 1991.

[46] M. Emery and W.A. Zheng. Fonctions convexes et semimartingales dans une variété. In *Séminaire de Probabilités XVIII, Lecture Notes in Mathematics*, volume 1059, pages 501–518. Springer-Verlag, New York-Heidelberg-Berlin, 1984.

[47] J. Fill. An interruptible algorithm for exact sampling via Markov Chains. *Annals of Applied Probability*, to appear, 1997. *WWW* URL: `www.mts.jhu.edu/~fill/`.

[48] R.A. Fisher. Dispersion on a sphere. *Proc. Roy. Soc. London*, A217:295–305, 1953.

[49] M. Fréchet. Les éléments aléatoires de nature quelconque dans un espace distancié. *Ann. Inst. Henri Poincaré*, 10:215–310, 1948.

[50] M. Fukushima and M. Okada. On Dirichlet forms for plurisubharmonic functions. *Acta Math.*, 159:171–214, 1984.

[51] R. Gangolli. On the construction of certain diffusions on a differentiable manifold. *Zeitschrift für Wahrscheinlichkeitstheorie und verwe Gebiete*, 2:406–419, 1964.

[52] S. Goldberg, T. Ishihara, and N.C. Petridis. Mappings of bounded dilatation of Riemannian manifolds. *J. Diff. Geom.*, 10:619–630, 1975.

[53] S. Goldberg and C. Mueller. Brownian motion, geometry, and generalizations of Picard's little theorem. *The Annals of Probability*, 11:833–846, 1983.

[54] R.E. Greene and H. Wu. *Function Theory on Manifolds Which Possess a Pole*, volume 699 of *Lecture Notes in Mathematics*. Springer-Verlag, New York, 1979.

[55] S. Hildebrandt. Harmonic mappings of Riemannian manifolds. In E. Giusti, editor, *Harmonic mappings and minimal immersions*, volume 1161 of *Lecture Notes in Mathematics*, pages 1–117, New York, 1985. Springer-Verlag.

[56] E.P. Hsu. Heat semigroup on a complete Riemannian manifold. *The Annals of Probability*, 17:1248–1254, 1989.

[57] E.P. Hsu and W.S. Kendall. Limiting angle of Brownian motion in certain two-dimensional Cartan-Hadamard manifolds. *Annales de la Faculté des Sciences de Toulouse Mathématiques (Série 6)*, 1:169–186, 1992.

[58] E.P. Hsu and P. March. The limiting angle of certain Riemannian Brownian motions. *Communications on Pure and Applied Mathematics*, 38:755–768, 1985.

[59] H. Huang and W.S. Kendall. Correction note to 'Martingales on manifolds and harmonic maps'. *Stochastics*, 37:253–257, 1991.

[60] N. Ikeda and S. Watanabe. *Stochastic differential equations and diffusion processes (First Edition)*. North-Holland / Kodansha, Amsterdam / Tokyo, 1981.

[61] K. Itô. Stochastic integral. *Proc. Imp. Acad. Tokyo*, 20:519–524, 1944.

[62] K. Itô. Stochastic differentials. *Applied Mathematics and Optimization*, 1:374–381, 1975.

[63] W. Jager and H. Jaul. Uniqueness and stability of harmonic maps and their Jacobi fields. *manuscript mathematica*, 28:269–291, 1979.

[64] E. Jørgensen. Construction of the Brownian motion and the Ornstein-Uhlenbeck process in a Riemannian manifold on basis of the Gangolli-McKean injection scheme. *Zeitschrift für Wahrscheinlichkeitstheorie und verwe Gebiete*, 44:71–87, 1978.

[65] J. Jost. Equilibrium maps between metric spaces. *Calc. Var.*, 2:173–204, 1994.

[66] J. Jost. *Riemannian Geometry and Geometric Analysis*. Springer-Verlag, New York-Heidelberg-Berlin, 1995.

[67] H.P. McKean Jr. *Stochastic Integrals*. Academic Press, New York, 1969.

[68] S. Kakutani. Markoff process and the Dirichlet problem. *Proceedings of the Japanese Academy. Series A. Mathematical Sciences*, 21:227–233, 1945.

[69] H. Kaneko. Liouville theorems based on symmetric diffusions. *Bulletin de la Société Mathématique de France*, 124:545–557, 1996.

[70] I. Karatzas and S.E. Shreve. *Brownian Motion and Stochastic Calculus*. Springer-Verlag, New York, 1988.

[71] H. Karcher. Riemannian centre of mass and mollifier smoothing. *Communications in Pure and Applied Mathematics*, 30:509–541, 1977.

[72] D.G. Kendall. The diffusion of shape. *Advances in Applied Probability*, 9:428–430, 1977.

[73] W.S. Kendall. Brownian motion, negative curvature, and harmonic maps. In D.Williams, editor, *Stochastic Integrals, Proceedings, LMS Durham Symposium, 1980*, volume 851 of *Lecture Notes in Mathematics*, pages 479–491, New York-Heidelberg-Berlin, 1981. Springer-Verlag.

[74] W.S. Kendall. Brownian motion and a generalised little Picard's theorem. *Transactions of the American Mathematical Society*, 275:751–760, 1983.

[75] W.S. Kendall. Brownian motion on 2-dimensional manifolds of negative curvature. *Lecture Notes in Mathematics*, 1059:70–76, 1984.

[76] W.S. Kendall. Nonnegative Ricci curvature and the Brownian coupling property. *Stochastics*, 19:111–129, 1986.

[77] W.S. Kendall. Stochastic differential geometry, a coupling property, and harmonic maps. *The Journal of the London Mathematical Society (Second Series)*, 33:554–566, 1986.

[78] W.S. Kendall. Stochastic differential geometry. In Yu.V. Prohorov and V.V. Sazonov, editors, *Proceedings, First World Congress of the Bernoulli Society*, volume 1, pages 515–524, Utrecht, 1987. VNU Press.

[79] W.S. Kendall. The radial part of Brownian motion on a manifold: Semimartingale properties. *The Annals of Probability*, 15:1491–1500, 1987.

[80] W.S. Kendall. Stochastic differential geometry: An introduction. *Acta Applicandae Mathematicae. An International Journal on Applying Mathematics and Mathematical Applications*, 9:29–60, 1987.

[81] W.S. Kendall. Martingales on manifolds and harmonic maps. In R. Durrett and M. Pinsky, editors, *The Geometry of Random Motion*, volume 73, pages 121–157, Providence, RI, 1988. American Mathematical Society.

[82] W.S. Kendall. Busemann functions and Brownian motion. In J. Norris, editor, *Stochastic Calculus in Applications, Symposium Proceedings, Cambridge 1986*, volume 197 of *Pitman Research Notes in Mathematics*, pages 63–68. Longman, 1988.

[83] W.S. Kendall. Symbolic computation and the diffusion of shapes of triads. *Advances in Applied Probability*, 20:775–797, 1988.

[84] W.S. Kendall. Coupled Brownian motions and partial domain monotonicity for the Neumann heat kernel. *Journal of Functional Analysis*, 86:226–236, 1989.

[85] W.S. Kendall. Probability, convexity, and harmonic maps with small image I: Uniqueness and fine existence. *Proceedings of the London Mathematical Society (Third Series)*, 61:371–406, 1990.

[86] W.S. Kendall. The Euclidean diffusion of shape. In D. Welsh and G. Grimmett, editors, *Disorder in Physical Systems*, pages 203–217, Oxford, 1990. Oxford University Press.

[87] W.S. Kendall. Symbolic Itô calculus: An overview. In N. Bouleau and D. Talay, editors, *Probabilités Numeriques*, pages 186–192, Rocquencourt, 1991. INRIA.

[88] W.S. Kendall. Convexity and the hemisphere. *The Journal of the London Mathematical Society (Second Series)*, 43:567–576, 1991.

[89] W.S. Kendall. Convex geometry and nonconfluent Γ-martingales I: Tightness and strict convexity. In M.T. Barlow and N.H. Bingham, editors, *Stochastic Analysis, Proceedings, LMS Durham Symposium, 11th - 21st July 1990*, pages 163–178, 1991.

[90] W.S. Kendall. The Propeller: A counterexample to a conjectured criterion for the existence of certain convex functions. *The Journal of the London Mathematical Society (Second Series)*, 46:364–374, 1992.

[91] W.S. Kendall. Convex geometry and nonconfluent Γ-martingales II: Well-posedness and Γ-martingale convergence. *Stochastics and Stochastic Reports*, 38:135–147, 1992.

[92] W.S. Kendall. Brownian motion and partial differential equations: From the heat equation to harmonic maps. In *Proc. ISI* 49th *session, Firenze (August 1993)*, volume 3, pages 85–101, 1993.

[93] W.S. Kendall. Doing stochastic calculus with *Mathematica*. In H. Varian, editor, *Economic and Financial Modeling with* Mathematica, pages 214–238. Springer-Verlag, New York, 1993.

[94] W.S. Kendall. Probability, convexity, and harmonic maps II: Smoothness via probabilistic gradient inequalities. *Journal of Functional Analysis*, 126:228–257, 1994.

[95] W.S. Kendall. The radial part of a Γ-martingale and a non-implosion theorem. *The Annals of Probability*, 23:479–500, 1995.

[96] W.S. Kendall. Perfect simulation for the area-interaction point process. In C.C. Heyde and L. Accardi, editors, *Probability Perspective*, Singapore, 1998. World Scientific Press.

[97] W.S. Kendall. A diffusion model for Bookstein triangle shape. *Advances in Applied Probability*, 30(2), 1998.

[98] Yu. Kifer. Brownian motion and harmonic functions on manifolds of negative curvature. *Theory of Probability and its Applications. An English translation of the Soviet journal Teoriya Veroyatnosteĭ i ee Primeneniya*, 21:81–95, 1976.

[99] Yu. Kifer. Spectrum, harmonic functions, and hyperbolic metric spaces. *Israel Journal of Math.*, 89:377–428, 1995.

[100] N.V. Krylov. *Controlled Diffusion Processes.* Springer-Verlag, New York, 1980.

[101] H.L. Le. Exit probability estimates for Γ-martingales on manifolds. *Stochastic and Stochastics Reports*, 54:199–209, 1995.

[102] H.L. Le. Mean size-and-shapes and mean shapes: a geometric point of view. *Advances in Applied Probability*, 27:44–55, 1995.

[103] H.L. Le. Limiting angle of Brownian motion on certain manifolds. *Probability theory and related fields*, 106:137–149, 1996.

[104] H.L. Le and D. Barden. Itô correction terms for the radial parts of semimartingales on manifolds. *Probability Theory and Related Fields*, 101:133–146, 1995.

[105] T.M. Liggett. *Interacting Particle Systems.* Springer-Verlag, New York, 1985.

[106] T. Lindvall. On coupling of Brownian motions. Research report, Department of Mathematics, Chalmers University of Technology & University of Göteborg, Göteborg, 1982.

[107] T. Lindvall. *Lectures on the Coupling Method.* John Wiley & Sons, Chichester and New York, 1992.

[108] T. Lyons and D. Sullivan. Function theory, random paths, and covering spaces. *Journal of Differential Geometry*, 19:299–323, 1984.

[109] P. Malliavin. *Géométrie Differentielle Stochastique.* Séminaire de Mathématiques Supérieures, Université de Montréal, 1978.

[110] P. March. Brownian motion and harmonic functions on rotationally symmetric manifolds. *Annals of Probability*, 14:793–801, 1986.

[111] D. Michel. Comparaison des notions de variétés faiblement et fortement harmoniques. *Comptes Rendus Acad. Sci. Paris*, 282:1007–1010, 1976.

[112] D. Mollison. Spatial contact models for ecological and epidemic spread (with discussion). *J. Roy. Statist. Soc.*, B39:283–326, 1977.

[113] B. Øksendal. *Stochastic Differential Equations: An Introduction with Applications.* Springer-Verlag, New York - Berlin, 1985.

[114] J.M. Oller and J.M. Corcuera. Intrinsic analysis of the statistical estimation. *The Annals of Statistics*, 23:1562–1581, 1995.

[115] E. Pardoux and S.G. Peng. Adapted solutions of a backward stochastic differential equation. *Systems & Control Letters*, 14:55–61, 1990.

[116] J. Picard. Calcul stochastique avec sauts sur une variété. In *Séminaire de Probabilités XXV, Lecture Notes in Mathematics*, volume 1485, pages 196–219. Springer-Verlag, New York-Heidelberg-Berlin, 1991.

[117] J. Picard. Martingales on Riemannian manifolds with prescribed limit. *Journal of Functional Analysis*, 99:223–261, 1991.

[118] J. Picard. Barycentres and martingales on a manifold. *Annales de l'institut Henri Poincare - probabilites et statistiques*, 30:647–702, 1994.

[119] M. Pinsky. Homogenization and stochastic parallel transport. In D. Williams, editor, *Stochastic Integrals*, volume 851 of *Lecture Notes in Mathematics*, pages 271–284, New York, 1981. Springer-Verlag.

[120] J.J. Prat. Étude asymptotique et convergence angulaire du mouvement brownien sur une variété à courbure négative. *C.R. Acad. Sciences Paris Séries A–B,*, 280A:1539–1542, 1975.

[121] G. Price and D. Williams. .Rolling with slipping I. *Séminaire de Probabilités*, XVII:194–197, 1983.

[122] J.G. Propp and D.B. Wilson. Exact sampling with coupled Markov chains and applications to statistical mechanics. *Random Structures and Algorithms*, 9:223–252, 1996.

[123] P.H. Roberts and H.D. Ursell. Random walk on a sphere and a Riemannian manifold. *Philosophical Transactions of the Royal Society of London. Series A. Mathematical and Physical Sciences*, 252:317–356, 1960.

[124] L.C.G. Rogers and D. Williams. *Diffusions, Markov Processes, and Martingales volume 2: Itô Calculus.* John Wiley & Sons, Chichester and New York, 1987.

[125] L. Schwartz. *Semi-martingales sur des varietes et martingales conformes sur des varietes analytiques complexes*, volume 780 of *Lecture Notes in Mathematics.* Springer-Verlag, New York-Heidelberg-Berlin, 1980.

[126] L. Schwartz. *Semimartingales and their stochastic calculus on Manifolds.* Les Presses de l'Université de Montréal, Montréal, 1984.

[127] S. Stafford. A probabilistic proof of S.Y. Cheng's theorem. *Annals of Probability*, 18:1816–1822, 1990.

[128] D. Stroock. On the growth of stochastic integrals. *Zeitschrift für Wahrscheinlichkeitstheorie und Verwandte Gebiete*, 18:340–344, 1971.

[129] D. Stroock and S.R.S. Varadhan. *Multidimensional diffusion processes.* Springer-Verlag, New York, 1979.

[130] D. Sullivan. The Dirichlet problem at infinity for a negatively curved manifold. *Journal of Differential Geometry*, 18:723–732, 1983.

[131] A. Thalmaier. Brownian motion and the formation of singularities in the heat flow for harmonic maps. *Probability theory and related fields*, 105:335–367, 1996.

[132] A. Thalmaier. On the differentiation of heat semigroups and Poisson integrals. *Stochastics and Stochastics Reports*, 61:297–321, 1997.

[133] A. Thalmaier and F.-Y. Wang. Gradient estimates for harmonic functions on regular domains in Riemannian manifolds. Research report 32, University of Warwick Department of Maths, 1996.

[134] M. van den Berg and J.T. Lewis. Brownian motion on a hypersurface. *Bull. London Math. Soc.*, 17:144–150, 1985.

[135] F.-Y. Wang. Brownian coupling property for negatively curved Riemannian manifolds. Research report, Department of Math., Beijing Normal University, Beijing, 1997.

[136] H. Wu. An elementary method in the study of nonnegative curvature. *Acta Mathematica*, 142:57–78, 1979.

[137] S.-T. Yau. Harmonic functions on complete Riemannian manifolds. *Communications on Pure and Applied Mathematics*, 28:201–228, 1975.

[138] W.A. Zheng. Sur le théorème de convergence des martingales dans une variété Riemannienne. *Zeitschrift für Wahrscheinlichkeitstheorie und Verwandte Gebiete*, 63:511–515, 1983.

AMS/IP Studies in Advanced Mathematics
Volume 8, 1998

Umberto Mosco: Dirichlet forms and self–similarity

3.1 Introduction

We will be concerned with *Dirichlet forms* and *fractals*. The two theories come naturally in connection when one studies dynamical properties of fractal structures. Both theories come out enriched by this interaction.

On the one hand, fractal theory finds in the Dirichlet forms the right tools for a variational approach to dynamics, based on a measure–valued Lagrangian formalism. On the other hand, the theory of Dirichlet forms, when applied to fractals, is opened to the concept of self-similarity, the study of functional invariance under *self–similar* transformations being itself a novel development in the calculus of variations.

The theory of Dirichlet forms — from the point of view of main concern here — can be seen as a development of the classic energy approach of Frostman, Cartan and others to potential theory. In its present axiomatic form, it was developed in the middle fifties by Beurling and Deny in two seminal papers, [3] and [4]. The first one of these papers is particularly inspiring from the point of view of fractals. In fact, the main features of Dirichlet forms are presented there in a finite dimensional setting, in connection with the theory of general electric networks. This has made their theory immediatly operative on typical fractal models — like the Sierpinski gasket — which occur as continuous limits of discrete *pre–fractals* and on which energy forms are constructed as limits of suitable renormalized discrete Dirichlet forms, following the so–called *decimation* procedure of physicists.

Fractals were introduced by Mandelbrot, as a general framework for the comprehension of a variety of natural phenomena. In the last two decades, various kinds of fractal models have received vast attention among physicists. An overview of physical applications can be found for example in the survey papers [20], [9], [40].

The mathematical theory is more recent. The study of dynamical properties of fractals meets two basic difficulties. Standard fractals — like the Sierpinski gaket — have a complicated fine structure that escapes any description of them as smooth, differentiable manifolds. This rules out standard classical partial differential equations. Second, measuring distances — a basic requirements of physical description — may be a cumbersome task on a fractal. Typical fractals, due to their highly fragmented and topological complicated structure, are purely unrectifiable sets in the sense of geometric measure theory.

The property enjoied by most common fractals and which makes it posssible to develop a fractal theory is *self–similarity*. From the physical point of view, the study of a fractal structure can be developed in three stages: the *geometry* of the structure, the distribution of mass on it (*statics*), the energy field (*dynamics*). In this physical context, self–similarity describes *invariance*.

The first general description of (self–similar) fractal statics in Euclidean spaces is the one given by Hutchinson [21]. More recent studies of self–similar measures in non–Euclidean spaces — like for example in certain homogeneous Lie groups of the kind of the Heisenberg group — are due to Strichartz [49], mostly in connection with Fourier transform theory and Olsen [41]. Extensions to metric spaces have been considered by Schief [47] and Marchi [31].

In our talks we will be mainly interested in dynamics. The dynamical properties of general self–similar measures being largely unexplored, we shall confine ourselves to homogeneous fractals. These are self–similar fractals for which the distribution of mass can be identified with the volume, the relevant volume measure being the Hausdorff measure in the — possibly fractional — dimension of the structure.

The first fundamental contributions to the mathematical theory of fractal dynamics have been of constructive type and have been given in the late eighties by Goldstein [19], Kusuoka [25], Barlow-Perkins [2], by probabilistic methods, with the construction of a diffusion process on the Sierpinski gasket. Lindstrøm [29] was able to identify an interesting class of fractals, so–called *nested fractals*, on which the construction of a diffusion process could be also carried on.

These "standard" fractal diffusions are constructed as the limit in law of suitably renormalized random walks on the pre–fractal approximations. The *generator* of the process is given the role of "standard Laplace operator" within the fractal. The process behaves like a *bona–fide* "Brownian motion" inside the structure, even if various basic scaling properties exhibit "anomalous" behavior — when compared with their classic Brownian analogues — as already observed by Alexander–Orbach [1] and Rammal–Toulouse [44].

The analytic approach was developed later, first for the Sierpinski gasket, by Kigami [23], Kusuoka [26], Fukushima-Shima [16], then extended also to nested fractals by Kusuoka [27], Fukushima[17].

Kigami constructed a "Laplacian" on the Sierpinski gasket, as the limit of suitable *finite–difference schemes*. Kusuoka produced a general framework for the construction of fractal Dirichlet forms. The paper by Fukushima–Shima — devoted as [44] to the spectral analysis of the Sierpinski gasket — went in the direction of interpreting the approximate finite–difference schemes of Kigami as discrete Dirichlet forms in the sense of Beurling-Deny. These discrete forms give in the limit a Dirichlet form of diffusion type on the continuos fractal. This form uniquely defines a self–adjoint operator in the natural L^2 Hilbert space of the fractal and this operator was shown to be the *Friedrichs extension* of the Laplace operator constructed by Kigami. Later this operator was seen to be the same — up to multiplicative constant — as the generator of the diffusion process obtained in the limit of random walks, as explained before. More general uniqueness and non–uniqueness results have been obtained by Metz [32], Sabot [46] and Peirone [42].

Finite–difference schemes and random walks — as we have seen, two parallel approachs to constructive fractal theory — were first brought together in the perspective of numerical analysis in the fundamental paper of Courant, Friedrichs and Lewy [12], devoted to the approximate solution of partial differential equations. The notion of Friedrichs extension and more generally the theory of self–adjoint operators of von Neumann and Friedrichs, also dates from those years, [15]. Generally speaking, all these notions — as well as that of Dirichlet form — have found wide applicability well beyond the classic framework of differential equations. As the fractal theory shows, they play also a basic role in dealing with non–smooth geometries and "irregular" variational problems, [35]. The lectures of J. Jost at this School open new interesting perspectives in this regard.

Let us give the plan of the paper. We shall first describe general *self–similar fractals*, section 1, and then *self–similar Lagrangians*, section 2. Following [34], a self–similar fractal possessing a non trivial self–similar Lagrangian will be called a *variational fractal*.

Self–similar *Lagrangian metrics* are described in section 3. We shall consider two cases, according to whether or not the local energy — the *Lagrangian* — is absolutely continous with respect to the underlying volume measure — a Hausdorff measure of possibly fractional dimension. In the absolutely continuous case, we describe the metric first introduced in [5] — an extension of the geodesic distance of geometric optics — and show that it inherits self–similarity from the Lagrangian. In the general case of a possibly singular Lagrangian — the one of interest in fractal theory — we define a (quasi–)metric, that reproduces Lagrangian self-similarity, again following [34]. We shall call it the *effective*

Lagrangian metric, or simply, the *intrinsic metric* of the given variational fractal.

In section 4 we describe the intrinsic scaling of the volume. We show that the volume of the intrinsic metric balls has polynomial growth in the radius. This property makes a variational fractal — with its intrinsic metric — a *space of homogeneous type*, briefly, a *homogeneous space*, in the sense of abstract harmonic analysis, [11], [48]. The exponent of the volume growth is the *homogeneous dimension* of the space. We call this dimension also the *intrinsic* (or *Lagrangian*) *dimension* of the fractal.

The intrinsic dimension plays a basic role in the theory and it is indeed the fundamental exponent of fractal dynamics. In section 5 we define the *field operators* — "Laplacians" — and show that the intrinsic dimension governs Weyl's asymptotics of their eigenvalues, [7], [39], [43]. An important dynamical notion — the so–called *spectral dimension* — is thus explained by metric "Lagrangian geometry". Local solutions and interior scaling of Laplacians are discussed in section 6.

In section 7 we describe a result from [36] — first stated in [34] — namely a family of *scaled Poincaré inequalities* on the intrinsic balls. These inequalities are the starting point of the variational theory developed in [5], ... , [8] for Dirichlet forms on homogeneous spaces, as a non–differentiable extension of the classic theory of De Giorgi–Nash–Moser for second order partial differential equations with measurable coefficients. Some important results, like *Nash inequalities* and *Morrey–Sobolev imbeddings*, can thus be extended to fractals, as described in section 8.

We conclude the paper with some remarks on the intrinsic diffusions. We refer to [37] [38] for further comments and examples and for some applications to physics.

3.2 Self–similar fractals

Self–similarity is a metric notion. From the point of view of geometry, it describes objects that — roughly speaking — are composed of a finite number of similar small copies of themselves.

By saying that B is a small similar copy of A we mean that A and B are both compact subsets of a given metric space $X \equiv (X, d_0)$, and $\psi(A) = B$, where $\psi : X \to X$ has the property $d_0\big(\psi(x),\, \psi(y)\big) = \alpha^{-1} d_0(x, y)$ for every $x,\, y \in X$, with $\alpha > 1$ some given constant (we say, that ψ is a *contractive similarity* of X, with contraction factor α^{-1}).

A (*self–similar*) *fractal* is then a subset K of a complete metric space $X \equiv$

(X, d_0), such that

$$K = \bigcup_{i=1}^{N} \psi_i(K) \ , \tag{3.2.1}$$

where $\psi_i : X \to X$, $i = 1, \ldots, N$, are $N \geq 1$ contractive similarities of X.

A theorem by Hutchinson [21] affirms that, given the set $\{\psi_1, \ldots, \psi_N\}$, there exists a unique non-empty compact set K in X that satisfies the self-similarity identity above. This is the *invariant set* associated with $\{\psi_1, \ldots, \psi_N\}$.

It should be noted that there may be different sets of mappings ψ_i's possessing the same invariant K. In the following, we shall assume that K is associated with a specific given set $\{\psi_1, \ldots, \psi_N\}$, $\psi_i : X \to X$, that satisfies

$$d_0\big(\psi_i(x), \psi_i(y)\big) = \alpha_i^{-1} d_0(x, y), \qquad \alpha_1 \geq \cdots \geq \alpha_N > 1 \ . \tag{\textbf{A1}}$$

Whenever in the following we shall describe self-similar properties of K, we will implicitly refer to this set $\{\psi_1, \ldots, \psi_N\}$. If in (A1) $\alpha_i \equiv \alpha$, we say that K is *isotropic*.

The exponent $d_f \geq 0$ uniquely defined by the identity

$$\sum_{i=1}^{N} \alpha_i^{-d_f} = 1 \tag{3.2.2}$$

is the *self-similar dimension* of K. If K is isotropic, then $d_f = \ln N / \ln \alpha$.

In the following we shall take the embedding space $X \equiv (X, d_0)$ to be the D-dimensional Euclidean space $X \equiv R^D$ with the usual Euclidean distance

$$d_0(x, y) \equiv |x - y| = \left(\sum_{h=1}^{D} |x_h - y_h|^2 \right)^{\frac{1}{2}} .$$

In [21] it is also proved that — given the set $\{\psi_1, \ldots, \psi_N\}$ — for every choice of constants $m_i > 0$, $\sum_{i=1}^{N} m_i = 1$, there exists a unique positive, regular Borel measure μ in X, with $\mu(X) = 1$, such that

$$\mu(\cdot) = \sum_{i=1}^{N} m_i \mu\big(\psi_i^{-1}(\cdot)\big) \ . \tag{3.2.3}$$

Moreover, μ is supported on K, the invariant set of $\{\psi_1, \ldots, \psi_N\}$.

The relation (1.3) can be equivalently written as

$$\int_K \varphi(y)d\mu = \sum_{i=1}^N m_i \int_K \varphi \circ \psi_i(x)d\mu$$

for every $\varphi \in C(K)$.

Any measure μ obtained in this way can be regarded as a — possibly non-homogeneous — invariant (i.e., self-similar) distribution of mass in K (a structure often referred to in the physical literature as a *multifractal*).

We will restrict our theory to homogeneous fractals, for which μ can be identified with the volume measure, *Vol.* In this case, μ denotes the invariant measure uniquely asssociated with the similitudes $\{\psi_1, \ldots, \psi_N\}$ by choosing

$$m_i \equiv \alpha_i^{-d_f} \ . \tag{3.2.4}$$

The pair $K = (K, \mu)$ — satisfying (1.1), (1.3), (1.4) — uniquely associated with a given set $\{\psi_1, \ldots, \psi_N\}$ will be called a *homogeneous fractal* if, in addition to (A1), $\psi_1, \ldots, \psi_n$ verify the following "open set condition", [21]:

(A2) $$\psi_i(O) \subset O, \ \psi_i(O) \cap \psi_j(O) = \oslash \forall i \neq j \in \{1, \ldots, N\},$$
for some nonempty open set O of X .

Roughly speaking, this condition implies that distinct copies $\psi_i(K)$, $\psi_j(K)$, $i \neq j$, either are disjoint (then K is totally disconnected, as the Cantor set), or they "just touch" without overlapping their interiors.

Any homogeneous fractal $K = (K, \mu)$ enjoies additional interesting metric properties, namely:

(i) $$0 < H^{d_f}(K) < \infty$$

where H^{d_f} denotes the d_f–dimensional Hausdorff measure in X. Hence d_f coincides with the Hausdorff dimension of K. d_f will be also called the *fractal dimension* of K. Moreover,

(ii) $$\mu = H^{d_f}(K)^{-1} H^{d_f} \llcorner K \ ,$$

that is, μ coincides with the normalized restriction to K of the d_f–dimensional Hausdorff measure H^{d_f} of X. This property allows us to consider μ — up to normalization — as the natural volume measure *within* K.

Before proceeding we introduce a few notations. For arbitrary indeces $i_1, \ldots, i_n \in \{1, \ldots, N\}$, $n = 1, 2, \ldots$, we define

$$\psi_{i_1 \ldots i_n} \equiv \psi_{i_1} \circ \cdots \circ \psi_{i_n} \ , \qquad A_{i_1 \ldots i_n} \equiv \psi_{i_1 \ldots i_n}(A)$$

The diameter of $K_{i_1\dots i_n}$ in X, denoted $\mathrm{diam}_X K_{i_1\dots i_n}$, satifies

$$\mathrm{diam}_X K_{i_1\dots i_n} = \alpha_{i_1}^{-1}\dots\alpha_{i_n}^{-1}\mathrm{diam}_X K\,. \tag{3.2.5}$$

We say that $K_{i_1\dots i_n}$ is *of size* $\ell > 0$ (in X) if

$$\alpha_1^{-1}\ell < \mathrm{diam}_X K_{i_1\dots i_n} \le \ell\,. \tag{3.2.6}$$

The set K can be obtained as the union of an increasing number of "copies" of smaller and smaller size. In fact, by iterating (1.1), for every $n = 1, 2, \dots$ we get

$$K = \bigcup_{i_1\dots i_n=1}^{N} K_{i_1\dots i_n}\,. \tag{3.2.7}$$

Moreover,

$$\mu(K_{i_1\dots i_n}\cap K_{j_1\dots j_n}) = 0 \quad \text{whenever} \quad \{i_1\dots i_n\}\ne\{j_1\dots j_n\}\,, \tag{3.2.8}$$

[21]. Any such fine subdivision of K obeys an important property, namely, the following *finite–overlapping property*, [34], [36] Theorem 2.1:

Lemma 3.2.1 *Under (A1), (A2), there exists an integer M, such that for every $\ell > 0$ the number of all $K_{i_1\dots i_n}$'s of size ℓ — $i_1\dots i_n\in\{1,\dots,N\}$, $n = 1, 2, \dots$ — that have non–empty intersection with a given ball B_ℓ^X of X is bounded above by M.*

The constant M depends only on D — the dimension of X — on α_1, $\mathrm{diam}_X K$ and on the radii of a ball contained in O and a ball containing O.

It follows that for every $\ell > 0$, we have

$$K\cap B_{\ell/2}^X \subset \bigcup_{\substack{K_{j_1\dots j_m}\text{ of size } \ell/2\\ K_{j_1\dots j_m}\cap B_{\ell/2}^X\ne\emptyset}} K_{j_1\dots j_m} \subset K\cap B_\ell^X \subset \bigcup_{\substack{K_{i_1\dots i_n}\text{ of size } \ell\\ K_{i_1\dots i_n}\cap B_\ell^X\ne 0}} K_{i_1\dots i_n} \tag{3.2.9}$$

where only a finite number ($\le M$) of sets contribute to the unions. This shows that every spherical cut $K\cap B_\ell^X$ of K in X, while not itself a self–similar copy of K, can be approximated both from outside and from inside by at most M exact self–similar copies of K, M being independent of ℓ as $\ell\downarrow 0$.

Remark 1.1 Lemma 1.1 is related to a density result in [21], Theorem 5.3 (1) (i). The basic property of X on which Lemma 1.1 relies is the following *homogeneity property*: there exists $\nu > 0$, such that every ball B_ℓ^X

contains at most $c\varepsilon^{-\nu}$ points with mutual distance $\geq \varepsilon\ell$. A (complete) quasi–metric space X enjoing this property is called a *space of homogeneous type*, [11]. Clearly, the D–dimensional Euclidean space possesses the homogeneity property, with $\nu = D$. The theory above can be extended to arbitrary (quasi–)metric spaces of homogeneous type [31]. Extensions of the open set condition to metric spaces have been considered also in [47].

Before concluding this section on fractal statics, let us give the (intrinsic) definition of the *boundary* Γ of K. According to [34], Γ is defined to be the set

$$\Gamma = \bigcup_{i \neq j} \psi_i^{-1}(K_i \cap K_j) \ . \tag{3.2.10}$$

Lemma 3.2.2 *Γ is a compact subset of $K \cap \partial O$ and $\mu(\Gamma) = 0$.*

Proof. By the open set condition, $K_i \subset \overline{O}_i$ for every $i = 1, \ldots, N$ and $K_i \cap O_j = 0$ if $i \neq j$. Therefore, $K_i \cap K_j = \partial O_i \cap \partial O_j$ and $\psi_i^{-1}(K_i \cap K_j) \subset K \cap \partial O$ for every $i \neq j$. Moreover, $\psi_i^{-1}(K_i \cap K_j)$ is compact. We have $\mu(\Gamma) \leq \sum_{i \neq j} \mu\left(\psi_i^{-1}(K_i \cap K_j)\right) = \sum_{i \neq j} \alpha_i^{d_f} \mu(K_i \cap K_j) = 0$.

It is easy to see that $K_i \cap K_j = \Gamma_i \cap \Gamma_j$, for $i \neq j$. Later on we shall assume that this relation scales down by self–similarity to arbitrary small $K_{i_1 \ldots i_n}$, that is, that the following property holds

(A3) $\quad K_{i_1 \ldots i_n} \cap K_{j_1 \ldots j_n} = \Gamma_{i_1 \ldots i_n} \cap \Gamma_{j_1 \ldots j_n}$ if $\{i_1, \ldots, i_n\} \neq \{j_1, \ldots, j_n\}$.

We recall that this is one of the properties defining *nested fractals*, for which $\Gamma = F$, F being the (finite) set of *essential fixed points* of the maps $\psi_1, \ldots, \psi_N$, [29].

3.3 Variational fractals

We now consider fractal dynamics. The Lagrangian approach to dynamics is based on the definition of a local energy L — the *Lagrangian* — which gives the total energy W after integration on the structure.

In the classic linear field theory, the Lagrangian is taken to be proportional to the square of the gradient of the potential field u,

$$\mathsf{L}\,[u] \propto \left|\operatorname{grad} u\right|^2 , \tag{3.3.1}$$

both having well defined pointwise values, that is, densities with respect to the underlying volume form *Vol* of the structure. Therefore, the total energy has the expression

$$W[u] \propto \int |\operatorname{grad} u|^2 d\,Vol\,. \tag{3.3.2}$$

In a typical fractal K — for example, the Sierpinki gasket — where no differentiable structure is available, both the square–of–the–field $|\operatorname{grad} u|^2$ and the Lagrangian $\mathsf{L}[u]$ cannot be expected to possess well defined pointwise values. We will assume that they are well defined as *measures* in K, giving the total energy after integration on K,

$$W[u] \propto \int_K d|\operatorname{grad} u|^2\,. \tag{3.3.3}$$

The theory of Dirichlet forms provides the technical tools for such a measure–valued Lagrangian formalism. Following [34], we give the following definition:

Definition 2.1 A (homogeneous) *variational fractal* — embedded in the space $X \equiv (X, d_0)$ — is a triple $K \equiv (K, \mu, W)$, where K is the invariant set of a given family $\{\psi_1, \ldots, \psi_N\}$ satisfying $(A1)$, $(A2)$; μ is the invariant measure on K; moreover,
(B1) W is a irreducible, strongly local, regular Dirichlet form in the Hilbert space $H = L^2(K, \mu)$ — with domain D_W — such that:

$$W[u] = \sum_{i=1}^{N} \rho_i W[u \circ \psi_i],\ u \in D_W\ , \tag{3.3.4}$$

where the constants $\rho_1, \ldots, \rho_N$ satisfy

$$\rho_i \equiv \mu(K_i)^\sigma\ , \tag{3.3.5}$$

for some constant $\sigma < 1$, independent of $i = 1, \ldots, N$.

By irreducible, we mean above that $W[u] = 0$ implies u =constant, that is, potentials with null energy are constants. This property rules out the trivial case $W \equiv 0$. Since $\mu(K_i) = \alpha_i^{-d_f}$, (2.5) is equivalent to $\rho_i \equiv \alpha_i^{-d_f\sigma}$.

As we have seen, an invariant measure μ exists on every self–similar fractal. Instead, there is no general result assuring the existence of a non–trivial self–similar energy on an arbitrary self–similar fractal. However, as mentioned in the Introduction, in addition to obvious classic Euclidean examples, invariant (self-similar) energy forms have been constructed on a variety of non-trivial fractals. Some basic examples of variational fractals are described for instance in [36] [37] [43].

Let us recall the definition of the local energy, from the general theory of Dirichlet forms [28], [18]. For bounded $u,\ v \in D_W$, the (signed) Borel measure $\mathsf{L}\,(u,v)$ is defined by the identity

$$\int_K \varphi d\mathsf{L}\,(u,v) \equiv \frac{1}{2}[W(\varphi u, v) + W(u, \varphi v) - W(\varphi, uv)]\ , \tag{3.3.6}$$

for $\varphi \in C_b \cap D_W$. We put $\mathsf{L}\,[u] \equiv \mathsf{L}\,(u,u)$. The definition of $\mathsf{L}\,[u]$ is extended to arbitrary $u \in D_W$ by taking limits of truncations and, by polarization, this gives $\mathsf{L}\,(u,v)$ on D_W.

We recall that the measures $\mathsf{L}\,(u,v)$ have a local character. The restriction of $\mathsf{L}\,(u,v)$ to any open subset U of X depends only on the restrictions of u and v to U.

The gradient-like character of L is reflected by the *Leibnitz rule* and *chain rule* satisfied by $\mathsf{L}\,(u,v)$. Roughly speaking, the usual derivation rules for products and composite functions now hold in the measure-valued sense, as $d\mathsf{L}\,(uv,z) = ud\mathsf{L}\,(v,z) + vd\mathsf{L}\,(u,z)$ and $d\mathsf{L}\,\big(\eta(u_1,\ldots,u_n),\ v\big) = \sum_{i=1}^{n} \eta_{x_i}(u_i,\ldots,u_n)d\mathsf{L}\,(u,v)$, [18], [33]. In the following, we identify $\mathsf{L}\,[u]$ with the (measure-valued) square-of-the-field, $|\operatorname{grad} u|^2$. The notation $|\operatorname{grad} u|^2$ reveals the quadratic gradient nature of $\mathsf{L}\,[u]$ more explicitly.

The Lagrangian can also be obtained from the transition function $p_t(x,dy)$ of the semigroup P_t by means of the formula

$$\int_K \varphi d\mathsf{L}\,[u] \equiv \lim_{t\downarrow 0} \frac{1}{2} \int\int_{K\times K-\mathrm{diag}} \varphi(x)\big(u(x)-u(y)\big)^2 p_t(x,dy)\mu(dx)\ , \tag{3.3.7}$$

which shows explicitly the positivity of $\mathsf{L}\,[u]$. For the relations between Dirichlet forms and Markov processes we refer to [18].

The Lagrangian inherits the self-similar invariance of W, in measure-valued sense:

Lemma 3.3.1 *For every $u,\ v \in D_W$, we have*

$$\int_K \varphi d\mathsf{L}\,(u,v) \equiv \sum_{i=1}^{N} \rho_i \int_K \varphi\circ\psi_i d\mathsf{L}\,(u\circ\psi_i,\ v\circ\psi_i)\ , \tag{3.3.8}$$

for all $\varphi \in C(K)$.

Proof. We do the proof for $\mathsf{L}[u]$. By (2.6), (2.4) and again (2.6), for bounded $u \in D_W$ and $\varphi \in D_W \cap C(K)$ we find

$$\int_K \varphi d\mathsf{L}[u] = W(\varphi u, u) - (1/2)W(\varphi, u^2) =$$

$$= \sum_{i=1}^{N} \rho_i \left[W\big((\varphi\circ\psi_i)(u\circ\psi_i),\, u\circ\psi_i\big) - (1/2)W\big(\varphi\circ\psi_i,\, (u\circ\psi_i)^2\big)\right] =$$

$$= \sum_{i=1}^{N} \rho_i \int_K \varphi \circ \psi_i d\mathsf{L}[u \circ \psi_i] .$$

This extends to all $\varphi \in C(K)$ by the regularity of the measures $\mathsf{L}[u]$ and to arbitrary $u \in D_W$ by approximation with truncated functions as before.

The local scaling of energy can be also obtained from the analogous property of the global energy W by a classical Fourier type argument, consisting in replacing rapidly oscillating functions $\varphi\exp(iu)$ into the identity (2.4), [33], Theorem 4.4.

The identity (2.8) extends also to characteristic functions $\varphi = 1_F$ of Borel sets F in K

$$\mathsf{L}[u](F) = \sum_{i=1}^{N} \rho_i \mathsf{L}[u \circ \psi_i]\big(\psi_i^{-1}(F)\big) , \qquad F \subset K , \tag{3.3.9}$$

which in standard measure notation can be equivalently written as

$$\mathsf{L}[u] = \sum_{i=1}^{N} \rho_i \psi_{i\#} \mathsf{L}[u \circ \psi_i] . \tag{3.3.10}$$

By iteration on the indices $i_1, \ldots, i_n \in \{1, \ldots, N\}$, we obtain the identity

$$\mathsf{L}[u] = \sum_{i_1 \ldots i_n = 1}^{N} \rho_{i_1} \ldots \rho_{i_n} \psi_{i_1 \ldots i_n \#} \mathsf{L}[u \circ \psi_{i_1 \ldots i_n}] . \tag{3.3.11}$$

The support of each measure $\psi_{i_1 \ldots i_n \#} \mathsf{L}[u \circ \psi_{i_1 \ldots i_n}]$ is the image under the map $\psi_{i_1 \ldots i_n}$ of the support of the measure $\mathsf{L}[u \circ \psi_{i_1 \ldots i_n}]$, hence

$$\operatorname{supp} \psi_{i_1 \ldots i_n \#} \mathsf{L}[u \circ \psi_{i_1 \ldots i_n}] \subset K_{i_1 \ldots i_n} \tag{3.3.12}$$

Therefore, for $(i_i, \ldots, i_n) \neq (j_i, \ldots, j_n)$, the supports of $\psi_{i_1 \ldots i_n \#} \mathsf{L}[u \circ \psi_{i_1 \ldots i_n}]$ and $\psi_{j_1 \ldots j_n \#} \mathsf{L}[u \circ \psi_{j_1 \ldots j_n}]$ intersect only on a subset of $K_{i_1 \ldots i_n} \cap K_{j_1 \ldots j_n}$. Even if we know, by (1.8), that $\mu(K_{i_1 \ldots i_n} \cap K_{j_1 \ldots j_n}) = 0$, we cannot say in general that $\psi_{j_1 \ldots j_n \#} \mathsf{L}[u \circ \psi_{j_1 \ldots j_n}]$ and $\psi_{i_1 \ldots i_n \#} \llcorner [u \circ \psi_{i_1 \ldots i_n}]$ also vanish on $K_{i_1 \ldots i_n} \cap K_{j_1 \ldots j_n}$. However, this is clearly the case if the measures $\mathsf{L}[u \circ \psi_{i_1 \ldots i_n}]$ and $\mathsf{L}[u \circ \psi_{j_1 \ldots j_n}]$ are *absolutely continuous* with respect to the measure μ. In fact, if $(i_1, \ldots, i_n) \neq (j_1, \ldots, j_n)$,

$$\mu(\psi^{-1}_{j_1 \ldots j_n}(K_{i_1 \ldots i_n} \cap K_{j_1 \ldots j_n})) = Lip(\psi^{-1}_{j_1 \ldots j_n})^{d_f} \mu(K_{i_1 \ldots i_n} \cap K_{j_1 \ldots j_n}) =$$

$$= \alpha_{j_1}^{d_f} \ldots \alpha_{j_n}^{d_f} \mu(K_{i_1 \ldots i_n} \cap K_{j_1 \ldots j_n}) = 0 \; .$$

Therefore, if $\mathsf{L}[u \circ \psi_{j_1 \ldots j_n}] \ll \mu$, then $\mathsf{L}[u \circ \psi_{j_1 \ldots j_n}](\psi^{-1}_{j_1 \ldots j_n}(K_{i_1 \ldots i_n} \cap K_{j_1 \ldots j_n})) = 0$, hence

$$\psi_{j_1 \ldots j_n \#} \mathsf{L}[u \circ \psi_{j_1 \ldots j_n}](K_{i_1 \ldots i_n} \cap K_{j_1 \ldots j_n}) =$$

$$= \mathsf{L}[u \circ \psi_{j_1 \ldots j_n}](\psi^{-1}_{j_1 \ldots j_n}(K_{i_1 \ldots i_n} \cap K_{j_1 \ldots j_n})) = 0$$

Lemma 3.3.2 *Let $u \in D_W$. Then $\mathsf{L}[u] \ll \mu$ if and only if $\mathsf{L}[u \circ \psi_i] \ll \mu$ for every $i = 1, \ldots, N$.*

Proof. Let $\mathsf{L}[u] \ll \mu$. We want to prove that if $\mu(E) = 0$, $E \subset K$, then $\mathsf{L}[u \circ \psi_i](E) = 0$ for every $i = 1, \ldots, N$. In fact, let $F = \psi_i(E)$. Then $\mu(F) = m_i \mu(E) = 0$. Since $\mathsf{L}[u] \ll \mu$, then $\mathsf{L}[u](F) = 0$. Therefore, by (2.9), $0 = \mathsf{L}[u](F) = \sum_{j=0}^{N} \rho_j \mathsf{L}[u \circ \psi_j](\psi_j^{-1}(F)) = \rho_i \mathsf{L}[u \circ \psi_i](E)$, hence $\mathsf{L}[u \circ \psi_i](E) = 0$ for each $i = 1, \ldots, N$. Now let $\mathsf{L}[u \circ \psi_i] \ll \mu$ for all $i = 1, \ldots, N$. Let $F \subset K$, $\mu(F) = 0$. If $E = \psi_i^{-1}(F)$, then $\mu(E) = m_i^{-1} \mu(F) = 0$. Since $\mathsf{L}[u \circ \psi_i] \ll \mu$, then $\mathsf{L}[u \circ \psi_i](E) = 0$. Therefore, $\mathsf{L}[u](F) = \sum_{i=1}^{N} \rho_i \mathsf{L}[u \circ \psi_i](\psi_1^{-1}(F)) = 0$. Thus, $\mathsf{L}[u] \ll \mu$.

By the previous lemma and preceding remarks, if $\mathsf{L}[u] \ll \mu$ we have

$$\psi_{j_1 \ldots j_n \#} \mathsf{L}[u \circ \psi_{j_1 \ldots j_n}](K_{i_1 \ldots i_n} \cap K_{j_1 \ldots j_n}) = 0 \tag{3.3.13}$$

if $(i_1, \ldots, i_n) \neq (j_1, \ldots, j_n)$.

By taking this into account in (2.11), for each fixed set of indeces $(i_1, \ldots, i_n)$ we find for arbitrary Borel sets $F = \psi_{i_1 \ldots i_n}(E) \subset K_{i_1 \ldots i_n}$

$$\mathsf{L}[u](F) \quad = \sum_{j_1 \ldots j_n = 1}^{N} \rho_{j_1} \ldots \rho_{j_n} \psi_{j_1 \ldots j_n \#} \mathsf{L}[u \circ \psi_{j_1 \ldots j_n}](F) =$$

$$= \rho_{i_1} \ldots \rho_{i_n} \psi_{i_1 \ldots i_n \#} \mathsf{L}[u \circ \psi_{j_1 \ldots j_n}](F) \; ,$$

that is

$$\mathsf{L}[u] \llcorner K_{i_1\ldots i_n} = \rho_{i_1}\ldots\rho_{i_n}\psi_{i_1\ldots i_n\#}\mathsf{L}[u\circ\psi_{i_1\ldots i_n}] \ . \tag{3.3.14}$$

Thus, the restriction of $\mathsf{L}[u]$ to $K_{i_1\ldots i_n}$ coincides with the renormalized "pushed–forward" image of the measure $\mathsf{L}[u\circ\psi_{i_1\ldots i_n}]$ by the map $y = \psi_{i_1\ldots i_n}(x)$ from K onto $K_{i_1\ldots i_n}$. Since $\psi_{i_1\ldots i_n\#}\mathsf{L}[u\circ\psi_{i_1\ldots i_n}]$ is supported in $K_{i_1\ldots i_n}$, in (2.14) we have identified $\psi_{i_1\ldots i_n\#}\mathsf{L}[u\circ\psi_{i_1\ldots i_n}]$ — the image under the map $\psi_{i_1\ldots i_n}$ from K into K — with its restriction to $K_{i_1\ldots i_n}$ — *i.e*, the image under the map $\psi_{i_1\ldots i_n}$ from K onto $K_{i_1\ldots i_n}$.

We can thus state the following transformation rule of the measure $\mathsf{L}[u]$ under the "change of variable" $y = \psi_{i_1\ldots i_n}(x)$ that carries K onto $K_{i_1\ldots i_n}$:

Lemma 3.3.3 *Let $u \in D_W$ be such that $\mathsf{L}[u] \ll \mu$. Then, for every $(i_1,\ldots,i_n)$,*

$$\mathsf{L}[u] \llcorner K_{i_1\ldots i_n} = \rho_{i_1}\ldots\rho_{i_n}\psi_{i_1\ldots i_n\#}\mathsf{L}[u\circ\psi_{i_1\ldots i_n}] \ , \tag{3.3.15}$$

that is,

$$\mathsf{L}[u](F) = \rho_{i_1}\ldots\rho_{i_n}\mathsf{L}[u\circ\psi_{i_1\ldots i_n}](E) \ ,$$

for arbitary Borel sets $F = \psi_{i_1\ldots i_n}(E)$, $E \subset K$, $F \subset K_{i_1\ldots i_n}$.

3.4 The Lagrangian metrics

We now face the problem of defining intrinsic fractal metrics. We take a different approach than in the usual fractal theory. There, metrics in K — for example the so–called *chemical distance* — are derived from the geometry of the fractal, as this is seen in the embedding space X. Here, we derive an effective metric from the intrinsic Lagrangian of K.

Before introducing this effective fractal metric, let us recall from [5] how the Lagrangian metrics are defined in the absolutely continuous case.

More precisely, let us assume that the Radon–Nikodym density $\mathsf{L}[\Phi](\cdot) = |\operatorname{grad}\Phi|^2(\cdot) = d\mathsf{L}[\Phi]/d\mu \in L^1_{\text{loc}}$ exists, that is

$$\mathsf{L}[\Phi] \ll \mu \ , \tag{3.4.1}$$

for every "test function" $\Phi \in D_W \cap C(K)$.

Following [5], we then define for every x, $y \in K$,

$$d_W(x,y) = \sup\{\Phi(x) - \Phi(y)\} \ , \tag{3.4.2}$$

where the supremum is over all $\Phi \in D_W \cap C(K)$ with $\mathsf{L}[\Phi] \le \mu$ on K.

The notation $\mathsf{L}[\Phi] \leq \mu$ on K is in the sense of measures, that is, $\mathsf{L}[\Phi](E) \leq \mu(E)$ for every Borel set $E \subset K$.

This defines a (possibly infinite–valued) distance $d_W(\cdot,\cdot)$ in K. It is then further assumed in [5] that the metric topology induced by d_W on K coincides with the initial topology of K (in particular, $d_W(\cdot,\cdot)$ is a distance).

The motivation for this definition comes from Riemannian geometry, where analogous "dual" definitions of geodesic distances are classic, see for example [22] and [35] where also *sub–Riemannian* metrics are described in this connection.

The relation with geodesic distances can be better understood by remarking that, at least formally, for each fixed $x \in K$, the function $\Phi(y) \equiv d_W(x,y)$ given by (3.2) is the *maximal subsolution* of the *eiconal equation* of geometric optics

$$(3.4.3) \qquad \left|\operatorname{grad}\Phi\right|^2 = 1, \quad \text{in } K\ ,$$

satisfying the "initial condition"

$$(3.4.4) \qquad \Phi(x) = 0\ .$$

Note that, in our case, the equation (3.3) has a meaning μ–a.e., $\left|\operatorname{grad}\Phi\right|^2(\cdot) \in \mathsf{L}^1(K,\mu)$ being the density of the Lagrangian $\mathsf{L}[\Phi]$ with respect to μ. A subsolution of (3.3) is any $\Phi \in D_W \cap C(K)$ such that $\left|\operatorname{grad}\Phi\right|^2(x) \leq 1$ μ–a.e. in K.

We now study the behavior of d_W under the transformations $y = \psi_{i_1\dots i_n}(x)$, $i_1,\dots,i_n \in \{1,\dots,N\}$, carrying K onto $K_{i_1\dots i_n}$. To this end, for every x, $y \in K$ we define for every $(i_1,\dots,i_n)$,

$$(3.4.5) \qquad d_W\big(\psi_{i_1\dots i_n}(x),\ \psi_{i_1\dots i_n}(y)\big) = \sup\{\Phi\circ\psi_{i_1\dots i_n}(x) - \Phi\circ\psi_{i_1\dots i_n}(y)\}\ ,$$

where now the supremum is over all $\Phi \in D_W \cap C(K)$ with $\mathsf{L}[\Phi] \leq \mu$ on $K_{i_1\dots i_n}$.

Here the notation $\mathsf{L}[\Phi] \leq \mu$ on $K_{i_1\dots i_n}$ is again in the sense of measures, that is $\mathsf{L}[\Phi](F) \leq \mu(F)$ for every Borel set $F \subset K_{i_1\dots i_n}$.

We shall now prove that d_W inherits invariance, *i.e.*, self–similarity, from the form W. More precisely, we have

Proposition 3.4.1 *Let* (3.1) *be satisfied. Then, the following identities hold for all x, $y \in K$:*

$$(3.4.6) \qquad d_W(x,y) = \left(\rho_{i_1}/m_{i_1}\right)^{1/2}\dots(\rho_{i_n}/m_{i_n})^{1/2} d_W\big(\psi_{i_1\dots i_n}(x),\ \psi_{i_1\dots i_n}(y)\big)$$

for every $(i_1,\dots,i_n)$, and

$$(3.4.7) \qquad d_W^2(x,y) = \sum_{i_1\dots i_n=1}^{N} \rho_{i_1}\dots\rho_{i_n} d_W^2\big(\psi_{i_1\dots i_n}(x),\ \psi_{i_1\dots i_n}(y)\big)\ .$$

the constants $\rho_1,\dots,\rho_N$ being those occurring in (2.9).

Proof. For arbitrary Borel sets $F = \psi_{i_1 \ldots i_n}(E)$, $E \subset K$, $F \subset K_{i_1 \ldots i_n}$, we have the transformation rule

$$\mu(F) = m_{i_1} \ldots m_{i_n} \mu(E)$$

and, by Lemma 2.3, for every $\Phi \in D_W \cap C(K)$,

$$\mathsf{L}[\Phi](F) = \rho_{i_1} \ldots \rho_{i_n} \mathsf{L}[\Phi \circ \psi_{i_1 \ldots i_n}](E) .$$

Therefore, the condition

$$\mathsf{L}[\Phi](F) \leq \mu(F)$$

is equivalent to the condition

$$\rho_{i_1} \ldots \rho_{i_n} \mathsf{L}[\Phi \circ \psi_{i_1 \ldots i_n}](E) \leq m_{i_1} \ldots m_{i_n} \mu(E) .$$

If we introduce the function $\tilde{\Phi} = \left(\rho_{i_1}/m_{i_1}\right)^{1/2} \ldots (\rho_{i_n}/m_{i_n})^{1/2} \Phi \circ \psi_{i_1 \ldots i_n}$, we have $\tilde{\Phi} \in D_W \cap C(K)$ and the latter condition can be equivalently written as

$$\mathsf{L}[\tilde{\Phi}](E) \leq \mu(E) .$$

Therefore, for x, $y \in K$,

$$\begin{aligned} d_W(x,y) \quad & \geq \tilde{\Phi}(x) - \tilde{\Phi}(y) = \left(\rho_{i_1}/m_{i_1}\right)^{1/2} \ldots \left(\rho_{i_n}/m_{i_n}\right)^{1/2} \\ & \left(\Phi \circ \psi_{i_1 \ldots i_n}(x) - \Phi \circ \psi_{i_1 \ldots i_n}(y)\right) , \end{aligned}$$

hence also

$$d_W(x,y) \geq \left(\rho_{i_1}/m_{i_1}\right)^{1/2} \ldots (\rho_{i_n}/m_{i_n})^{1/2} d_W\left(\psi_{i_1 \ldots i_n}(x) \ , \ \psi_{i_1 \ldots i_n}(y)\right) .$$

On the other hand,

$$\begin{aligned} d_W\left(\psi_{i_1 \ldots i_n}(x), \ \psi_{i_1 \ldots i_n}(y)\right) \ & \geq \ \Phi \circ \psi_{i_1 \ldots i_n}(x) - \Phi \circ \psi_{i_1 \ldots i_n}(y) = \\ & = \ \left(\rho_{i_1}/m_{i_1}\right)^{-1/2} \ldots (\rho_{i_n}/m_{i_n})^{-1/2} \left(\tilde{\Phi}(x) - \tilde{\Phi}(y)\right) \end{aligned}$$

hence

$$d_W\left(\psi_{i_1 \ldots i_n}(x), \ \psi_{i_1 \ldots i_n}(y)\right) \geq \left(\rho_{i_1}/m_{i_1}\right)^{-1/2} \ldots (\rho_{i_n}/m_{i_n})^{-1/2} d_W(x,y) .$$

Therefore,

$$d_W(x,y) = \left(\rho_{i_1}/m_{i_1}\right)^{1/2} \ldots \left(\rho_{i_n}/m_{i_n}\right)^{1/2} d_W\left(\psi_{i_1 \ldots i_n}(x), \ \psi_{i_1 \ldots i_n}(y)\right)$$

and this proves (3.6). We now rewrite the identity above as

$$m_{i_1} \dots m_{i_n} d_W^2(x,y) = \rho_{i_1} \dots \rho_{i_n} d_W^2\big(\psi_{i_1 \dots i_n}(x),\ \psi_{i_1 \dots i_n}(y)\big)$$

and sum the two sides over the indeces $i_1, \dots, i_n \in \{i, \dots, N\}$. Since

$$\sum_{i_1 \dots i_n = 1}^{N} m_{i_1} \dots m_{i_n} = 1$$

we obtain (3.7) and this concludes the proof of the Proposition.

We now come to the general case, by allowing the Lagrangian to be possibly singular with respect to the measure μ. In this case the test functions occurring in the definition of d_W may be only the constants, leading to a useless $d_W(x,y) \equiv 0$. In this case we cannot follow our previous procedure.

An alternative approach has been given in [34]. The idea comes from the theory of *subelliptic operators* of Hörmander's type. In this theory, as shown in [45] and [14], the "geodesic" distance associated with the given (smooth) vector fields — which is the same as the distance d_W of section 4, [22] — can be estimated by $d_W(x,y) \leq c|x-y|^{\varepsilon}$, where $\varepsilon > 0$ is the relevant subelliptic exponent, that is, the exponent occurring in the subelliptic estimate $\|u\|_{\varepsilon}^2 \leq c\left[W(u,u) + \|u\|_{L^2}\right]$. Here $\|\cdot\|_{\varepsilon}$ is the norm of the fractional Sobolev space H^{ε}. The exponent ε is also equal to the reciprocal of the length of the commutators in Hörmander's condition. Further comments on these metrics and on the role they play in sub-elliptic extensions of DeGiorgi–Nash–Moser theory can be found in [35].

Given a variational fractal, we now look for a *quasi–distance* d of the type

$$d(x,y) = d_0(x,y)^{\delta}\,, \qquad x,\ y \in K\,, \tag{3.4.8}$$

where $\delta > 0$ is a parameter to be chosen conveniently. By "quasi–distance" we mean that the triangle inequality is satisfied up to a multiplicative constant, denoted c_T in the following.

It turns out that such a δ is uniquely determined if we impose the additional condition that d — as the distance d_W in (3.7) — obeys the identity

$$d^2(x,y) = \sum_{i=1}^{N} \rho_i d^2\big(\psi_i(x),\ \psi_i(y)\big), \qquad x,\ y \in K\,, \tag{3.4.9}$$

where $\rho_i, \dots, \rho_N$ are the constants of the energy scaling (2.4).

In fact, we can easily prove the following Lemma:

Lemma 3.4.2 *There exists a unique $\delta \geq 0$ such that both* (3.8), (3.9) *hold. Such a δ is the unique solution of the equation*

(3.4.10)
$$\sum_{i=1}^{N} \rho_i \alpha_i^{-2\delta} = 1 \ ;$$

moreover,

(3.4.11)
$$\delta = d_f(1-\sigma)/2 \ .$$

Proof. By replacing (3.8) in (3.9) and using (A1) we obtain

$$d_0^{2\delta}(x,y) = \sum_{i=1}^{N} \rho_i \alpha_i^{-2\delta} d_0^{2\delta}(x,y) \ .$$

Therefore (3.10) holds — we are implicitly assuming that K is not a singleton — and this determines $\delta > 0$. By (2.5) $\rho_i = \alpha_i^{-d_f\sigma}$, thus (3.10) can be written as

$$\sum_{i=1}^{N} \alpha_i^{-(d_f\sigma+2\delta)} = 1 \ .$$

By definition of d_f this implies $d_f\sigma + 2\delta = d_f$, hence δ is given by (3.11).

If K is isotropic — $\alpha_i \equiv \alpha$, $\rho_i \equiv \rho$ — then (3.11) gives

(3.4.12)
$$\delta = \frac{1}{2}\frac{\log(N\rho)}{\log \alpha} \ .$$

This is the same as

(3.4.13)
$$N\rho = \alpha^{2\delta} \ ,$$

which is the fractal analogue of the Euclidean relation $N\rho = \alpha^2$. Note that $\alpha^{-\delta}$ is the contraction factor of the self–similarities ψ_i's in the new metric d.

The quasi–distance d given by (3.8), (3.10) is the *effective Lagrangian metric* of K. We also call d the *intrinsic metric* of K. The (quasi–)balls in the distance d,

(3.4.14)
$$B_R(x) \equiv B(x,R) = \{y \in K : d(x,y) < R\}, \qquad x \in K, \ R > 0 \ ,$$

are the *intrinsic balls* within K. For every $x \in K$ and $R > 0$, we have

(3.4.15)
$$B(x,R) \equiv K \cap B^X(x, R^{1/\delta}) \ ,$$

where $B^X(x,\ell)$, $\ell > 0$, denotes, as in section 1, the ball of radius ℓ in the embedding space X.

3.5 The intrinsic homogeneous structure

From now on, $K \equiv (K, \mu, W)$ is a given variational fractal — according to Definition 2.1 — d is the intrinsic metric within K, given by (3.8), (3.11), $B_R = B_R(x)$ are the intrinsic balls (3.14).

The diameter of $E \subset K$ in the metric d will be denoted by $\operatorname{diam} E$. With notation from section 1, $\operatorname{diam} E = (\operatorname{diam}_X E)^\delta$.

We study the volume growth of the intrinsic balls within K.

Theorem 3.5.1 *For every $x \in K$ and every $0 < r \leq R \leq \operatorname{diam} K$ we have*

$$(3.5.1) \qquad M^{-1}\alpha_1^{-1}\mu(B(x,R))\left(\frac{r}{R}\right)^\nu \leq \mu(B(x,r)) \leq M\alpha_1\mu(B(x,R))\left(\frac{r}{R}\right)^\nu,$$

where

$$(3.5.2) \qquad \nu = \frac{d_f}{\delta},$$

M the constant in Lemma 1.1.

From (4.2), by taking (3.11) into account, we get the

Corollary *The volume growth constant ν in* (4.1) *and the constant σ of the energy scaling* (2.4), (2.5) *are mutually related by*

$$(3.5.3) \qquad \nu = \frac{2}{1-\sigma}, \quad \sigma = \frac{\nu - 2}{\nu}.$$

The proof of the theorem is based on the finite–overlapping property of section 1, Lemma 1.1. We first use the approximation (1.9) of the Euclidean spherical cuts of the fractal to get an analogous approximation of the intrinsic balls within the fractal and then we exploit exact piece–wise self–similarity of the approximating sets. For details we refer to [36], Theorem 4.2.

It can be seen, as a consequence of the volume lower bound in (4.1), that the intrinsic balls B_r of K possess the homogeneity property mentioned in Remark 1.1, namely, every ball B_r contains at most $\varepsilon^{-\nu}$ disjoint balls $B_{\varepsilon r}$. Since the upper bound in (4.1) also holds, ν is the least (positive) constant for which this property holds.

The fractal K — with the intrinsic quasi–distance d — is thus a *space of homogeneous type* [11] and ν is the *homogeneous dimension* of K. We also call ν the *intrinsic dimension* of K, and, occasionally, also the Lagrangian dimension of K to stress the Lagrangian nature of d. As we shall see below, both d and ν play an important role in bringing into focus the main dynamical scaling laws.

The contraction factors of the mappings ψ_i in the intrinsic metric are

$$r_i = \alpha_i^{-\delta}, \qquad i = 1, \dots, N, \tag{3.5.4}$$

that is,

$$d(\psi_i(x,),\ \psi_i(y)) = r_i d(x,y), \qquad i = 1, \dots, N. \tag{3.5.5}$$

From Theorem 4.1 we easily get the following

Theorem 3.5.2 *For every $u \in D_W$ we have*

$$W[u] = \sum_{i=1}^{N} r_i^{\nu-2} W[u \circ \psi_i]. \tag{3.5.6}$$

Proof. By (4.2), (4.3), we have $\rho_i = \mu(K_i)^\sigma = \alpha_i^{-d_f\sigma} = r_i^{\nu-2}$ for every $i = 1, \dots, N$, therefore (4.6) follows from (2.4), (2.5).

By taking Lemma 2.1, (2.12) and Lemma 2.3 into account, we derive from Theorem 4.2 the intrinsic scaling of the Lagrangian:

Theorem 3.5.3 *For every u, $v \in D_W$, we have*

$$\int_K \varphi d\mathsf{L}(u,v) \equiv \sum_{i=1}^{N} r_i^{\nu-2} \int_K \varphi \circ \psi_i d\mathsf{L}(u \circ \psi_i,\ v \circ \psi_i), \tag{3.5.7}$$

for all $\varphi \in C(K)$. Moreover,

$$\mathsf{L}[u] = \sum_{i_1 \dots i_n = 1}^{N} (r_{i_i} \dots r_{i_n})^{\nu-2} \psi_{i_1 \dots i_n \#} \mathsf{L}[u \circ \psi_{i_1 \dots i_n}], \tag{3.5.8}$$

with supp $\psi_{i_1 \dots i_n \#}\mathsf{L}[u \circ \psi_{i_1 \dots i_n}] \subset K_{i_1 \dots i_n}$.

If $\mathsf{L}[u] \ll \mu$, then for every $(i_1, \dots, i_n)$,

$$\mathsf{L}[u](F) = (r_{i_i} \dots r_{i_n})^{\nu-2} \mathsf{L}[u \circ \psi_{i_1 \dots i_n}](E), \tag{3.5.9}$$

for arbitary Borel sets $F = \psi_{i_1 \dots i_n}(E)$, $E \subset K$.

The diameter of $K_{i_1 \ldots i_n}$ in the intrinsic metric is

$$\text{diam} K_{i_1 \ldots i_n} = r_{i_i} \ldots r_{i_n} \text{diam} K \ , \tag{3.5.10}$$

therefore the scaling factors in (4.8), (4.9) can be equivalently written as

$$(r_{i_i} \ldots r_{i_n})^{\nu - 2} \equiv (\text{diam} K_{i_1 \ldots i_n} / \text{diam} K)^{\nu - 2} \ .$$

If $K_{i_1 \ldots i_n}$ is of (intrinsic) size $r > 0$, *i.e.*,

$$r_i \, r < \text{diam} K_{i_1 \ldots i_n} \leq r \ ,$$

then

$$(r_{i_i} \ldots r_{i_n})^{\nu - 2} \approx (r / \text{diam} K)^{\nu - 2} \ .$$

Therefore, if $\mathsf{L}[u] \ll \mu$, the transformation rule (4.9) of the Lagrangian — under the "change of variable" $y = \psi_{i_1 \ldots i_n}(x)$ that carries K onto $K_{i_1 \ldots i_n}$ of size r, — can be written in loose notation as

$$\mathsf{L}[u](dy) \approx (r / \text{diam} K)^{\nu - 2} \psi_{i_1 \ldots i_n \#} \mathsf{L}[u \circ \psi_{i_1 \ldots i_n}](dx) \ , \tag{3.5.11}$$

for $y = \psi_{i_1 \ldots i_n}(x)$, $\text{diam} K_{i_1 \ldots i_n} \approx r$.

Theorems 4.2 and 4.3 show that energy and Lagrangian scale in the intrinsic metric as their classic analogues, now with the intrinsic dimension ν in the role of the Euclidean dimension D. In particular, (4.11) reproduces the transformation of the classic (measure-valued) Euclidean Lagrangian $\mathsf{L}[u] = \left| \text{grad}\, u \right|^2 dy$ under homotheties $y = rx$. The "correct" scaling of the energy has far reaching influence on all intrinsic dynamical scalings, as we shall see in the next sections.

3.6 Field operators and spectral asymptotics

Uniquely associated with the form W, there is a non-positive, self-adjoint operator Δ in the space $H = L^2(K, \mu)$ — with domain D_Δ — such that

$$W(u, v) = - \int_K (\Delta u) v \, d\mu \quad , \quad u \in D_\Delta \, , \ v \in D_W \ . \tag{3.6.1}$$

In the present Lagrangian formalism, this is the *field operator*, with "Neumann boundary condition", uniquely determined (up to multiplicative constant) by the given Lagrangian. In the following, we call Δ the *Laplacian*, with Neumann boundary condition on K.

The diffusion process X_t, $t > 0$, generated *within* K by the operator Δ is the *Brownian motion* in K "reflected on the boundary". The semigroup $P_t u = E^x\big(u(X_t)\big)$, $X_0 = x$, $u(0,x) = u$ defined by this process is the contraction semigroup $P_t u = \exp(t\Delta)u$ generated by Δ in H.

We define the form W_0 to be the closure in $L^2(K,\mu)$ of the restriction of W to the subspace $D_W \cap C_0(K-\Gamma)$, $C_0(K-\Gamma)$ being the space of continuous functions with compact support in $K - \Gamma$. W_0 is a closed, regular form in $L^2(K,\mu)$, with $D_W \cap C_0(K)$ dense both in D_{W_0} for the norm $\left(\|u\|^2_{L^2(K,\mu)} + W[u]\right)^{1/2}$ and in $C_0(K-\Gamma)$ for the uniform norm. Moreover, W_0 is strongly local [18] Th 3.1.2. The self adjoint operator Δ_0 with domain D_{Δ_0} in $L^2(K,\mu)$ is defined by the identity

$$(3.6.2) \qquad W_0(u,v) = -\int_K (\Delta_0 u)v \, d\mu, \quad u \in D_{\Delta_0}, \quad v \in D_{W_0}.$$

Δ_0 is the *Laplacian* with Dirichlet boundary condition on K. The diffusion process X_t, $t > 0$, generated within K by Δ_0 is the *Brownian motion* in K, with "absorption on the boundary".

The operators Δ, Δ_0 have a global character and differ only on the boundary. They act in the same way on functions with bounded support in $K-\Gamma$. We shall study the local behavior of Δ in section 6, together with its scaling properties.

Remark 5.1 The operators Δ and Δ_0 depend jointly on the energy form W and on the measure μ. If the restriction of W to $D_W \cap C(K)$ turns out to be closable in the space $L^2(K,\tilde{\mu})$ for some other measure $\tilde{\mu}$ on K, by the usual closure procedure we would get a closed form $\tilde{W}$ in the space $L^2(K,\tilde{\mu})$ and another pair of self–adjoint operators $\tilde{\Delta}$, $\tilde{\Delta}_0$ in the space $L^2(K,\tilde{\mu})$ will be uniquely associated with $\tilde{W}$. For example, on the Sierpinski gaket, a measure $\tilde{\mu}$ can be found such that the local energy measures of $\tilde{W}$ become absolutely continuous with respect to $\tilde{\mu}$, [26]. However, such a $\tilde{\mu}$ turns out to be *orthogonal* with respect to the natural Hausdorff volume measure μ in the dimension of the fractal. The operators $\tilde{\Delta}$, $\tilde{\Delta}_0$ have in this case quite a different nature than the field operators Δ, Δ_0 considered above.

In order to study the spectral behavior of the self–adjoint operators Δ, Δ_0 associated with the forms W, W_0, we shall further assume that the domain D_W of W is *compactly imbedded* in $L^2(K,\mu)$. Both operators $-\Delta$, $-\Delta_0$ possess then a discrete set of non–negative eigenvalues λ_n, λ_n^0 which tend to $+\infty$ as $n \to +\infty$.

The *integrated density of states* of $-\Delta$, $-\Delta_0$, are defined to be the functions $r(x)$, $r_0(x)$, that for each positive real $x \in R$ count the eigenvalues of $-\Delta$, $-\Delta_0$ — taken with their multiplicity — which are lesser or equal to x.

For the classic Laplace operator in a bounded open domain Ω of R^D, both functions $r(x)$, $r_0(x)$ — according to a well known result of H. Weyl [50] — behave like $c|\Omega|x^{D/2} + o(x^{D/2})$ as $x \to \infty$, were $|\Omega|$ denotes the D-dimensional Lebesgue measure of Ω.

An analogous asymptotic behavior of $r(x)$, $r_0(x)$,

$$c^{-1}x^{d_s/2} \leq r_0(x) \leq r(x) \leq cx^{d_s/2}$$

as $x \to \infty$ has been found to hold also on standard fractals, first by Alexander–Orbach [1], Rammal–Toulouse [44] and by Barlow–Perkins [2] and Fukushima–Shima [16] for the Sierpinski gasket, then also for general nested fractals and more general p.c.f. fractals by Fukushima [17], Kigami-Lapidus [24] and others.

However, the exponent d_s that takes the role of the Euclidean dimension D in the fractal spectral asymptotics has always been found to be different not only from the dimension D of the embedding Euclidean space, but also from the fractal (volume) dimension d_f itself, $d_s < d_f < D$. This is somewhat surprising and indeed the conjecture $d_s = d_f$ appears in early physical literature on the subject.

Despite the fact that $d_s \neq d_f$, we still expect d_s to reflect some suitable volume behavior of the structure also in the fractal case.

Such a "geometric" interpretation of d_s was first given for the Sierpinski gasket K in [7], [34], [39], where it was observed that d_s coincides with the dimension ν of K. As seen in section 3, the homogeneous dimension of K is related to the growth rate of the volume of concentric balls in the intrinsic metric of K. Since the intrinsic metric reflects Lagrangian scaling, the equality $d_s = \nu$ provides in some sense a Lagrangian "geometric" interpretation of d_s, as mentioned in the Introduction.

A fractal version of Weyl's result has been recently obtained by G. Posta in the general framework of variational fractals.

Theorem 3.6.1 [43]. *Let $K \equiv (K, \mu, W)$ be a variational fractal and let D_W be compactly injected in $L^2(K, \mu)$. Then, there exist constants $c > 0$, $\tilde{x} > 0$ such that*

$$c^{-1}x^{\nu/2} \leq r_0(x) \leq r(x) \leq cx^{\nu/2}$$

for all $x \geq \tilde{x}$, where ν is the Lagrangian dimension of K.

The proof is similar to the proof of the classic Weyl's result — as given for example in Courant–Hilbert [13] — and is based on the variational mini–max characterization of the eigenvalues and on the intrinsic scaling of the energies described in section 4.

As shown in [43], this result covers at the same time the classsic case, as well as most of known fractal examples, like nested fractals, p.c.f. fractals with regular harmonic structures in the sense of Kigami–Lapidus, the Sierpinski carpet in the two–dimensional Euclidean space.

3.7 Local solutions

The domains of the (global) self–adjoint operators Δ, Δ_0 depend on boundary conditions and are not stable under the transformations ψ_i. A function $u \circ \psi_i$ may not satisfy the same boundary condition satisfied by u.

We now define the *local* operator Δ_{loc} in $K - \Gamma$ and show how Δ_{loc} transforms under the maps ψ_i. The definition of Δ_{loc} is based on the notion of *local (weak) solutions* in $K - \Gamma$.

We define $D[W]_{\text{loc}}$ to be the space of all (μ-measurable) function u on K with the following property: for any relatively compact open subset U in $K - \Gamma$, there exists a function $w \in D[W]$ such that $u = w$ μ–*a.e.* on U. Here we denote D_W also by $D[W]$. Given $u \in D[W]_{\text{loc}}$, the measure $\mathsf{L}[u]$ is well defined in $K - \Gamma$ by putting $\mathsf{L}[u] \llcorner U$, $\mathsf{L}[w] \llcorner U$ for arbitrary U, w as above. The measure $L(u, v)$ in $K - \Gamma$, for u, $v \in D[W]_{\text{loc}}$, is defined by polarization.

Lemma 3.7.1 *Let $u \in D[W]_{loc}$. Then $u \circ \psi_i \in D[W]_{loc}$ for every $i = 1, \ldots, N$.*

Proof. Let $u \in D[W]_{\text{loc}}$ and $i = 1, \ldots, N$. Let V be a relatively compact open subset of $K - \Gamma$. Then $U = \psi_i(V)$ is a relatively compact open subset in $K_i - \Gamma_i$. Since $u \in D[W]_{\text{loc}}$, there exists a function $w \in D[W]$ such that $w = u$ μ–*a.e.* in U. Let $v = w \circ \psi_i$. Then, $v \in D[W]$ and $v = w \circ \psi_i = u \circ \psi_i$ μ–*a.e.* in V.
If $f \in L^2_{\text{loc}}(K, \mu)$ and $\lambda \geq 0$, we say that $u \in L^2_{\text{loc}}(K, \mu) \cap D[W]_{\text{loc}}$ is a *weak local solution* of the (formal) equation

$$-\Delta u + \lambda u = f \quad \text{in } K - \Gamma \,, \tag{3.7.1}$$

if

$$\int_{K-\Gamma} d\mathsf{L}\,(u, \varphi) + \lambda \int_{K-\Gamma} u\varphi \, d\mu = \int_{K-\Gamma} f\varphi \, du \tag{3.7.2}$$

for every $\varphi \in D[W] \cap C_0(K - \Gamma)$.

Given $f \in H = L^2(K, \mu)$ and $\lambda \geq 0$, the set of all weak solutions $u \in L^2(K, \mu) \cap D[W]_{\text{loc}}$ of (6.1)will be denoted by $R_\lambda[\Delta]_{\text{loc}}(f)$, that is,

$$R_\lambda[\Delta]_{\text{loc}}(f) = \{u : u \in L^2(K, \mu) \cap D[W]_{\text{loc}},\ u \text{ solution of } (6.2)\} \,. \tag{3.7.3}$$

This define $R_\lambda[\Delta]_{\rm loc}$ as a multivalued operator from H in H.

We denote by $(\lambda-\Delta)_{\rm loc}$ the inverse of $R_\lambda[\Delta]_{\rm loc}$, in the multivalued sense, that is, $(\lambda-\Delta)_{\rm loc}$ is the operator defined on the domain $D\left[(\lambda-\Delta)_{\rm loc}\right] \equiv R_\lambda[\Delta]_{\rm loc}(H)$ by setting $f \in (\lambda-\Delta)_{\rm loc}(u)$, for $u \in D[\lambda-\Delta]_{\rm loc}$, if and only if $u \in R_\lambda[\Delta]_{\rm loc}(f)$ for $f \in H$.

We remark that for every $u \in D[\lambda-\Delta]_{\rm loc}$, the set of functions $f \in (\lambda-\Delta)_{\rm loc}(u)$ consists of a single function $f \in H$. In fact, if f_1, $f_2 \in (\lambda-\Delta)_{\rm loc}(u)$, then we have both $u \in R_\lambda[\Delta]_{\rm loc}(f_1)$, $u \in R_\lambda[\Delta]_{\rm loc}(f_2)$, f_1, $f_2 \in H$, therefore we get from (6.2)

$$\int_{K-\Gamma} f_1 \varphi \, d\mu = \int_{K-\Gamma} f_2 \varphi \, d\mu .$$

for every $\varphi \in D[W] \cap C_0(K-\Gamma)$, hence, by Lemma 1.2 and the regularity of the form, $f_1 = f_2$ μ–$a.e.$in K. Therefore, $(\lambda-\Delta)_{\rm loc}$ is single-valued on its domain, and we write $f = (\lambda-\Delta)_{\rm loc}(u)$ if $u \in D[\lambda-\Delta]_{\rm loc}$.

In particular, for $\lambda = 0$ we get the operator $(-\Delta)_{\rm loc} \equiv -\Delta_{\rm loc}$, with domain $D[\Delta]_{\rm loc} = R_0[\Delta]_{\rm loc}$. That is, the functions $u \in D[\Delta]_{\rm loc}$ are the weak solutions $u \in L^2(K,\mu) \cap D[W]_{\rm loc}$ of the equation

$$\int_{K-\Gamma} d\mathsf{L}\,(u,\varphi) = \int_{K-\Gamma} f\varphi \, d\mu , \quad \forall \varphi \in D[W] \cap C_0(K-\Gamma) ,$$

for some $f \in H$, and $-\Delta_{\rm loc} u = f$.

It is easily verified that for every $\lambda \geq 0$ we have $D[(\lambda-\Delta)_{\rm loc}] = D[\Delta_{\rm loc}]$ and

$$(\lambda-\Delta)_{\rm loc}(u) = \lambda u - \Delta_{\rm loc} u$$

for every $u \in D[(\lambda-\Delta)_{\rm loc}] \equiv D[\Delta_{\rm loc}]$.

This shows that $R_\lambda[\Delta]_{\rm loc}$ is the inverse operator of $\lambda - \Delta_{\rm loc}$ in the multivalued sense,

$$R_\lambda[\Delta]_{\rm loc} \equiv R_\lambda[\Delta_{\rm loc}] \equiv (\lambda-\Delta_{\rm loc})^{-1} .$$

We also have $\Delta_{\rm loc} u = \lambda u - R_\lambda[\Delta_{\rm loc}]^{-1}(u)$, for every $u \in D[\Delta_{\rm loc}] \equiv (\lambda-\Delta_{\rm loc})^{-1}(H)$.

The transformation rules of $\Delta_{\rm loc}$ and $R_\lambda[\Delta_{\rm loc}] \equiv (\lambda-\Delta_{\rm loc})^{-1}$ under the mappings ψ_i's are described in the following theorem

Theorem 3.7.1 *If $u \in D[\Delta_{\rm loc}]$, then for every $\lambda \geq 0$ and for every $i = 1,\ldots,N$, we have $u \circ \psi_i \in D[\Delta_{\rm loc}]$ and*

$$(r_i^2\lambda - \Delta_{\rm loc})(u \circ \psi_i) = r_i^2\big((\lambda-\Delta_{\rm loc})u\big) \circ \psi_i ,$$

Proof. Let $u \in D[\Delta_{loc}]$, $i \in \{1, \dots, N\}$. For every $\lambda \geq 0$, $u \in D[\Delta_{loc}] \equiv (\lambda - \Delta_{loc})^{-1}(H)$. Therefore $u \in \mathsf{L}^2(K, \mu) \cap D[W]_{loc}$ and there exists $f \in \mathsf{L}^2(K, \mu)$ such that

$$\int_{K-\Gamma} d\mathsf{L}\,(u, \varphi) + \lambda \int_{K-\Gamma} u\varphi\, d\mu = \int_{K-\Gamma} f\varphi\, d\mu$$

for every $\varphi \in D[W] \cap C_0(K - \Gamma)$. We have $f = \lambda u - \Delta_{loc} u$.

Let $i \in \{1, \dots, N\}$. By Lemma 6.1, $u \circ \psi_i \in \mathsf{L}^2(K, \mu) \cap D[W]_{loc}$.

Let $g := f \circ \psi_i$. Then $g \in \mathsf{L}^2(K, \mu)$. Let us consider arbitrary functions $\varphi \in D[W] \cap C_0(K - \Gamma)$, $\Phi \in D[W] \cap C_0(K_i - \Gamma_i)$, $\varphi = \Phi \circ \psi_i$, $\Phi = \varphi \circ \psi_i^{-1}$, Φ extended to be $= 0$ outside $K_i - \Gamma_i$. We have

$$\int_{K-\Gamma} \Phi \circ \psi_i \big((f - \lambda u) \circ \psi_i\big) d\mu + \lambda \int_{K-\Gamma} \Phi \circ \psi_i (u \circ \psi_i) d\mu = \tag{3.7.4}$$

$$= \int_{K-\Gamma} \Phi \circ \psi_i g\, d\mu\ .$$

Since supp $\Phi \subset K_i - \Gamma_i$ and $\mu(K_i \cap K_j) = 0$ for $i \neq j$, we have by the self–similarity of μ,

$$\begin{aligned} r_i^\nu \int_K \Phi \circ \psi_i \big((f - \lambda u) \circ \psi_i\big) du &= \sum_{j=1}^N r_j^\nu \int_K \big(\Phi(f - \lambda u)\big) \circ \psi_j\, du = \\ &= \int_K \Phi(f - \lambda u) d\mu\ . \end{aligned}$$

On the other hand, by (6.2) and Theorem 4.3 — which extends in $K - \Gamma$ to arbitrary u, $v \in D[W]_{loc}$ — we have

$$\int_{K-\Gamma} \Phi(f - \lambda u) d\mu = \int_{K-\Gamma} d\mathsf{L}\,(u, \Phi) = \sum_{j=1}^N r_j^{\nu-2} \int_{K-\Gamma} d\mathsf{L}\,(u \circ \psi_j, \Phi \circ \psi_j)\ .$$

By the strong locality of the form W, since supp $\Phi \subset K_i - \Gamma_i$

$$\sum_{j=1}^N r_j^{\nu-2} \int_{K-\Gamma} d\mathsf{L}\,(u \circ \psi_j, \Phi \circ \psi_j) = r_i^{\nu-2} \int_{K-\Gamma} d\mathsf{L}\,\big((u \circ \psi_i, \Phi \circ \psi_i)\big)$$

and by collecting the previous identities, we get

$$r_i^\nu \int_K \Phi \circ \psi_i \big((f - \lambda u) \circ \psi_i\big) d\mu = r_i^{\nu-2} \int_{K-\Gamma} d\mathsf{L}\,\big((u \circ \psi_i, \Phi \circ \psi_i)\big)\ .$$

Therefore, we get from (6.4)

$$r_i^{-2}\int_{K-\Gamma} d\mathsf{L}\left((u\circ\psi_i,\,\Phi\circ\psi_i)\right)+\lambda\int_{K-\Gamma}\Phi\circ\psi_i(u\circ\psi_i)d\mu=\int_{K-\Gamma}\Phi\circ\psi_i\,g\,d\mu\;,$$

thus

$$\int_{K-\Gamma} d\mathsf{L}\left((u\circ\psi_i,\,\varphi)\right)+\lambda r_i^2\int_{K-\Gamma}\varphi(u\circ\psi_i)d\mu=r_i^2\int_{K-\Gamma}\varphi(f\circ\psi_i)d\mu\;,$$

for all $\varphi\in D_W\cap C_0(K-\Gamma)$.

We have shown that $u\circ\psi_i\in L^2(K,\mu)\cap D[W]_{\mathrm{loc}}$, $u\circ\psi_i\in D[\Delta_{\mathrm{loc}}]$ and

$$\lambda r_i^2 u\circ\psi_i-\Delta_{\mathrm{loc}}(u\circ\psi_i)=r_i^2 f\circ\psi_i\in L^2(K,\mu)\;,$$

therefore, by the expression of f,

$$(r_i^2\lambda-\Delta_{\mathrm{loc}})(u\circ\psi_i)=r_i^2\left((\lambda-\Delta_{\mathrm{loc}})u\right)\circ\psi_i$$

and this concludes the proof.

Corollary 3.7.2 *If $u\in D[\Delta_{loc}]$, then $u\circ\psi_i\in D[\Delta_{loc}]$ for every $i=1,\dots,N$ and*

$$r_i^{-2}\Delta_{loc}(u\circ\psi_i)=(\Delta_{loc}u)\circ\psi_i\;,$$

Any solution $u\in L^2(K,\mu)\cap D[W]_{\mathrm{loc}}$ of (6.2) with $f\equiv 0$ and $\lambda=0$ is said to be "*locally harmonic*" in $K-\Gamma$. Thus, u is locally harmonic in $K-\Gamma$ if $u\in D[\Delta_{\mathrm{loc}}]$ and $\Delta_{\mathrm{loc}}u=0$. From the previous corollary it follows

Corollary 3.7.3 *If $u\in L^2(K,\mu)\cap D[W]_{loc}$ is locally harmonic in $K-\Gamma$ then $u\circ\psi_i\in L^2(K,\mu)\cap D[W_{loc}]$ is locally harmonic in $K-\Gamma$ for every $i=1,\dots,N$.*

From Theorem 6.1 we also get the following scaling of resolvent operators.

Corollary 3.7.4 *For every $f\in L^2(K,\mu)$ we have $f\circ\psi_i\in L^2(K,\mu)$ for every $i=1,\dots,N$ and*

$$R_\lambda[\Delta_{loc}](f)\circ\psi_i\subset r_i^2 R_{\lambda r_i^2}[\Delta_{loc}](f\circ\psi_i)\;,$$

for $\lambda\geq 0$ in multivalued sense.

Similar formulas as in Theorem 6.1 and its corollaries hold if we replace ψ_i by $\psi_{i_1 \dots i_n}$, where $i_1, \dots, i_n \in \{1, \dots, N\}$, $n \geq 1$. Then, the factor r_i^2 will be replaced everywhere by the factor $(r_{i_1} \dots r_{i_n})^2$. Note that, if $K_{i_1 \dots i_n}$ is of size r, then $(r_{i_1 \dots i_n})^2 \approx (r/\text{diam } K)^2$, as in section 4.

If Δ is the classic Laplace operator, or any uniformly elliptic operator with *smooth coefficients*, then — by the H^2 regularity theory of elliptic equations — $D[\Delta_{\text{loc}}] = H^2_{\text{loc}}$. However, if Δ is a uniformly elliptic operator with *discontinuous* bounded measurable coefficients, then local harmonics — after the theory of De Giorgi and Nash — are only known to be Hölder continuous, but no further information on the weak derivatives is available in general.

De Giorgi–Nash Hölderianity theory has been extended to Dirichlet forms in [5], [6], [7], under the assumption that the Lagrangian measures $\mathsf{L}[u]$ are absolutely continuous with respect to the measure μ. The starting point of this theory is a family of *local scaled Poincaré inequalities* on the balls of the intrinsic eiconal metric d_W. The theory has a structural character, all relevant estimates — like Harnack inequalities — being uniform for a family of forms mutually equivalent up to given fixed constants, $c^{-1}W \leq W' \leq cW$.

In the following section we derive these scaled local Poincaré inequalities in a variational fractal from a single global Poincaré inequality. This shows that in presence of self–similarity — up to form equivalence — we can free the theory from its local assumptions.

3.8 Scaled Poincaré inequalities

The theory of Dirichlet forms on homogeneous spaces developed in [6], [7] is based on the assumption that the metric of the space and the Lagrangian are further related one another by a family of *local scaled Poincaré inequalities* on the metric balls of the space.

In this section we show that if the structure enjoys self–similar invariance — as any variational fractal does — then a much simpler starting point can be given to the whole theory. This is a single *global Poincaré inequality*, itself a consequence, for example, of a *Rellich property* — *i.e.*, the L^2 compact imbedding of the space of the functions of finite energy. This brings new light also into the classic theory itself.

In a variational fractal $K = (K, \mu, W)$, the local Poincaré inequalities mentioned before can be written as

$$\int_{B_r} \left|u - u_{B_r}\right|^2 d\mu \leq c \left(\frac{r}{\text{diam}K}\right)^2 \int_{B_{qr}} d\mathsf{L}[u] \tag{3.8.1}$$

for every $u \in D_W$ and every $0 < r \leq \text{diam}K$.

Here $B_r \equiv B(x,r)$ denotes as usual a ball within K for its intrinsic metric d and $c > 0$ is a structural constant, independent of x and r, that will be better specified later on, $q \geq 1$. Moreover, $u_{B_r} := \int_{B_r} u\, d\mu$.

If inequality (7.1) holds, then any potential u with vanishing Lagrangian $\mathsf{L}[u]$ in the ball $B_r \equiv B(x,r)$ must be constant, *i.e.*, (7.1) implies *local* irreducibility of the form. This property clearly demands some kind of *local connectivity* of K. In fact — by the local character of $\mathsf{L}[u]$ — if $B_r \equiv B(x,r)$ is composed, for example, by two disjoint open sets, then any function u equal to a constant on each of these two sets would violate (7.1).

A variational fractal is composed locally of pieces $K_{i_1 \dots i_n}$ mutually intersecting on sets of μ-volume null, therefore it is disconnected in measure sense. In order the inequalities (7.1) to hold, K must then possess some kind of finer local connectivity property. In fact, we shall assume below that K is locally connected in a *capacity sense*, the capacity referred to here being the capacity defined within K by the form W.

In addition to (A1), (A2), (A3) and (B1) — that define a variational fractal $K = (K, \mu, W)$ — we now assume that the following properties (B2) and (B3) also hold:

(B2) The form $\int_{K-\Gamma} d\mathsf{L}[u]$, with domain D_W, is closed in $L^2(K,\mu)$, moreover, there exists a constant $c_p > 0$ such that

$$\int_K \left|u - u_K\right|^2 d\mu \leq c_p \int_{K-\Gamma} d\mathsf{L}[u] \tag{3.8.2}$$

for every $u \in D_W$, where $u_K = \int_K u\, d\mu$ is the average of u on K and Γ is the boundary of K.

By the closed graph theorem, the closure property in (B2) is equivalent to the condition

$$\int_K d\mathsf{L}[u] \leq c\Big(\int_{K-\Gamma} d\mathsf{L}[u] + \int_K u^2 d\mu\Big)\,, \quad u \in D_W\,, \tag{3.8.3}$$

for some constant $c > 0$. If this condition is satisfied, a sufficient condition for the inequality (7.1) to hold is the compact imbedding of D_W in the space $L^2(K,\mu)$.

By Lemma 1.1, for every $x \in K$ and every $0 < \ell \leq \operatorname{diam}_X K$, the family of all $K_{i_1 \dots i_n}$ — $i_1, \dots, i_n \in \{1, \dots, N\}$, $n = 1, 2, \dots$ — of size ℓ which have non–empty intersection with the ball $B^X_\ell = B^X(x,\ell)$ is finite, and bounded by the constant M, and

$$K \cap B^X_\ell \subset \bigcup_{\substack{K_{i_1 \dots i_n} \text{ of size } \ell \\ K_{i_1 \dots i_n} \cap B^X_\ell \neq \emptyset}} K_{i_1 \dots i_n}\,. \tag{3.8.4}$$

We assume that these $K_{i_1\dots i_n}$ — suitably ordered — form a chain which is connected in the capacity sense.

More precisely, we assume that (B3) below holds:

(B3) K is *locally connected in the capacity sense*, as expressed by the following condition:

For every $x \in K$ and every $0 < \ell \leq \operatorname{diam}_X K$, the sets $K_{i_1\dots i_n}$'s of (Euclidean) size ℓ with non–empty intersection with the ball $B^X(x,\ell)$ — the total number of which is bounded by M — can be so ordered that two contiguous $K_{i_1\dots i_m}$, $K_{j_1\dots j_n}$, $(i_1,\dots,i_m) \neq (j_1,\dots,j_n)$, verify the following condition:

$$K_{i_1\dots i_m} \cap K_{j_1\dots j_n} = \psi_{i_1\dots i_m} T = \psi_{j_1\dots j_n} T' \tag{3.8.5}$$

where T, $T' \in \Gamma$ are such that

$$\operatorname{cap} T \geq k \quad \operatorname{cap} T' \geq k \ , \tag{3.8.6}$$

$k \operatorname{cap} S \leq \operatorname{cap} S' \leq (1/k) \operatorname{cap} S$ whenever $S' = \psi^{-1}_{j_i\dots j_n} \circ \psi_{i_i\dots i_m}$, $S \subset T$, for some constant $k > 0$.

Then, the following theorem holds, [36] Theorem 5.1:

Theorem 3.8.1 *Let $K = (K, \mu, W)$ be a variational fractal connected in the capacity sense, according to (A1), (A2), (A3), (B1), (B2), (B3). Then, there exist two constants $c > 0$ and $q \geq 1$, such that for every $u \in D_W$ and every $0 < r \leq \operatorname{diam} K$ the inequalities (7.1) hold. The constant c depends only on α_1, c_p, k and on the constant M in Lemma 1.1, and $q = 2^\delta$.*

The main steps of the proof are the following.

1. First we prove that for every $u \in D[E]$ and for every $i_1,\dots,i_n$, $n \geq 0$, we have the scaled Poincaré inequality

$$\int_{K_{i_1\dots i_n}} \left|u - u_{K_{i_1\dots i_n}}\right|^2 d\mu \leq \leq c_p(\operatorname{diam} K_{i_1\dots i_n}/\operatorname{diam} K)^2 \int_{K_{i_1\dots i_n} - \Gamma_{i_1\dots i_n}} d\mathsf{L}\,[u] \ . \tag{3.8.7}$$

We simply use the intrinsic scaling of μ and $\mathsf{L}\,[u]$.

2. The most difficult step is to extend inequality (7.7) to two contiguous sets $K_{i_i\dots i_m}$, $K_{j_i\dots j_n}$, $m \geq 1$ $n \geq 1$, $(i_i \dots i_m) \neq (j_i \dots j_n) \in \{1,\dots,N\}$. To this end we rely on capacity methods based on the property that the two sets are

connected in capacity sense according to (7.5), (7.6) for some constant $k > 0$. We prove that there exists a constant $c = c(c_p, k)$, such that for every $u \in D[E]$

$$\int_Q |u - u_Q|^2 d\mu \leq$$

$$\leq c \left\{ \begin{array}{c} (\mathrm{diam} K_{i_i \dots i_m} / \mathrm{diam} K)^2 \int_{K_{i_i \dots i_m} - \Gamma_{i_1 \dots i_m}} d\mathsf{L}\,[u] + \\ + (\mathrm{diam} K_{j_i \dots j_n} / \mathrm{diam} K)^2 \int_{K_{j_i \dots j_n} - \Gamma_{j_1 \dots j_n}} d\mathsf{L}\,[u] \end{array} \right\},$$

where $Q = K_{i_i \dots i_m} \cup K_{j_1 \dots j_n}$.

3. In the third step we develop an easier chaining argument on successive contiguous pairs as in step 2 to propagate Poincaré inequality accross the finite family of n_s–complexes $K_{i_1^s \dots i_{n_s}^s}$, $s = 1, \dots, M$ of size ℓ, covering a given spherical cut as in (7.4). If $Q = \bigcup_{s=1}^{M} K_{i_1^s \dots i_{n_s}^s}$ and, for each $s = 1, \dots, M-1$, $Q_s = K_{i_1^s \dots i_{n_s}^s} \cup K_{i_1^{s+1} \dots i_{n_s}^{s+1}}$, we first prove the inequality

$$\int_Q |u - u_Q|^2 d\mu \leq \left[4 \frac{\mu(Q)}{\min\limits_{s=1,\dots,M-2} \mu(K_{i_1^{s+1} \dots i_{n_s}^{s+1}})} \right]^{M-2}$$

$$\max_{s=1,\dots,M-1} \int_{Q_s} |u - u_{Q_s}|^2 d\mu\,.$$

Together with Poincaré inequalities of step 2, this gives

$$\int_Q |u - u_Q|^2 d\mu \leq c \left[4 \frac{\mu(Q)}{\min\limits_{s=2,\dots,M-1} \mu(K_{i_1^s \dots i_{n_s}^s})} \right]^{M-2}$$

$$\max_{s=1,\dots,M} \left(\frac{\mathrm{diam} K_{i_1^s \dots i_{n_s}^s}}{\mathrm{diam} K} \right)^2 \int_{K_{i_1^s \dots i_{n_s}^s} - \Gamma_{i_1^s \dots i_{n_s}^s}} d\mathsf{L}\,[u]\,,$$

where $Q = \bigcup_{s=1}^{M} K_{i_1^s \dots i_{n_s}^s}$.

4. The final step consists in rescaling the inequalities to the intrinsic metric to get

$$\int_{B(x,r)} |u(y) - u_{B(x,r)}|^2 d\mu \leq$$

$$\leq c \left(4 M \alpha_1^{d_f}\right)^M \left(\frac{R}{\mathrm{diam} K} \right)^2 \int_{B(x, 2^\delta R)} d\mathsf{L}\,[u]$$

with $c = c(c_p, k)$.

For the details of the proof we refer to [36].

3.9 Nash inequalities and Morrey–Sobolev imbeddings

The scaled Poincaré inequalities of Theorem 7.1 allow us to apply the imbedding theory for Dirichlet forms on spaces of homogeneous type, developed in [7], [8]. We point out the *local* character of the inequalities stated below, in particular their explicit scaling with respect to the size of the (intrinsic) balls.

As in section 7, we consider a variational fractal $K = (K, \mu, E)$, satisfying in addition (B.2) (B.3). We recall, in particular, that K, with the intrinsic metric, is a space of homogeneous type of dimension ν. Below by c we denote a constant that depends only on the constants N, α, ν, and on the constants c_p and k.

The first important scaled inequalities — the so–called *Nash inequalities* — are proved by relying on a finite–overlapping property of homogeneous spaces, see [8].

Theorem 3.9.1 *Let $\nu > 0$ be the intrinsic dimension of K. There exists a constant $c > 0$, such that for every $x \in K$ and every $0 < R \leq diamK$,*

$$(3.9.1) \quad \|u\|^{2+\frac{4}{\nu}}_{L^2(B(x,R),\mu)} \leq \left[R^2 \int_{B\left(x, c_T(q+1)R\right)} d\mathsf{L}[u] + \int_{B(x,R)} |u|^2 d\mu\right] \mu\big(B(x,R)\big)^{-\frac{2}{\nu}} \left[\int_{B(x,R)} |u| d\mu\right]^{\frac{4}{\nu}} .$$

for every $u \in D_W\big[B\big(x, c_T(q+1)R\big)\big]$.

Proof. It suffices to prove the normalized inequality

$$\|u\|^{2+\frac{4}{\nu}}_{L^2\left(B(x,R),\mu\right)} \leq c\left[R^2 \int_{B\left(x, c_T(q+1)R\right)} d\mathsf{L}[u] + \int_{B(x,R)} |u|^2 d\mu\right]$$

for every $u \in D_W\big[B\big(x, c_T(q+1)R\big)\big]$ such that

$$\mu\big(B(x,R)\big)^{-\frac{1}{2}} \int_{B(x,R)} |u| d\mu = 1 .$$

Given arbitrary $0 < r = 4c_T s \leq R$, the ball $B_R = B(x,R)$ can be covered by the union of a sequence of balls $B_{i,r} \equiv B(y_i, r)$, $y_i \in B(x,R)$, with $B(y_i, s)$, and $B(y_j, s)$ disjoint if $i \neq j$, such that every $y \in K$ belongs to at most P balls $B(y_i, \sigma s)$ for arbitrary $\sigma > 0$, $0 < \sigma s = R \leq \operatorname{diam} K$, where $P < \infty$ is independent of s. We then estimate

$$\int_{B_R} |u|^2 d\mu \leq \sum_i \int_{B(y_i,r)} |u|^2 d\mu \leq$$

$$\leq c \sum_i \left\{ \int_{B_{i,r}} |u - u_{i,r}|^2 d\mu + |u_{i,r} - u_{i,s}|^2 \mu(B_{i,r}) + |u_{i,s}|^2 \mu(B_{i,r}) \right\} \leq$$

$$\leq c \left\{ s^2 \sum_i \int_{B_{i,r}} d\mathsf{L}\,[u] + \left(\frac{R}{s}\right)^\nu \mu(B_R)^{-1} \sum_i \left(\int_{B_{i,r}} |u| d\mu \right)^2 \right\}$$

and by the finite overlapping of the covering and the normalization we get

$$\left(\int_{B_r} |u|^2 d\mu \right)^{1/2} \leq c \left\{ s \left(\int_{B_{c_T(q+1)R}} d\mathsf{L}\,[u] \right)^{1/2} + \left(\frac{R}{s}\right)^{\nu/2} \right\} .$$

If

$$s = \frac{1}{4c_T} \left[\left(\int_{B_{c_T(q+1)R}} d\mathsf{L}\,[u] \right)^{1/2} + R^{-1} \left(\int_{B_R} |u|^2 d\mu \right)^{1/2} \right]^{-2/(2+\nu)}$$

then $r = 4c_T s \leq R$ and the previous inequality gives

$$\|u\|_{L^2(B_R)}^{2+\frac{4}{\nu}} \leq c \left\{ R^2 \int_{B_{c_T(q+1)R}} d\mathsf{L}\,[u] + \int_{B_R} |u|^2 d\mu \right\} .$$

De Giorgi's classic truncation method is the main tool used in [8] to deduce *scaled Sobolev inequalities* — when the dimension of K is $\nu > 2$ — from the scaled Poincaré inequalities of section 7 and from the scaled Nash inequalities above.

Theorem 3.9.2 *Let $\nu > 2$. Then, there exists a constant $c > 0$, such that for every $x \in K$ and every $R > 0$, we have*

$$\left(\frac{1}{\mu(B(x,R))} \int_{B(x,R)} |u|^{2^*} d\mu \right)^{1/2^*} \leq \tag{3.9.2}$$

$$\leq c\left[\frac{R^2}{\mu\big(B(x,R)\big)}\int_{B\left(x,c_T(q+1)R\right)} d\mathsf{L}\,[u] + \frac{1}{\mu\big(B(x,R)\big)}\int_{B(x,R)}\left|u\right|^2 d\mu\right]^{1/2}$$

with $2^* = 2\nu/(\nu-2)$.

These inequalities are proved by an iteration argument on the summability exponent, we refer to [8] for more details.

If the dimension of K is $\nu = 2$, we have "conformal invariance" of the energy, according to the scaling results of section 3. In this case, we obtain scaled inequalities of John–Nirenberg type involving intrinsic BMO norms and related exponential integrability properties of the functions of finite energy. We refer to [7], [8].

We now consider fractals of dimension $\nu < 2$ and state below the *Morrey inequalities* from [7]. This case is of particular relevance because it includes some of the most interesting examples of fractals constructed so far, like for example the so-called nested fractals. The main result is that on *all* variational fractals with dimension $\nu < 2$ — under the assumption of section 7 — the functions of finite energy are *Hölder continuous*. In the classic Euclidean theory in R^D — that is, for the classic Laplace operator — this property holds only in dimension $D = 1$.

Theorem 3.9.3 *Let* $\nu < 2$. *Then, for every* $x \in K$ *and every* $R > 0$,

$$|u(y) - u(z)| \leq c\left[\frac{R^2}{\mu\big(B(x,R)\big)}\int_{B\left((x,c_T(q+1)R)\right)} d\mathsf{L}\,[u]\right]^{1/2} d(y,z)^{1-\frac{\nu}{2}}\ ,$$

$$|u(y)| \leq c\left[\frac{R^2}{\mu\big(B(x,R)\big)}\int_{B\left((x,c_T(q+1)R)\right)} d\mathsf{L}\,[u] + \right.$$

$$\left. + \frac{1}{\mu\big(B(x,R)\big)}\int_{B(x,R)}\left|u\right|^2 d\mu\right]^{1/2},$$

for all y, $z \in B(x,R)$.

The intrinsic Morrey's imbedding $D_W \subset C^\beta$ of Theorem 8.3, $\beta = 1 - (\nu/2)$, clearly implies the Euclidean imbedding $D_W \subset C^{\beta_e}_{\text{eucl}}$, where now $C^{\beta_e}_{\text{eucl}}$ is the space of Hölder continuous functions with Hölder exponent $\beta_e = \delta\beta$ in the Euclidean metric of K. For the Sierpinski gasket, this estimate was first obtained by S. Kozlov, by direct calculations. The Euclidean Hölder exponent β_e, however, could not be explained as an exponent of Morrey's type, because it violates the expected Morrey's type relation $\beta_e = 1-(\tilde{d}/2)$ for all relevant fractal dimensions $\tilde{d} = D$, d_f, ν.

In order to explain how Theorem 8.3 follows from the scaled Poincaré inequalities of section 7, let us show how Kozlov's estimate can be derived from these inequalities. If $\nu < 2$, from the scaled Poincaré inequalities and the estimate $\frac{\mu(B_r)}{\mu(B_R)} \approx \left(\frac{r}{R}\right)^\nu$ we derive the estimate

$$\sup_B \left[\frac{1}{\mu(B)} \int_B \left|u - u_B\right|^2 d\mu\right]^{1/2} \leq C\mu(B)^\gamma ,$$

for every ball $B \equiv B_r$ in K, where $\gamma = \frac{1}{\nu} - \frac{1}{2}$ and $C \equiv C(u) = c_1\sqrt{W[u]}$. By a general result of the theory of homogeneous spaces, due to Macias–Segovia [30], this estimate implies that u — up to μ–a.e. equivalence — satisfies the estimate

$$|u(x) - u(y)| \leq c_2 C(u)\mu(B)^\gamma$$

in K for any ball B containing x and y. This easily gives the inequality

$$\sup_{x,y\in K} \frac{|u(x) - u(y)|}{d(x,y)^\beta} \leq c\sqrt{W(u)} \ .$$

with the exponent $\beta = 1 - \frac{\nu}{2}$.

Let us conclude with some remarks on the transition function $p_t(x,y)$ of the diffusion associated with the form W, by disregarding possible boundary effects.

It is well known from the work of Nash and Carlen–Kusuoka–Stroock, [10], that Nash inequality can be used to derive on diagonal estimates of $p_t(x,y)$. In particular, if the semigroup of W is a Feller semigroup, the estimate $p_t(x,x) \leq c t^{-\nu/2}$ for every $t \in (0,+\infty)$ follows from (8.1), as shown in [10]. By Theorem 8.3 above, the semigroup is of Feller type whenever the intrinsic dimension of K is $\nu < 2$.

On the other hand, the classic Gaussian behavior of $p_t(x,y) \equiv p_t(r)$ in the range $r^2/t \gg 1$ — where $r = d(x,y)$ — does not extend to fractals, as the example of the Sierpinski gasket shows, see [9], [2]. In our present framework, as shown in [38], $p_t(r)$ can be estimated in the intrinsic metric as

$$p_t(r) \approx t^{-\frac{\nu}{2}} \exp\left(-c\frac{r^2}{t}\right)^{\frac{\nu_{\min}}{2-\nu_{\min}}} ,$$

where $\nu_{\min}$ denotes the minimum *intrinsic* fractal dimension of the paths connecting two sites x, y within the fractal. We refer to [38] for further comments on fractal transport properties.

Acknowledgements. The author wishes to thank the Alexander von Humboldt Stiftung for its support and the Institute of Applied Mathematics at the University of Bonn for its hospitality.

References

[1] ALEXANDER S., ORBACH R., *Density of states on fractals:"fractons"*, J. Physique Lett. 43 (1982) L–625.

[2] BARLOW M.T., PERKINS E.A., *Brownian motion on the Sierpinski gasket*, Prob. Th. Rel. Fields 79 (1988) 543–624.

[3] BEURLING A., DENY J., *Espaces de Dirichlet, I. Le cas élémentaire*, Acta Math. 99 (1958) 203–224.

[4] BEURLING A., DENY J., *Dirichlet spaces*, Proc. Nat. Acad. Sci. U.S.A. 45 (1959) 208–215.

[5] BIROLI M, MOSCO U., *Formes de Dirichlet et estimations structurelles dans les milieux discontinus*, C. R. Acad. Sci. Paris Série I, t.313 (1991) 593–598.

[6] BIROLI M, MOSCO U., *A Saint–Venant principle for Dirichlet forms on discontinuous media*, Preprint Series no. 224, SFB 256, Universität Bonn, pp. 1–50, 1992; *A Saint–Venant principle type for Dirichlet forms on discontinuous media*, Ann. Mat. Pura Appl. (IV)CLXIX (1995) 125–181.

[7] BIROLI M., MOSCO U., *Sobolev and isoperimetric inequalities for Dirichlet forms on discontinuous media*, Rend. Mat. Acc. Lincei s. 9, vol. 6 (1995), 37–44.

[8] BIROLI M., MOSCO U., *Sobolev inequalities for Dirichlet forms on homogeneous spaces* , Publ. Lab. Anal. Num., Univ. Paris VI, n. R93004, 1993, and in " Boundary values for partial differential equations and applications ", C. Baiocchi and J.L. Lions eds., Research Notes in Applied Mathematics, Masson, 1993; *Sobolev inequalities on homogeneous spaces*, Potential Anal. 4 (4) (1995) 311–324.

[9] BUNDE A., HAVLIN S., *Fractals and Disordered Systems*, Springer V. 1991.

[10] CARLEN E.A., KUSUOKA S, STROOCK D.W., *Upper bounds for symmetric Markov transition functions*, Ann. Inst. H. Poincaré 2 (1987) 245–287.

[11] COIFMAN R.R, WEISS G., *Analyse harmonique sur certaines éspaces homogenes*, Lect. Notes in Math. 242, Springer V., 1971.

[12] COURANT R., FRIEDRICHS K.O, LEWY H., *Über partielle Differenzengleichungen der mathematischen Physik*, Math. Ann. 100 (1928) 32–74.

[13] Courant R., Hilbert D., *Methods of Mathematical Physics*, vol. 1, J. Wiley & Sons, 1966.

[14] Fefferman C.L., Phong D. H., *Sub-elliptic eigenvalue problems*, in "Harmonic Analysis", Wadsworth Math. Series 2, 1983, 590–606.

[15] Friedrichs K.O., *Spektraltheorie halbbeschränkter Operatoren und Anwendung auf die spektralzerlegung von Differentialoperatoren*, Math. Ann. I. 109 (1934) 465–487, Berichtigung 110 (1935) 777–779.

[16] Fukushima M., Shima T., *On a spectral analysis for the Sierpinski gasket*, Potential Anal. l:1 35 (1992) 1–35.

[17] Fukushima M., *Dirichlet forms, diffusion processes and spectral dimension for nested fractals*, in "Ideas and Methods in Mathematical Analysis, Stochastics and Applications", S. Albeverio and al. eds., Cambridge U. Press, 1992, 151–161.

[18] Fukushima M., Oshima Y. Takeda M., *Dirichlet forms and Symmetric Markov Processes*, Walter De Gruyter Co., 1995.

[19] Goldstein S., *Random walks and diffusions on fractals*, in "Percolation theory and ergodic theory of infinite particle systems", IMA Vol. Math. Appl. 8, Springer V., 1987.

[20] Havlin S., Ben-Avraham, *Diffusions in disordered media*, Adv. Phys, 36 (1987) 695–799.

[21] Hutchinson J. E., *Fractals and selfsimilarity*, Indiana U. Math. J. 30 (1981) 713–747.

[22] Jerison D., Sánchez Calle A., *Subelliptic, second order differential operators*, Springer V. Lect. Notes in Math. 1277 (1987) 46–77.

[23] Kigami J., *A harmonic calculus on the Sierpinski spaces*, Japan J. Appl. Math. 6 (1989) 259–290.

[24] Kigami J., Lapidus M.L., *Weyl's problem for the spectral distribution of Laplacians onp.c.f. self-similar fractals*, Commun. Math. Phys. 158 (1993) 93–125.

[25] Kusuoka S., *A diffusion process on a fractal*, in "Probabilistic methods in Mathematical Physics", Proc. Taniguchi Int. Symp., Katata and Tokyo, 1985, K. Ito and N. Ikeda eds., Kinokunya, Tokyo, 1987, 251–274.

[26] Kusuoka S., *Dirichlet forms on fractals and product of random matrices*, Publ. RIMS Kyoto U. 25 (1989) 659–680.

[27] Kusuoka S., *Diffusion processes in nested fractals*, Lect. Notes in Math. 1567, Springer V., 1993.

[28] LeJan Y., *Mesures associées à une forme de Dirichlet Applications*, Bull. Soc. Math. France 106 (1978) 61–112.

[29] Lindstrøm T., *Brownian motion on nested fractals*, Memoirs AMS 420, 83, 1990.

[30] Macias, R.A, Segovia, C., *Lipschitz functions on spaces of homogeneous type*, Adv. in Math. 33 (1979) 257–270.

[31] Marchi M.V., *Self-similarity in quasi-metric spaces*, in preparation.

[32] Metz V., *How many diffusions exist on the Viscek snowflake*, Acta Appl. Math. 32 (1993) 227–241.

[33] Mosco U., *Composite media and asymptotic Dirichlet forms*, J. Funct. Anal. 123 n. 2 (1994) 368–421.

[34] Mosco U., *Variational metrics on self-similar fractals*, C.R. Acad. Sci. Paris, t.321 Série I (1995) 715–720.

[35] Mosco U., *Variations and Irregularities*, in "Second Topological Analysis Workshop on Degree, Singularity and Variations: Developments of the Last 25 Years", M. Matzeu and A. Vignoli eds., Progress in Nonlinear Diff. Eqs. and their Appl., vol. 27, Birkhäuser, 1997, pp. 273–313.

[36] Mosco U., *Variational fractals*, Ann. Scuola Norm. Sup. Pisa, Special volume in Memory of E. De Giorgi, to appear.

[37] Mosco U., *Lagrangian metrics on fractals*, Proc. Symp. Appl. Math, 54, Amer. Math. Soc., R. Spigler and S. Venakides eds., 1998, 301–323.

[38] Mosco U., *Invariant field metrics and dynamical scalings on fractals*, Phys. Rev. Lett., vol. 79, n. 21, 24. Nov. 1997, p. 4067.

[39] Mosco U., Notarantonio L., *Homogeneous fractal spaces*, in "Variational methods for discontinuous structures", Villa Olmo, Como 1994, R. Serapioni and F. Tomarelli eds., Birkäuser, 1996.

[40] Nakayama T., Orbach R.L., Yakubo K., *Dynamical properties of fractal networks: Scaling, numerical solutions, and physical realizations*, Rev. Modern Phys., 66 no. 2, (1994) 381–443.

[41] Olsen L., *A multifractal formalism*, Adv. in Math. 116 (1995) 82–195.

[42] Peirone R., *Homogenization of functionals on fractals*, preprint, Dipartimento Mat., Univ. "Tor Vergata" di Roma, 1996.

[43] Posta G., *Spectral asymptotics for variational fractals*, to appear in: ZAA, Leipzig.

[44] Rammal R, Toulouse G., *Random walks on fractal structures and percolation clusters*, J. Physique Lett. 44 (1983) L13–L22.

[45] Rothschild L.P., Stein E.M., *Hypoelliptic differential operators and nilpotent groups*, Acta Math. 137 (1976) 247–320.

[46] Sabot C., *Diffusions sur les espaces fractals*, Thèse Doct. Univ. Paris VI, 1995.

[47] Schief A., *Self-similar sets in complete metric spaces*, Proc. Amer. Math. Soc. 124, n. 2 (1996) 481–490.

[48] Stein E. M., *Harmonic analysis*, Princeton Univ. Series, 1994.

[49] Strichartz R.S., *Self-similar measures and their Fourier transforms III*, Indiana U. Math. J. 42, 2 (1993) 367–411.

[50] Weyl H., *Das asymptotische Verteilungsgesetz der Eigenwerte der linearer partieller Differentialgleichungen (mit einer Anwendung auf die Theorie der Hohlraumstrahlung)*, Math. Ann. 71 (1912) 441–479.

AMS/IP Studies in Advanced Mathematics
Volume 8, 1998

Michael Röckner: Stochastic analysis on configuration spaces: basic ideas and recent results

4.1 Introduction

The purpose of this paper is on one hand to provide sufficiently many details to complement lectures given in Anogia at the Euro–Conference on "Dirichlet forms and their Applications in Geometry and Stochastics" in June 1997 and at the Mathematical Sciences Research Institute in Berkeley in October/November 1997 within the "MSRI–Year in Stochastic Analysis". On the other hand the aim is to provide a both comprehensive and summarizing "pedagogic" account of the quite substantial material recently published on analysis and geometry on configuration spaces.

In addition, this work might serve as a guide for the reader to the following papers on the subject: [AKR97a], [AKR97b], (see also [AKR96a], [AKR96b]), [MR97], [RS97], [RSch97], [dSKR98]. We also include some new results and a number of new resp. modified proofs for crucial results in those articles, as well as describe some very recent developments, not covered by the lectures in Anogia and Berkeley.

For an overview of the single topics treated in these notes we refer to the list of contents which we hope is descriptive enough. Here we only would like to point out the following parts of this paper to which we put special emphasis resp. which are complements to the above articles:

(1) We describe the lifting of the geometry of the underlying manifold X to the configuration space Γ_X in Section 2 in more detail than it was done in [AKR97a].

(2) In Subsection 3.2 (see also Proposition 4.4.6) we give a particularly detailed account of the (Markov–, strong, L^1–) uniqueness results for the "mixed Poisson

case" obtained in [AKR97a]. We complement [AKR97a] by additionally treating the *non–conservative case* (cf. Remark 4.3.8 (i)) in Subsection 3.2. In Subsection 3.3 (among other things also) the consequences of the uniqueness results for the corresponding diffusion processes on Γ_X are discussed.

(3) In Subsection 8.2 in case the underlying measure μ on Γ_X is a Gibbs (Ruelle) measure coming from a pair potential ϕ, we include an explanation in what sense our diffusion process solves the *heuristic* stochastic differential equation

$$dX_t^i = dW_t^i + \sum_{j:j\neq i} \nabla\phi(X_t^i - X_t^j)\,dt, \; i \in \mathbb{N}.$$

(4) In Subsection 6.3 we give an essentially self–contained proof of the closability results recently obtained in [dSKR98].

(5) In Section 5 we include the joint results with Alexander Schied (who presented a part of them himself in Anogia and Berkeley) from [RSch97]. In particular, we discuss the Dirichlet forms $(\mathcal{E}_\mu^\Gamma, \mathcal{F})$ and $(\mathcal{E}_\mu^\Gamma, \mathcal{F}^{(c)})$ introduced in that reference (see Proposition 4.4.9 below), and additionally prove that the latter has the local property (cf. Proposition 4.5.6 below).

(6) The recent results in [RS97], not particularly mentioned in Anogia, are presented in Theorems 4.3.16 and 4.8.2 (iv).

(7) Most of the main results from [AKR97b] are discussed in these notes for simplicity under the additional *finite range* condition (C) (cf. Subsection 7.2). This is supposed to ease the reading for non–experts.

We emphasize, however, that the quite large number of consequences for mathematical physics is not discussed in sufficient detail in this paper. We refer instead to [AKR97c] and [KLRRSh97].

For the reader, who is only interested in the (much simpler) "free case", i.e., where the underlying measure on Γ_X is a mixed Poisson measure, we tried to keep this part completely independent and have devoted Sections 2 and 3 entirely to this case. (The price we had to pay in turn is however, that we have to be a bit repetitious in the more general "Gibbsian case" discussed in Sections 6 – 9). Also for simplicity, in Sections 2 and 3 we mainly consider the situation where the intensity measure σ of the mixed Poisson measures is just the volume element m of the manifold X and leave the more general "σ–case" to respective remarks.

As far as prerequisites in general are concerned, we would like to point out that our framework is described in detail in Section 2. Some a-priori knowledge on Dirichlet forms is, however, advisable, e.g. Chapter I in [MR92] (though a *generator of a Dirichlet form* in this paper is (-1) times the generator of a Dirichlet form in [MR92]).

W.r.t. the (anyway quite substantial) length of this paper we did not include all proofs of the results stated here. Main references for the single sections (apart from the modifications and complements mentioned above which are all completely proved below) are as follows: Sections 2 and 3: [AKR97a], [RS97]; Section 4: [AKR97b], [MR97]; Section 5: [RSch97]; Section 6: [AKR97b], [dSKR98]; Section 7–9: [AKR97b].

Finally, it is a great pleasure to thank all my colleagues and friends with whom I shared the exciting recent work on analysis and geometry on configuration spaces, in particular, Sergio Albeverio, Yuri Kondratiev, Zhi–Ming Ma, Jiagang Ren, Alexander Schied, Byron Schmuland and Jose Luis da Silva. I would also like to thank all of them for the permission to report here on very recent results, only just submitted for publication. I also thank the organizers of the above mentioned two conferences, in particular, Susanna Papadopoulou and Karl Theodor Sturm resp. Steve Evans and Ruth Williams for very nice stays in Anogia resp. Berkeley and for very stimulating meetings.

4.2 Lifting the geometry from the base manifold to the configuration space

In this paper let X be a connected, oriented C^∞ Riemannian manifold. For each point $x \in X$, the tangent space to X at x will be denoted by T_xX; and the tangent bundle endowed with its natural differentiable structure will be denoted $TX = \cup_{x\in X} T_xX$. The Riemannian metric on X associates to each $x \in X$ an inner product on T_xX, which we denote by $\langle \cdot, \cdot \rangle_{TX}$. The associated norm will be denoted by $|\cdot|_{TX}$. Let m denote the volume element.

$\mathcal{O}(X)$ is defined as the family of all open sets of X and $\mathcal{B}(X)$ denotes the corresponding Borel σ–algebra. $\mathcal{O}_c(X)$ and $\mathcal{B}_c(X)$ denote the systems of all elements in $\mathcal{O}(X)$, $\mathcal{B}(X)$ respectively, which have compact closures.

The *configuration space* Γ_X over the manifold X is defined as the set of all locally finite subsets (configurations) in X:

$$\Gamma_X := \{\gamma \subset X \mid |\gamma \cap K| < \infty \text{ for any compact } K \subset X\}.$$

Here $|A|$ denotes the cardinality of a set A. For $\Lambda \subset X$ we sometimes use the shorthand γ_Λ for $\gamma \cap \Lambda$ and define

$$\Gamma_\Lambda := \{\gamma \in \Gamma_X \mid \gamma \cap (X \setminus \Lambda) = \emptyset\}.$$

The reader should watch, however, that if Λ is a sub–manifold not equal to X there is an ambiguity in the notation (consider e.g. the case where Λ is relatively compact). As in the standard literature we shall, however, never consider sub–manifolds of X and always understand Γ_Λ in the above sense.

We can identify any $\gamma \in \Gamma_X$ with a positive integer–valued Radon measure on $(X, \mathcal{B}(X))$, i.e.,

$$\gamma \equiv \sum_{x \in \gamma} \varepsilon_x \ ,$$

where $\sum_{x\in\emptyset} \varepsilon_x :=$ zero measure. The space Γ_X can hence be endowed with the vague topology, i.e., the weakest topology on Γ_X such that all maps

$$\Gamma_X \ni \gamma \mapsto \langle f, \gamma \rangle := \int_X f(x)\gamma(dx) = \sum_{x \in \gamma} f(x) \tag{4.2.1}$$

are continuous. Here $f \in C_0(X)$ (:= the set of all continuous functions on X with compact support). Let $\mathcal{B}(X)$ denote the corresponding Borel σ–algebra.

For later use we recall the "localized" description of Γ_X: for $\Lambda \in \mathcal{B}_c(X)$ define

$$\Gamma_\Lambda \ := \ \{\gamma \in \Gamma_X \mid \gamma(X \setminus \Lambda) = 0\},$$

and for $n \in \mathbb{Z}_+$

$$\Gamma_\Lambda^{(n)} \ := \ \{\gamma \in \Gamma_\Lambda \mid \gamma(\Lambda) = n\}.$$

There is a bijection

$$\tilde{\Lambda}^n / S_n \ \to \Gamma_\Lambda^{(n)},$$

where

$$\tilde{\Lambda}^n \ := \ \{(x_1, \dots, x_n) \in \Lambda^n | x_k \neq x_j, \ j \neq k\}, \tag{4.2.2}$$

and S_n is the permutation group over $\{1, \dots, n\}$. If $\Lambda \in \mathcal{O}_c(X)$, this bijection defines a locally compact metrizably Hausdorff topology on $\Gamma_\Lambda^{(n)}$, hence a corresponding (sum) topology on

$$\Gamma_\Lambda \ = \ \bigcup_{n=0}^{\infty} \Gamma_\Lambda^{(n)}. \tag{4.2.3}$$

If Γ_Λ is equipped with the associated Borel σ– algebra then $(\Gamma_X, \mathcal{B}_X(\Gamma))$ is the projective limit of the measurable spaces $(\Gamma_\Lambda, \mathcal{B}(\Gamma_\Lambda))$ as $\Lambda \nearrow X$ (cf. e.g. [MR97]).

In this section we shall "lift the geometry" of X onto Γ_X.

4.2.1 Test functions, flows, directional derivatives

We already know how to lift functions in the test function space $\mathcal{D} = C_0^\infty(X)$ ($:=$ the set of all C^∞–functions on X with compact support) onto Γ_X, namely according to (4.2.1) we obtain for each $f \in \mathcal{D}$ the function

$$\gamma \mapsto \langle f, \gamma \rangle$$

on Γ_X. We would like, however, an algebra of *test functions on* Γ_X. So, we define

$$\mathcal{F}C_b^\infty(\mathcal{D}, \Gamma_X) := \{g(\langle f_1, \cdot\rangle, \dots, \langle f_N, \cdot\rangle) \mid N \in \mathbb{N},\ g \in C_b^\infty(\mathbb{R}^N); \quad f_1, \dots, f_N \in \mathcal{D}\} \tag{4.2.4}$$

and sometimes set for simplicity $\mathcal{F}C_b^\infty := \mathcal{F}C_b^\infty(\mathcal{D}, \Gamma_X)$. Elements in $\mathcal{F}C_b^\infty$ are called *finitely based*, since they only depend on finitely many points in γ.

After lifting the test functions, we want to lift flows. Let us first consider the group of diffeomorphisms $\mathrm{Diff}_0(X)$ on X which are equal to the identity outside some compact set. Any $\psi \in \mathrm{Diff}_0(X)$ defines a transformation

$$\Gamma_X \ni \gamma \mapsto \psi(\gamma) = \{\psi(x) \mid x \in \gamma\} = \sum_{x \in \gamma} \varepsilon_{\psi(x)} \in \Gamma\,. \tag{4.2.5}$$

Correspondingly, if $V_0(X)$ denotes the set of smooth vector fields on X with compact support and if $v \in V_0(X)$ with corresponding flow ψ_t^v, $t \in \mathbb{R}$ (i.e., $\frac{d}{dt}\psi_t^v(x) = v(\psi_t^v(x))$, $\psi_0^v(x) = x$), then ψ_t^v, $t \in \mathbb{R}$, lifts to a *flow on* Γ_X via (4.2.3). Immediately, we now gain *directional derivatives on* Γ_X for any $F \in \mathcal{F}C_b^\infty$ by defining for $v \in V_0(X)$

$$\left(\nabla_v^\Gamma F\right)(\gamma) := \frac{d}{dt}\, F(\psi_t^v(\gamma))|_{t=0}\,. \tag{4.2.6}$$

4.2.2 Tangent bundle, gradient, vector fields and divergence

The next step is to define a gradient for functions in $\mathcal{F}C_b^\infty$ which corresponds to the directional derivatives in (4.2.6). At the same time we shall find the appropriate "tangent bundle" over Γ_X.

So let $F = g(\langle f_1, \cdot\rangle, \dots, \langle f_N, \cdot\rangle) \in \mathcal{F}C_b^\infty$, $v \in V_0(X)$ and $\gamma \in \Gamma_X$. By (4.2.6) and the chain rule we have with $\nabla_v^X f_i := \langle \nabla^X f_i, v\rangle_{TX}$

$$\begin{aligned}\nabla_v^\Gamma F(\gamma) &= \sum_{i=1}^N \partial_i g(\langle f_1,\gamma\rangle,\dots,\langle f_N,\gamma\rangle)\ \langle \nabla_v^X f_i,\gamma\rangle \\ &= \int \left\langle \sum_{i=1}^N \partial_i g(\langle f_1,\gamma\rangle,\dots,\langle f_N,\gamma\rangle)\ \nabla^X f_i, v\right\rangle_{TX} d\gamma \\ &= \langle \nabla^\Gamma F(\gamma), v\rangle_{L^2(X\to TX;\gamma)}, \qquad (4.2.7)\end{aligned}$$

where the *gradient* ∇^Γ *on* Γ_X is defined by

$$\nabla^\Gamma F(\gamma) := \sum_{i=1}^N \partial_i g(\langle f_1,\gamma\rangle,\dots,\langle f_N,\gamma\rangle)\ \nabla^X f_i \in C_0^\infty(X), \qquad (4.2.8)$$

∇^X denotes the gradient on X, ∂_i directional derivative w.r.t. the i–th coordinate and $L^2(X\to TX;\gamma)$ the space of γ–square integrable vector fields on X. We note that here and henceforth we identify every v in $V_0(X)$ with the corresponding class in $L^2(X\to TX;\gamma)$ though this is, of course, not a one–to–one operation. In particular, $\nabla^\Gamma F(\gamma)\in L^2(X\to TX;\gamma)$ and by (4.2.7) this class is independent of the representation of F in (4.2.8).

Equation (4.2.7) immediately leads to the appropriate *tangent bundle on* Γ_X, namely

$$T_\gamma\Gamma_X := L^2(X\to TX;\gamma),\ \ \gamma\in\Gamma_X, \qquad (4.2.9)$$

equipped with the usual L^2–inner product

$$\langle\ ,\ \rangle_{T_\gamma\Gamma_X} := \langle\ ,\ \rangle_{L^2(X\to TX;\gamma)},\ \ \gamma\in\Gamma_X. \qquad (4.2.10)$$

Corresponding, *finitely based vector fields* on $(\Gamma_X, T\Gamma_X)$ can be defined as follows:

$$\Gamma_X\ni\gamma\mapsto\sum_{i=1}^N F_i(\gamma)\ v_i\in C_0^\infty(X), \qquad (4.2.11)$$

where $F_1,\dots,F_N\in\mathcal{F}C_b^\infty$; $v_1,\dots,v_N\in V_0(X)$. Let $\mathcal{VF}C_b^\infty := \mathcal{VF}C_b^\infty(\mathcal{D},\Gamma_X)$ be the set of all such maps. We note that, of course, $\nabla^\Gamma F\in\mathcal{VF}C_b^\infty$ for all $F\in\mathcal{F}C_b^\infty$ and that each $v\in V_0(X)$ is identified with the vector field $\gamma\mapsto v$ in $T\Gamma_X$ which is constant modulo taking γ–classes. Its norm function $\gamma\mapsto\|v\|_{T_\gamma\Gamma_X} = (\int\|v\|^2_{T_xX}\gamma(dx))^{1/2}$ is, however, not bounded.

We can now lift the divergence div^X on X. For $v \in V_0(X)$ by our rule to lift functions we must define

$$\mathrm{div}^\Gamma v(\gamma) := \langle \mathrm{div}^X v, \gamma \rangle, \ \gamma \in \Gamma_X \ . \tag{4.2.12}$$

Requiring the usual product rule to hold we must define for $V := \sum_{i=1}^N F_i \, v_i \in \mathcal{VFC}_b^\infty$

$$\mathrm{div}^\Gamma V := \sum_{i=1}^N \left(\langle \nabla^\Gamma F_i, v_i \rangle_{T\Gamma_X} + F_i \, \mathrm{div}^\Gamma v_i \right) \ . \tag{4.2.13}$$

Also this will turn out to be a definition independent of the representation of V (see Remark 2.3 (ii) below).

The next question that now arises is whether ∇^Γ and div^Γ are "connected" through a volume element on Γ_X as ∇^X and div^X are through the volume element m on X. Since Γ_X is an infinite dimensional space, one might be sceptical whether such volume elements exist. But, in fact they do, if one uses the appropriate definition of "volume element" on Γ_X. Let us first recall the following well–known characterization of the volume element m on X (cf. e.g. [Ch84]):

m is up to a constant the unique *Radon* measure σ on $(X, \mathcal{B}(X))$ such that $(\nabla^X, C_0^\infty(X))$ and $(\mathrm{div}^X, V_0(X))$ are dual operators on $L^2(X;\sigma)$ w.r.t. $\langle \ , \ \rangle_{TX}$, in the sense that

$$\int_X \langle v, \nabla^X f \rangle_{TX} \, d\sigma = -\int \mathrm{div}^X v \ f \ d\sigma \text{ for all } f \in C_0^\infty(X), \ v \in V_0(X).$$

Definition 4.2.1 *A probability measure μ on $(\Gamma_X, \mathcal{B}(\Gamma_X))$ with Radon mean (i.e., $\int_{\Gamma_X} \gamma(K)\mu(d\gamma) < \infty$ for all compact $K \subset X$) is called a* volume element *on Γ_X, if*

$$\int \langle V, \nabla^\Gamma F \rangle_{T\Gamma_X} \, d\mu = -\int \mathrm{div}^\Gamma V \ F \, d\mu \text{ for all } F \in \mathcal{FC}_b^\infty, V \in \mathcal{VFC}_b^\infty. \tag{4.2.14}$$

4.2.3 Characterization of the volume elements as the mixed Poisson measures

We have a complete characterization of the volume elements on Γ_X. Let us first recall some standard definitions.

For any Radon measure σ on X the (*pure*) *Poisson measure with intensity* σ is the unique measure $\psi_{i_1 \dots i_n \sigma}$ on $(\Gamma_X, \mathcal{B}(\Gamma_X))$ with Laplace transform given by

(4.2.15) $$\int_{\Gamma_X} e^{\langle f, \gamma \rangle} \, \psi_{i_1 \dots i_n \sigma}(d\gamma) = e^{\int (e^f - 1) \, d\sigma} \quad \text{for all } f \in C_0^\infty(X).$$

The respective "local" description of $\psi_{i_1 \dots i_n \sigma}$ is as follows: for $\Lambda \in \mathcal{O}_c(X)$ and $n \in \mathbb{N}$ the product measure $\sigma^{\otimes n}$ can be considered as a (finite) measure on $\tilde{\Lambda}_n$. Let

(4.2.16) $$\sigma_{\Lambda,n} := \sigma^{\otimes n} \circ (s_\Lambda^n)^{-1}$$

be the corresponding measure on $\Gamma_\Lambda^{(n)}$ where $s_\Lambda^n : \tilde{\Lambda}^n \to \Gamma_\Lambda^{(n)}$, $s_\Lambda^n((x_1, \dots, x_n)) := \sum_{i=1}^n \varepsilon_{x_i}$. Define $p_\Lambda : \Gamma_X \to \Gamma_\Lambda$ by

$$p_\Lambda(\gamma) := \gamma_\Lambda$$

Then

(4.2.17) $$\psi_{i_1 \dots i_n \sigma} \circ p_\Lambda^{-1} = e^{-\sigma(\Lambda)} \sum_{n=0}^\infty \frac{1}{n!} \sigma_{\Lambda,n},$$

where $\sigma_{\Lambda,0} := \varepsilon_\emptyset$ on $\Gamma_\Lambda^{(0)} = \{\emptyset\}$.

(4.2.17) also leads to an existence proof of $\psi_{i_1 \dots i_n \sigma}$ as follows: Defining probability measures μ_Λ on $(\Gamma_\Lambda, \mathcal{B}(\Gamma_\Lambda))$ by the right hand side of (4.2.17), a version of Kolmogorov's theorem (cf. e.g. [Pa67]) implies that there exists a unique probability measure $\psi_{i_1 \dots i_n \sigma}$ on $(\Gamma_X, \mathcal{B}(\Gamma_X))$ such that

$$\psi_{i_1 \dots i_n \sigma} \circ p_\Lambda^{-1} = \mu_\Lambda \text{ for all } \Lambda \in \mathcal{O}_c(X).$$

It is easy to check that $\psi_{i_1 \dots i_n \sigma}$ indeed satisfies (4.2.15).

E.g. by (4.2.17) it is easy to check that

(4.2.18) $$\int_{\Gamma_X} \langle f, \gamma \rangle \, \psi_{i_1 \dots i_n \sigma}(d\gamma) = \int_X f \, d\sigma \text{ for all } f \in L^1(X; \sigma)$$

and

(4.2.19)
$$\int_{\Gamma_X} \langle f, \gamma \rangle^2 \, \psi_{i_1 \dots i_n \sigma}(d\gamma) = \int_X f^2 \, d\sigma + \left(\int_X f \, d\sigma \right)^2 \text{ for all } f \in L^1(X; \sigma) \cap L^2(X; \sigma)$$

We note that for $\sigma \equiv 0$, $\psi_{i_1 \dots i_n \sigma}$ is just the Dirac measure on $(\Gamma_X, \mathcal{B}(\Gamma_X))$ with total mass in the empty configuration $\gamma \equiv \emptyset$ $(\subset X)$.

For a probability measure λ on $(\mathbb{R}_+, \mathcal{B}(\mathbb{R}_+))$ the measure $\mu_{\lambda,\sigma}$ on $(\Gamma_X, \mathcal{B}(\Gamma_X))$ defined by

$$\mu_{\lambda,\sigma} := \int_{\mathbb{R}_+} \psi_{i_1 \dots i_n z \cdot \sigma} \ \lambda(dz),$$

is called *mixed Poisson measure.*

Theorem 4.2.2 *Suppose* $m(X) = \infty$ *and let* μ *be a probability measure on* $(\Gamma_X, \mathcal{B}(\Gamma_X))$ *with Radon mean. Then the following assertions are equivalent:*

(i) μ *is a volume element on* Γ_X.

(ii) μ *is a mixed Poisson measure* $\mu_{\lambda,m}$ *for some probability measure* λ *on* $(\mathbb{R}_+, \mathcal{B}(\mathbb{R}_+))$.

Proof. (ii) $\Rightarrow$ (i): (Here the assumption $m(X) = \infty$ is not needed.) We shall first prove that $\psi_{i_1 \dots i_n m}$ is a volume element on Γ_X.

Let $F = g_F(\langle f_1, \cdot \rangle, \dots, \langle f_N, \cdot \rangle) \in \mathcal{F}C_b^\infty$, $V \in \mathcal{V}\mathcal{F}C_b^\infty$. By linearity we may assume that $V(\gamma) = G(\gamma) v$ for all $\gamma \in \Gamma_X$ for some $G = g_G(\langle g_1, \cdot \rangle, \dots, \langle g_N, \cdot \rangle)$, $v \in V_0(X)$. Let $\Lambda \in \mathcal{O}_c(X)$ such that $\operatorname{supp} f_i$, $\operatorname{supp} g_i$, $\operatorname{supp} v \subset \Lambda$, $1 \le i \le N$. Then by (4.2.6), (4.2.7), (4.2.16), and (4.2.17)

$$\begin{aligned} \int_{\Gamma_X} \langle V, \nabla^\Gamma F \rangle_{T\Gamma_X} \, d\psi_{i_1 \dots i_n m} &= \int_{\Gamma_X} G \ \nabla_v^\Gamma F \ d\psi_{i_1 \dots i_n m} \\ &= \int_{\Gamma_X} G(\gamma_\Lambda) \ \nabla_v^\Gamma F(\gamma_\Lambda) \ \psi_{i_1 \dots i_n m}(d\gamma) \\ &= e^{-\sigma(\Lambda)} \sum_{n=0}^{\infty} \frac{1}{n!} \ A_{\Lambda,n} \end{aligned} \tag{4.2.20}$$

where

$$\begin{aligned}
A_{\Lambda,n} &:= \int_{\Lambda^n} g_G\left(\sum_{j=1}^n g_1(x_j),\dots,\sum_{j=1}^n g_N(x_j)\right)\cdot \\
&\quad \sum_{i=1}^N\left(\partial_i g_F\left(\sum_{j=1}^n f_1(x_j),\dots,\sum_{j=1}^n f_N(x_j)\right)\sum_{k=1}^n \nabla_v^X f_i(x_k)\right) \\
&\quad m(dx_1)\dots m(dx_n) \\
&= \sum_{k=1}^n \int_{\Lambda^n} g_G\left(\sum_{j=1}^n g_1(x_j),\dots,\sum_{j=1}^n g_N(x_j)\right)\cdot \\
&\quad \left\langle \nabla_{x_k}^X g_F\left(\sum_{j=1}^n f_1(x_j),\dots,\sum_{j=1}^n f_N(x_j)\right), v(x_k)\right\rangle_{T_{x_k}X} \\
&\quad m(dx_1)\dots m(dx_n)
\end{aligned}$$

where $\nabla_{x_k}^X$ denotes the gradient on X acting w.r.t. the variable x_k. Integrating by parts w.r.t. x_k we obtain that

$$\begin{aligned}
A_{\Lambda,n} &= -\sum_{k=1}^n \int_{\Lambda^n}\Bigg[\left\langle \nabla_{x_k}^X g_G\left(\sum_{j=1}^n g_1(x_j),\dots,\sum_{j=1}^n g_N(x_j)\right), v(x_k)\right\rangle_{T_{x_k}X} \\
&\quad + g_G\left(\sum_{j=1}^n g_1(x_j),\dots,\sum_{j=1}^n g_N(x_j)\right)\ \mathrm{div}^X v(x_k)\Bigg] \\
&\quad \cdot g_F\left(\sum_{j=1}^n f_1(x_j),\dots,\sum_{j=1}^n f_N(x_j)\right)\ m(dx_1)\dots m(dx_n) \\
&= -\int_{\Lambda^n}\Bigg[\sum_{i=1}^N \partial g_G\left(\sum_{j=1}^n g_1(x_j),\dots,\sum_{j=1}^n g_N(x_j)\right)\sum_{k=1}^n \nabla_v^X g_i(x_k) \\
&\quad + g_G\left(\sum_{j=1}^n g_1(x_j),\dots,\sum_{j=1}^n g_N(x_j)\right)\sum_{k=1}^n \mathrm{div}^X v(x_k)\Bigg] \\
&\quad g_F\left(\sum_{j=1}^n f_1(x_j),\dots,\sum_{j=1}^n f_N(x_j)\right)\ m(dx_1)\dots m(dx_N).
\end{aligned}$$

Substituting back into (4.2.20) by (4.2.16), (4.2.17) yields

$$\int \langle V, \nabla^\Gamma F\rangle_{T\Gamma_X}\ d\psi_{i_1\dots i_n m} = -\int \mathrm{div}^\Gamma V\ F\ d\psi_{i_1\dots i_n m}\,.$$

Since we have not used a specific normalization of m, this proves that any $z \cdot m$, $z \in]0, \infty[$, is a volume element on Γ_X. It is trivial to check that this is also true for $z = 0$, since $\psi_{i_1 \dots i_n 0 \cdot m}$ is the Dirac measure on $(\Gamma_X, \mathcal{B}(\Gamma_X))$ with total mass in the empty configuration $\emptyset \subset X$. Hence by Fubini's theorem this is also true for $\mu_{\lambda,m}$ as in the assertion, since integrability is ensured because $\mu_{\lambda,m}$ has a Radon mean (hence λ has finite first moments).

For the proof of the converse (i) $\Rightarrow$ (ii) (which is a little harder) we refer to the detailed expositions in [AKR97a] and [AKR97b], more precisely Theorem 3.2, 4.1 respectively ibidem. (The first is analytic, the latter is purely probabilistic and works in more general situations, see the following remark.) □

Remark 4.2.3 (i) ("*σ–case*") Let $\rho \in L^1_{loc}(X;m)$ be such that $\rho^{\frac{1}{2}} \in H^{1,2}_{loc}(X;m)$, i.e., $\rho^{\frac{1}{2}}$ has locally m–square integrable weak derivatives. Define

$$\sigma := \rho \cdot m \ .$$

the *logarithmic derivative* of the measure σ is given by the vector field

$$\beta^\sigma := \frac{\nabla^X \rho}{\rho} \tag{4.2.21}$$

where $\beta^\sigma := 0$ on $\{\rho = 0\}$. For every $v \in V_0(X)$ we set

$$\mathrm{div}^X_\sigma v := \langle \beta^\sigma, v\rangle_{TX} + \mathrm{div}^X v \ . \tag{4.2.22}$$

Note that $\langle \beta^\sigma, v\rangle \in L^2(X;\sigma) \cap L^1(X;\sigma)$ since it has compact support, hence the same is true for $\mathrm{div}^X_\sigma v$; consequently by (4.2.19)

$$\langle \mathrm{div}^X_\sigma v, \cdot \rangle \in L^2(\Gamma_X; \psi_{i_1 \dots i_n \sigma}) \ .$$

It is now easy to check that in the proof of Theorem 4.2.2, (ii) $\Rightarrow$ (i), m can be replaced by σ if in Definition 4.2.1 div^X is replaced by div^X_σ. If $\sigma(X) = \infty$ the same is true for the converse (i) $\Rightarrow$ (ii), if, in addition, e.g. $|\beta^\sigma|_{TX} \in L^1_{loc}(X;m)$. We refer to [AKR97b, Theorem 4.1 and Remark 4.1] for details.

(ii) Since the left hand side of (4.2.14) is independent of the representation of $V \in \mathcal{VFC}^\infty_b$ as $V = \sum_{i=1}^N F_i \cdot v_i$; $F_1, \dots, F_N \in \mathcal{FC}^\infty_b$, $v_1, \dots, v_N \in V_0(X)$, so is $\mathrm{div}^\Gamma V \in L^1(\Gamma_X;\mu)$ for any volume element μ on Γ_X, hence for any mixed Poisson measure $\mu_{\lambda,m}$ as in Theorem 4.2.2.

(iii) A proof of Theorem 4.2.2, (ii) $\Rightarrow$ (i), can also be derived from the well–known Mecke identity for $\psi_{i_1 \dots i_n \sigma}$ (cf. [Mec67, Satz 3.1]). This was observed and pointed out to us by V. Liebscher (private communication).

(iv) Theorem 4.2.2, (ii) $\Rightarrow$ (i), extends to so–called compound Poisson measures (cf. [dSKSt97]). The proof is similar.

4.2.4 Laplacian and classical pre–Dirichlet form

Now we proceed with our lifting operations. Since we have gradient and divergence on Γ_X, it is clear how to define the corresponding *Laplacian on* Γ_X, namely for $F \in \mathcal{F}C_b^\infty$

$$\Delta^\Gamma F := \operatorname{div}^\Gamma \nabla^\Gamma F \,. \tag{4.2.23}$$

(We here note that $\nabla^\Gamma F \in \mathcal{V}\mathcal{F}C_b^\infty$).

By (4.2.8), (4.2.13) it follows that for all $F = g(\langle f_1, \cdot\rangle, \ldots, \langle f_N, \cdot\rangle) \in \mathcal{F}C_b^\infty$

$$\begin{aligned} \Delta^\Gamma F(\gamma) &= \sum_{i,j=1}^N \partial_i\partial_j g(\langle f_1,\gamma\rangle,\ldots,\langle f_N,\gamma\rangle)\ \langle\langle \nabla^X f_i, \nabla^X f_j\rangle_{TX}, \gamma\rangle \\ &\quad + \sum_{i=1}^N \partial_i g(\langle f_1,\gamma\rangle,\ldots,\langle f_N,\gamma\rangle)\ \langle \Delta^X f_i, \gamma\rangle \,. \end{aligned} \tag{4.2.24}$$

Here $\Delta^X := \operatorname{div}^X \nabla^X$ is the Laplacian on X.

For any volume element $\mu_{\lambda,m}$ (cf. Theorem 4.2.2) we obtain the *classical pre–Dirichlet form on* Γ_X as

$$\mathcal{E}^\Gamma_{\mu_{\lambda,m}}(F,G) := \int \langle \nabla^\Gamma F, \nabla^\Gamma G\rangle_{T\Gamma_X}\, d\mu_{\lambda,m};\ F, G \in \mathcal{F}C_b^\infty. \tag{4.2.25}$$

"pre" is added because the form is not yet closed on $L^2(\Gamma_X;\mu_{\lambda,m})$. We come to the "problem of closability" in detail in Subsections 4.1 and 6.3 below (and also in the following subsection 3.1). We only emphasize here that only by passing to the closure of $(\mathcal{E}^\Gamma_{\mu_{\lambda,m}}, \mathcal{F}C_b^\infty)$ on $L^2(\Gamma_X;\mu_{\lambda,m})$ we start performing real infinite dimensional analysis on Γ_X. So far (apart from the construction of Poisson measures) we have been doing only essentially finite dimensional analysis, since our test function and vector fields on Γ_X are finitely based, hence depend only on a finite part of each configuration $\gamma \in \Gamma_X$. We note that until now we also never used the full tangent space $T_\gamma\Gamma_X = L^2(X \to TX;\gamma)$ at $\gamma \in \Gamma_X$, but rather only $V_0(X)$ with the inherited inner product.

Remark 4.2.4 The pre–Dirichlet form (4.2.25) seems to have appeared first in [BiGJ87] in the special case where X is an open subset of $\mathbb{R}^d$ and with $\psi_{i_1\ldots i_n\sigma}$ replacing $\mu_{\lambda,m}$. But $T\Gamma_X$ and ∇^Γ were neither identified nor used there. Instead, $\langle \nabla^\Gamma F, \nabla^\Gamma G\rangle_{T\Gamma_X}$ was replaced by the corresponding formula obtained by means of (4.2.8). ∇^Γ and $T\Gamma_X$ at least in the special case $X = \mathbb{R}^1$ seem to have first been introduced in [Sm88]. We thank F. Hirsch resp. V.I. Bogachev for pointing this out to us.

4.3 Infinite dimensional analysis and Brownian motion on configuration spaces

In this section we shall develop a "real" infinite dimensional analysis on Γ_X and discuss some applications. For the rest of this paper we shall use all terminology and notations from Section 2 without further notice. In particular, we fix $\mu_{\lambda,m}$ as in Theorem 4.2.2. It can be shown that $\mu_{\lambda,m}$ has full support on Γ_X, i.e., $\mu_{\lambda,m}(U) > 0$ for every open subset U of Γ_X (cf. [RSch97]). Therefore, we do not distinguish between $\mathcal{F}C_b^\infty$ and the corresponding $\mu_{\lambda,m}$ classes, since each of the latter has thus a *unique* representative in $\mathcal{F}C_b^\infty$. We additionally assume that

$$\int_{\mathbb{R}_+} z^2\, \lambda(dz) < \infty\,. \tag{4.3.1}$$

It follows by (4.2.18), (4.2.19) that

$$\int_{\Gamma_X} \langle f,\gamma\rangle\, \mu_{\lambda,m}(d\gamma) = \int_{\mathbb{R}_+} z\, \lambda(dz) \int_X f\, dm \quad \text{for all } f \in L^1(X;m). \tag{4.3.2}$$

and

$$\int_{\Gamma_X} \langle f,\gamma\rangle^2\, \mu_{\lambda,m}(d\gamma) = \int_{\mathbb{R}_+} z\, \lambda(dz) \int_X f^2\, dm + \int_{\mathbb{R}_+} z^2\, \lambda(dz) \left(\int_X f\, dm\right)^2 \quad \text{for all } f \in L^1(X;m) \cap L^2(X;m). \tag{4.3.3}$$

4.3.1 Dirichlet forms and operators

For the definitions of closability and closure we refer to Subsection 4.1 below.

Proposition 4.3.1 *For all $F, G \in \mathcal{F}C_b^\infty$*

$$\mathcal{E}^\Gamma_{\mu_{\lambda,m}}(F,G) = -\int \Delta^\Gamma F\; G\; d\mu_{\lambda,m}\,. \tag{4.3.4}$$

In particular, $(\mathcal{E}^\Gamma_{\mu_{\lambda,m}}, \mathcal{F}C_b^\infty)$ is closable on $L^2(\Gamma_X;\mu_{\lambda,m})$.

Proof. (4.3.4) is immediate because $\mu_{\lambda,m}$ is a volume element on Γ_X by Theorem 4.2.2. The closability follows then as a special case from [MR92, Chap. I, Proposition 3.3], since $\Delta^\Gamma F \in L^2(\Gamma_X;\mu_{\lambda,m})$ by (4.3.3) and (4.2.24). $\square$

Let $(\mathcal{E}^\Gamma_{\mu_{\lambda,m}}, D(\mathcal{E}^\Gamma_{\mu_{\lambda,m}}))$ be the closure of $(\mathcal{E}^\Gamma_{\mu_{\lambda,m}}, \mathcal{F}C_b^\infty)$ on $L^2(\Gamma_X;\mu_{\lambda,m})$. $D(\mathcal{E}^\Gamma_{\mu_{\lambda,m}})$ is the analogue of a first order Sobolev space on Γ_X. We, therefore, set

$$H_0^{1,2}(\Gamma_X;\mu_{\lambda,m}) := D(\mathcal{E}^\Gamma_{\mu_{\lambda,m}}) \tag{4.3.5}$$

with norm

$$\|\cdot\|_{H_0^{1,2}(\Gamma_X;\mu_{\lambda,m})} := \left(\mathcal{E}^\Gamma_{\mu_{\lambda,m}}(\cdot,\cdot) + (\cdot,\cdot)_{L^2(\Gamma_X;\mu_{\lambda,m})}\right)^{1/2}. \tag{4.3.6}$$

$(\mathcal{E}_{\mu_{\lambda,m}}, D(\mathcal{E}^\Gamma_{\mu_{\lambda,m}}))$ is a closed, positive definite, symmetric bilinear form on $L^2(\Gamma_X;\mu_{\lambda,m})$, hence it is associated with a unique positive definite self–adjoint operator $(H^\Gamma_{\mu_{\lambda,m}}, D(H^\Gamma_{\mu_{\lambda,m}}))$ (called its *generator*) on $L^2(\Gamma_X;\mu_{\lambda,m})$, i.e.,

$$\begin{aligned} D\left(\sqrt{H^\Gamma_{\mu_{\lambda,m}}}\right) &= D(\mathcal{E}^\Gamma_{\mu_{\lambda,m}}),\ \mathcal{E}^\Gamma_{\mu_{\lambda,m}}(F,G) = \left(\sqrt{H^\Gamma_{\mu_{\lambda,m}}}F, \sqrt{H^\Gamma_{\mu_{\lambda,m}}}G\right)_{L^2(\Gamma_X;\mu_{\lambda,m})} \\ &\text{for all } F,G \in D(\mathcal{E}^\Gamma_{\mu_{\lambda,m}}) \end{aligned} \tag{4.3.7}$$

(cf. e.g. [MR92, Chap. I] for details, except that the generator there is (-1) times our generator here).

Clearly, by (4.3.4)

$$H^\Gamma_{\mu_{\lambda,m}} F = -\Delta^\Gamma F \quad \text{for all } F \in \mathcal{F}C_b^\infty,$$

i.e., $(H^\Gamma_{\mu_{\lambda,m}}, D(H^\Gamma_{\mu_{\lambda,m}}))$ extends $(-\Delta^\Gamma, \mathcal{F}C_b^\infty)$. It is nothing but the Friedrichs' extension of $(-\Delta^\Gamma, \mathcal{F}C_b^\infty)$ on $L^2(\Gamma_X;\mu_{\lambda,m})$ and is also called *Dirichlet operator.* In fact it is also a Dirichlet operator in the sense of [MR92, Chap. I, Definition 4.1]. This follows from [MR92, Chap. I, Theorem 4.4, Proposition 4.3] and the following result:

Proposition 4.3.2 $(\mathcal{E}^\Gamma_{\mu_{\lambda,m}}, D(\mathcal{E}^\Gamma_{\mu_{\lambda,m}}))$ *is a symmetric* Dirichlet form *on* $L^2(\Gamma_X;\mu_{\lambda,m})$, *i.e.,*

$$\begin{aligned} F \in D(\mathcal{E}^\Gamma_{\mu_{\lambda,m}}) \ \Rightarrow\ & F^\# := F^+ \wedge 1 \in D(\mathcal{E}^\Gamma_{\mu_{\lambda,m}}) \text{ and} \\ & \mathcal{E}^\Gamma_{\mu_{\lambda,m}}(F^\#, F^\#) \le \mathcal{E}^\Gamma_{\mu_{\lambda,m}}(F,F)\,. \end{aligned} \tag{4.3.8}$$

Proof. The argument is entirely standard, since ∇^Γ satisfies the chain rule. (One proceeds exactly in the same way as e.g. in [MR92, Chap. II, Subsection 2c)].) □

E.g. by the spectral theorem $(H^{\Gamma}_{\mu_{\lambda,m}}, D(H^{\Gamma}_{\mu_{\lambda,m}}))$ generates a strongly continuous contraction semigroup $T^{\Gamma}_{\mu_{\lambda,m}}(t) := e^{-tH^{\Gamma}_{\mu_{\lambda,m}}}$, $t > 0$, on $L^2(\Gamma_X; \mu_{\lambda,m})$. Since (4.3.8) is equivalent to the *sub–Markov property* of $(T^{\Gamma}_{\mu_{\lambda,m}}(t))_{t>0}$ (i.e., $F \in L^2(\Gamma_X; \mu_{\lambda,m})$, $0 \leq F \leq 1 \Rightarrow 0 \leq T^{\Gamma}_{\mu_{\lambda,m}}(t)F \leq 1$ for all $t \geq 0$, cf. [MR92, Chap. II, Sect. 4]), it follows that $(T^{\Gamma}_{\mu_{\lambda,m}}(t))_{t>0}$ is a strongly continuous contraction semigroup on every $L^p(\Gamma_X; \mu_{\lambda,m})$, $p \in [1, \infty[$ (cf. [RS75, Theorem X.59]), which we shall denote by the same symbol. The corresponding generators we denote by $(H^{\Gamma}_{\mu_{\lambda,m},p}, D(H^{\Gamma}_{\mu_{\lambda,m},p}))$. We have in fact that $(T^{\Gamma}_{\mu_{\lambda,m}}(t))_{t\geq 0}$ is *Markovian*, i.e., in addition, it holds that $T^{\Gamma}_{\mu_{\lambda,m}}(t)\,1 = 1$ for all $t > 0$ (which follows from the fact that $1 \in \mathcal{F}C_b^{\infty} \subset D(H^{\Gamma}_{\mu_{\lambda,m}})$ and $H^{\Gamma}_{\mu_{\lambda,m}}\,1 = \Delta^{\Gamma}\,1 = 0$).

4.3.2 Markov uniqueness, essential self–adjointness and heat semigroup

We emphasize that in general the Friedrichs' extension is not the only self–adjoint extension of a symmetric operator on L^2. In Section 5 we shall consider Dirichlet forms $(\mathcal{E}, D(\mathcal{E}))$ on $L^2(\Gamma_X; \mu_{\lambda,m})$ which are a–priori extensions of $(\mathcal{E}^{\Gamma}_{\mu_{\lambda,m}}, D(\mathcal{E}^{\Gamma}_{\mu_{\lambda,m}}))$ and whose generators also extend $(-\Delta^{\Gamma}, \mathcal{F}C_b^{\infty})$. So, it is a non-trivial question whether these extensions are *strict* extensions or indeed coincide with $(\mathcal{E}^{\Gamma}_{\mu_{\lambda,m}}, D(\mathcal{E}^{\Gamma}_{\mu_{\lambda,m}}))$. Below we shall give a complete answer to this question and prove that, in fact, they all coincide (at least under conditions (A), (B) specified in Theorem 4.3.3). To this end we introduce the set $\underline{\underline{\mathcal{E}}}$ of all Dirichlet forms $(\mathcal{E}, D(\mathcal{E}))$ on $L^2(\Gamma_X; \mu_{\lambda,m})$ whose respective generators $(L(\mathcal{E}), D(L(\mathcal{E})))$ extend $(-\Delta^{\Gamma}, \mathcal{F}C_b^{\infty})$. We note that $(\mathcal{E}^{\Gamma}_{\mu_{\lambda,m}}, D(\mathcal{E}^{\Gamma}_{\mu_{\lambda,m}})) \in \underline{\underline{\mathcal{E}}}$ and that any $(\mathcal{E}, D(\mathcal{E})) \in \underline{\underline{\mathcal{E}}}$ then automatically extends $(\mathcal{E}^{\Gamma}_{\mu_{\lambda,m}}, D(\mathcal{E}^{\Gamma}_{\mu_{\lambda,m}}))$, since for all $F, G \in \mathcal{F}C_b^{\infty}$ $(\subset D(L(\mathcal{E})))$

$$\begin{aligned} \mathcal{E}(F,G) &= \int L(\mathcal{E})F\, G\, d\mu_{\lambda,m} = -\int \Delta^{\Gamma} F\, G\, d\mu_{\lambda,m} \\ &\underset{(4.3.4)}{=} \mathcal{E}^{\Gamma}_{\mu_{\lambda,m}}(F,G), \end{aligned}$$

hence this equality extends to the closure. In this sense $(\mathcal{E}^{\Gamma}_{\mu_{\lambda,m}}, D(\mathcal{E}^{\Gamma}_{\mu_{\lambda,m}}))$ is minimal in $\underline{\underline{\mathcal{E}}}$ and the extensions mentioned above to be studied in Section 5 are in $\underline{\underline{\mathcal{E}}}$. If $\underline{\underline{\mathcal{E}}} = \{(\mathcal{E}^{\Gamma}_{\mu_{\lambda,m}}, D(\mathcal{E}^{\Gamma}_{\mu_{\lambda,m}}))\}$ one says that *Markov uniqueness* holds which we shall prove to be the case in many situations. Before we state the corresponding theorem and give a complete proof, we recall that a symmetric operator on a Hilbert space is called *essentially self–adjoint* if its closure is self–adjoint. Let $(H^X_m, D(H^X_m))$ be the Friedrichs' extension of $(-\Delta^X, \mathcal{D})$ on $L^2(X; m)$.

Theorem 4.3.3 *Assume that the following conditions hold:*

(A) $(\Delta^X, \mathcal{D})$ *is essentially self-adjoint on* $L^2(X;m)$ *(which, as is well-known, is e.g. the case, if* X *is complete w.r.t. the Riemannian metric).*

(B) $\lambda = \varepsilon_z$ *for some* $z \in [0,\infty[$ *or*

$$\int H_m^X f \, dm = 0 \qquad \text{for all } f \in D(H_m^X) \cap L^1(X;m)$$
$$\text{such that } H_m^X f \in L^1(X;m).$$

(The latter just means that $(H_m^X, D(H_m^X))$ *is* conservative*).*

Then

$$\underline{\underline{\mathcal{E}}} = \{(\mathcal{E}^\Gamma_{\mu_{\lambda,m}}, D(\mathcal{E}^\Gamma_{\mu_{\lambda,m}}))\},$$

i.e., Markov uniqueness holds.

Remark 4.3.4 *("σ–case")* Let $\sigma = \rho \cdot m$ be as specified at the beginning of Remark 4.2.3 (i). Then Theorem 4.3.3 (as well as all subsequent results in Subsections 3.2, 3.3) extend to the case where m is replaced by σ. The proof is word for word the same as the one given below. We only note here that condition (A) for σ is e.g. fulfilled if X is complete and if $|\beta^\sigma|_{TX} \in L^p_{loc}(X;m)$ for some $p > \dim(X)$, where $\beta^\sigma := \frac{\nabla \rho}{\rho}$. This immediately follows from the proof of Theorem 1 and Remark 4 (iii) in [BKR97] and Gaffney's Lemma (cf. e.g. [Ba87]).

Theorem 4.3.3 is a consequence of the following result:

Theorem 4.3.5 *Suppose that conditions (A), (B) in Theorem 4.3.3 hold. Then* $\mathcal{F}C_b^\infty$ *is dense in* $D(H^\Gamma_{\mu_{\lambda,m},1})$ *w.r.t. the graph norm*
$\|\cdot\|_1 := \|H_{\mu_{\lambda,m},1} \cdot \|_{L^1(\Gamma_X;\mu_{\lambda,m})} + \|\cdot\|_{L^1(X;\mu_{\lambda,m})}$.

Before we prove Theorem 4.3.5, we show that it implies Theorem 4.3.3:

Proof of Theorem 4.3.3. Let $(\mathcal{E}, D(\mathcal{E})) \in \underline{\underline{\mathcal{E}}}$ with corresponding generator $(L, D(L))$. Since $T_t := e^{-tL}$, $t > 0$, is Markovian, it extends to a strongly continuous contraction semigroup on $L^1(\Gamma_X;\mu_{\lambda,m})$. Let $(L_1, D(L_1))$ denote the corresponding generator. Since $\mu_{\lambda,m}$ has finite total mass, also $(L_1, D(L_1))$ extends $(-\Delta, \mathcal{F}C_b^\infty)$. Therefore, it extends its L^1–closure, which coincides by Theorem 4.3.5 with $(H^\Gamma_{\mu_{\lambda,m},1}, D(H^\Gamma_{\mu_{\lambda,m},1}))$. But one generator of a strongly continuous semigroup cannot strictly extend another, because this would contradict the fact that by the Hille–Yosida theorem both $1 + L_1$ and $1 + H^\Gamma_{\mu_{\lambda,m},1}$ have range all of $L^1(\Gamma_X;\mu_{\lambda,m})$. Thus, $(L_1, D(L_1)) = (H^\Gamma_{\mu_{\lambda,m},1}, D(H^\Gamma_{\mu_{\lambda,m},1}))$, hence $e^{-tL} = e^{-tH^\Gamma_{\mu_{\lambda,m}}}$ for all $t \geq 0$, consequently $(\mathcal{E}, D(\mathcal{E})) = (\mathcal{E}^\Gamma_{\mu_{\lambda,m}}, D(\mathcal{E}^\Gamma_{\mu_{\lambda,m}}))$. □

We shall prove Theorem 4.3.5 along with two other related results. To formulate them we need some preparations.

Consider the classical Dirichlet form $(\mathcal{E}_m^X, D(\mathcal{E}_m^X))$ on X defined as the closure of

$$\begin{aligned} \mathcal{E}_m^X(f,g) &:= \int_X (\nabla^X f, \nabla^X g)_{TX}\, dm \\ &= -\int_X \Delta^X f\, g\, dm \quad ;\ f,g \in \mathcal{D}. \end{aligned} \tag{4.3.9}$$

Let $(H_m^X, D(H_m^X))$ be the corresponding generator (i.e., the Friedrichs' extension of $(-\Delta^X, \mathcal{D})$ on $L^2(X;m)$ mentioned before). We note that if condition (A) holds $(H_m^X, D(H_m^X))$ coincides with the closure of $(-\Delta^X, \mathcal{D})$ since one self–adjoint operator cannot strictly extend another. Since also $(e^{-tH_m^X})_{t>0}$ is sub–Markovian, it is a strongly continuous contraction semigroup on every $L^p(X;m)$, $p \in [1,\infty[$ (cf. [Da89, Theorem 1.4.1]). Define

$$\begin{aligned} \mathcal{D}_1 := \{f \in D(H_m^X) \ \cap\ & L^1(X;m) \mid H_m^X f \in L^1(X;m) \text{ and} \\ & -\delta \le f \le 0 \text{ for some } \delta \in]0,1[\} \end{aligned} \tag{4.3.10}$$

and for $\mathcal{D}_0 \subset \mathcal{D}_1$

$$E(\mathcal{D}_0) := \text{linear hull of} \{\exp(\langle \log(1+f), \cdot\rangle) \mid f \in \mathcal{D}_0\}. \tag{4.3.11}$$

Remark 4.3.6 (i) It is an easy exercise to show that if $f \in D(H_m^X) \cap L^1(X;m)$ and $H_m^X f \in L^1(X;m)$, then $f \in D(H_{m,1}^X)$ and $H_{m,1}^X f = H_m^X f$ (cf. the remark following the proof of Theorem 1.4.1 in [Da89]). In particular, $\mathcal{D}_1 \subset D(H_{m,1}^X)$.

(ii) We note that obviously $e^{-tH_m^X}\mathcal{D}_1 \subset \mathcal{D}_1$.

Proposition 4.3.7 *(i) Suppose condition (A) holds and let $t \ge 0$, $f \in L^1(X;m)$, $-1 < f \le 0$.*
If $\lambda = \varepsilon_z$ for some $z \in [0,\infty[$, then

$$\begin{aligned} & T^{\Gamma}_{\psi_{i_1\dots i_n}\, z\cdot m}(t)\,(\exp(\langle \log(1+f), \cdot\rangle)) \\ = \ & \exp\left(\langle \log(1+e^{-tH_m^X} f), \cdot\rangle - z\int \left(e^{-tH_m^X} f - f\right) dm\right). \end{aligned}$$

If $(H_m^X, D(H_m^X))$ is conservative, then

$$T^{\Gamma}_{\mu_{\lambda,m}}(t)\,(\exp\langle \log(1+f), \cdot\rangle)) = \exp\left(\langle \log(1+e^{-tH_m^X} f, \cdot\rangle\right).$$

(ii) Suppose conditions (A) and (B) hold. Then $E(\mathcal{D}_1)$ is a dense subset of both $D(H^\Gamma_{\mu_{\lambda,m}})$ and $D(H^\Gamma_{\mu_{\lambda,m},1})$ w.r.t. the respective graph norms $\|\cdot\|_2 := \left(\|H^\Gamma_{\mu_{\lambda,m}}\cdot\|^2_{L^2(\Gamma_X;\mu_{\lambda,m})} + \|\cdot\|^2_{L(\Gamma_X;\mu_{\lambda,m})}\right)^{1/2}$ (i.e., $(H^\Gamma_{\mu_{\lambda,m}}, E(\mathcal{D}_1))$ is essentially self-adjoint on $L^2(\Gamma_X;\mu_{\lambda,m})$) and $\|\cdot\|_1$.

(iii) Suppose conditions (A) and (B) hold. Then any function in $E(\mathcal{D}_1)$ can be approximated by a uniformly bounded sequence of functions in $E(\mathcal{D}_1 \cap C_0^\infty(X))$ $(\subset \mathcal{F}C_b^\infty)$ w.r.t. $\|\cdot\|_1 + \|\cdot\|_{H_0^{1,2}(\Gamma_X;m)}$.

Clearly, Theorem 4.3.5 is an immediate consequence of Proposition 4.3.7 (ii) and (iii). so, it remains to prove 4.3.7 whose parts (ii), (iii) are essentially consequences of its first part (i).

Remark 4.3.8 (i) In the preliminary preprint version of [AKR97a] it was claimed that assertions (i) and (ii) of Proposition 4.3.7 also hold if $(H^X_m, D(H^X_m))$ is not conservative and $\mu_{\lambda,m}$ is not (necessarily) a pure Poisson measure. However, there was a gap in the proof. This case is so far unproved.

(ii) In fact if (B) holds, the two formulae in Proposition 4.3.7 (i) coincide, because (B) implies that

$$\int \left(e^{-tH^X_m} f - f\right) dm = 0 \quad \text{for all } t > 0, f \in L^1(X;m).$$

(iii) If $\lambda = \varepsilon_z$ for some $z \in [0,\infty[$, then it can be shown using second quantization that $\mathcal{F}C_b^\infty$ is also dense in $D(H^\Gamma_{\psi_{i_1 \dots i_n z \cdot m}})$ w.r.t. $\|\cdot\|_2$ (cf. [AKR97a, Subsection 5.4]).

For the proof of Proposition 4.3.7 we need a number of lemmas. We first recall the following well–known result:

Lemma 4.3.9 *Let μ be a σ–finite measure on a measurable space $(E,\mathcal{B})$ and $p \in [1,\infty)$. Let $\chi \in C_b^1(\mathbb{R})$ with $\chi(0) = 0$, if $\mu(E) = \infty$. Let $\mathbb{R}_+ \ni t \mapsto f_t \in L^p(E;\mu)$ be a continuously differentiable map. Then so is $\mathbb{R}_+ \ni t \mapsto \chi \circ f_t \in L^p(E;\mu)$ and $\frac{d(\chi\circ f_t)}{dt} = \chi'(f_t)\frac{df_t}{dt}$.*

Proof. Since $L^p(E;\mu) \ni g \mapsto \int h\, \chi(g)\, d\mu$ is differentiable for all $h \in L^q(E;\mu)$, $q \in (1,\infty]$, $p^{-1} + q^{-1} = 1$, the assertion follows immediately by the fundamental theorem of calculus. □

For $f \in \mathcal{D}_1$ (cf. (4.3.10)) define

$$f_t := e^{-H_m^X} f, \quad t \geq 0.$$

Lemma 4.3.10 *Let $f \in \mathcal{D}_1$. Then $\mathbb{R}_+ \ni t \mapsto \langle \log(1+f_t), \cdot\rangle \in L^2(\Gamma_X; \mu_{\lambda,m})$ is differentiable with derivative equal to*

$$-\langle (1+f_t)^{-1} H_m^X f_t, \cdot \rangle, t \in \mathbb{R}_+.$$

Proof. Let $t \in \mathbb{R}_+$, $\varepsilon \in \mathbb{R}$ with $\varepsilon > 0$ if $t = 0$. Then by (4.3.3) (since $H_m^X f_t \in L^1(X;m) \cap L^2(X;m)$)

$$\int \left(\frac{1}{\varepsilon}(\langle \log(1+f_{t+\varepsilon}), \gamma\rangle - \langle \log(1+f_t), \gamma\rangle) + \langle (1+f_t)^{-1} H_m^X f_t, \gamma\rangle \right)^2 \mu_{\lambda,m}(d\gamma)$$

$$= \int z\lambda(dz) \int A(\varepsilon,t)^2 \, dm + \int z^2 \lambda(dz) \left(\int A(\varepsilon,t)\, dm \right)^2,$$

where

$$A(\varepsilon,t) := \frac{1}{\varepsilon}(\log(1+f_{t+\varepsilon}) - \log(1+f_t)) + (1+f_t)^{-1} H_m^X f_t \, .$$

Now the assertion follows by Remark 4.3.6(i) and Lemma 4.3.9 □

Note that clearly the measure

$$\mathcal{B}(\Gamma_X) \ni B \mapsto \int_{\mathbb{R}_+} \int_{\Gamma_X} 1_B(\gamma) \; \psi_{i_1 \dots i_n \, z \cdot m}(d\gamma) \; z \; \lambda(dz)$$

is absolutely continuous w.r.t. $\mu_{\lambda,m}$. Let $\rho_{\lambda,m}$ denote the corresponding Radon–Nikodym derivative. Since by Hölder's inequality

$$\int \rho_{\lambda,m}^2 \, d\mu_{\lambda,m} = \iint \rho_{\lambda,m}(\gamma) \, z \, \psi_{i_1 \dots i_n \, z \cdot m}(d\gamma) \, \lambda(dz) \leq \left(\int \rho_{\lambda,m}^2 \, d\mu_{\lambda,m} \right)^{1/2} \left(\int z^2 \lambda(dz) \right)^{1/2},$$

it follows that

$$\int \rho_{\lambda,m}^2 \, d\mu_{\lambda,m} \leq \int z^2 \, \lambda(dz) \, . \tag{4.3.12}$$

Lemma 4.3.11 *Suppose condition (A) holds and let $f \in \mathcal{D}_1$. Then:*

(i) $\exp(\langle\log(1+f),\cdot\rangle) \in D(H^\Gamma_{\mu_{\lambda,m}})$
and

(4.3.13)
$$H^\Gamma_{\mu_{\lambda,m}} \exp(\langle\log(1+f),\cdot\rangle)=\Big(\langle\frac{H^X_m f}{1+f},\cdot\rangle - \rho_{\lambda,m}\int H^X_m f\, dm\Big) \exp(\langle\log(1+f),\cdot\rangle)\ .$$

(ii) For every $F \in E(\mathcal{D}_1)$ *there exist* $F_n \in E(\mathcal{D}_1 \cap C_0^\infty(X))$ $(\subset \mathcal{F}C_b^\infty)$, $n \in \mathbb{N}$, *such that* $\sup_n \|F_n\|_\infty < \infty$ *and* $F_n \to F$ *as* $n \to \infty$ *w.r.t.* $\|\cdot\|_1$.

Proof. Since $\exp(\langle\log(1+f),\cdot\rangle)$ and the right hand side of (4.3.13) are in $L^2(\Gamma_X;\mu_{\lambda,m})$, to show (i) it suffices to prove that $\exp(\langle\log(1+f),\cdot\rangle) \in D(H^\Gamma_{\mu_{\lambda,m},1})$ and that (4.3.13) holds with $H^\Gamma_{\mu_{\lambda,m},1}$ replacing $H^\Gamma_{\mu_{\lambda,m}}$.

By condition (A) there exist $\psi_n \in \mathcal{D}$, $n \in \mathbb{N}$, such that

(4.3.14) $$\psi_n \to f,\quad -\Delta^X\psi_n \to H^X_m f \text{ in } L^2(X;m) \text{ as } n\to\infty.$$

Let $\delta,\delta' \in (0,1)$ such that $-1 < -\delta' < -\delta \le f$, and let $\chi \in C_0^\infty(\mathbb{R})$ such that $-\delta' \le \chi$ and $\chi(s) = s$ for all $s \in [-\delta, 1]$. Then for $n \in \mathbb{N}$ we have that $f_n := \chi\circ\psi_n \in \mathcal{D}$, hence $\log(1+f_n) \in \mathcal{D}$. Clearly, by (4.3.2)

$$\begin{aligned}&\int \langle(\log(1+f_n) - \log(1+f))^2,\gamma\rangle\, \mu_{\lambda,m}(d\gamma)\\ =\ &\int (\log(1+f_n) - \log(1+f))^2 dm \int z\, \lambda(dz) \underset{n\to\infty}{\longrightarrow} 0.\end{aligned}$$

Hence for any $\xi \in C^\infty(\mathbb{R})$ such that $\xi(s) = s$ for all $s \in]-\infty,0]$ and $\operatorname{supp}\xi\cap[0,\infty[$ is compact, it easily follows that

(4.3.15) $$\exp\circ\xi(\langle\log(1+f_n),\cdot\rangle) \to \exp(\langle\log(1+f),\cdot\rangle)$$

in $L^1(\Gamma_X;\mu_{\lambda,m})$.

Furthermore, by (4.2.24) (since $H^\Gamma_{\mu_{\lambda,m},1}$ extends $H^\Gamma_{\mu_{\lambda,m}}$, hence Δ^Γ)

$$\begin{aligned}&H^\Gamma_{\mu_{\lambda,m},1} \exp\circ\xi(\langle\log(1+f_n),\cdot\rangle)\\ =\ &\Big(\langle\frac{\Delta^X f_n}{1+f_n},\cdot\rangle\cdot\xi'(\langle\log(1+f_n),\cdot\rangle)\\ &+\big[\xi'(\langle\log(1+f_n),\cdot\rangle) - \xi''(\langle\log(1+f_n),\cdot\rangle)\\ &-(\xi')^2(\langle\log(1+f_n),\cdot\rangle)\big]\cdot\langle\frac{|\nabla^X f_n|^2_{TX}}{(1+f_n)^2},\cdot\rangle\Big)\\ &\exp\circ\xi(\langle\log(1+f_n),\cdot\rangle).\end{aligned} \tag{4.3.16}$$

We have that for all $n \in \mathbb{N}$

$$\Delta^X f_n = \chi'(\psi_n)\Delta^X \psi_n + \chi''(\psi_n)|\nabla^X \psi_n|_{TX}^2. \tag{4.3.17}$$

Since $\chi''(f) = 0$, (4.3.14) implies that

$$\frac{\chi''(\psi_n)}{1+f_n}|\nabla^X \psi_n|_{TX}^2 \to 0 \text{ in } L^1(X;m) \text{ as } n \to \infty.$$

Hence by (4.3.2)

$$\langle \frac{\chi''(\psi_n)}{1+f_n}|\nabla^X \psi_n|_{TX}^2, \cdot\rangle \to 0 \text{ in } L^1(\Gamma_X;\mu_{\lambda,m}) \text{ as } n \to \infty. \tag{4.3.18}$$

Furthermore, by (4.3.12) and (4.3.3) resp. (4.2.18)

$$\begin{aligned}
&\int \left(\langle \frac{H_m^X f}{1+f}, \gamma\rangle - \rho_{\lambda,m} \cdot \int H_m^X f\, dm - \langle \frac{\chi'(\psi_n)}{1+f_n}\, \Delta^X \psi_n, \gamma\rangle\right)^2 \mu_{\lambda,m}(d\gamma) \\
= &\int \rho_{\lambda,m}^2\, d\mu_{\lambda,m} \left(\int H_m^X f\, dm\right)^2 + \int \langle \frac{H_m^X f}{1+f} - \frac{\chi'(\psi_n)}{1+f_n}\Delta^X \psi_n, \gamma\rangle^2 \mu_{\lambda,m}(d\gamma) \\
&\quad -2\int H_m^X f\, dm \int\int \langle \frac{H_m^X f}{1+f} - \frac{\chi'(\psi_n)}{1+f_n}\Delta^X \psi_n, \gamma\rangle\ \psi_{i_1\ldots i_n\, z\cdot m}(d\gamma)\, z\, \lambda(dz) \\
\leq &\int z^2\lambda(dz) \left(\int H_m^X f\, dm\right)^2 + \int z\ \lambda(dz) \int (\frac{H_m^X f}{1+f} - \frac{\chi'(\psi_n)}{1+\chi(\psi_n)}\, \Delta^X \psi_n)^2\, dm \\
&\quad + \int z^2\lambda(dz) \left(\int (\frac{H_m^X f}{1+f} - \frac{\chi'(\psi_n)}{1+\chi(\psi_n)}\Delta^X \psi_n)\, dm\right)^2 \\
&\quad -2\int z^2\lambda(dz)\ \int H_m^X f\, dm \int (\frac{H_m^X f}{1+f} - \frac{\chi'(\psi_n)}{1+\chi(\psi_n)}\Delta^X \psi_n)\, dm\ .
\end{aligned}$$

But

$$\begin{aligned}
&\int \frac{\chi'(\psi_n)}{1+\chi(\psi_n)}\Delta^X \psi_n\, dm \\
= &\int \frac{\chi''(\psi_n)(1+\chi(\psi_n)) - \chi'(\psi_n)^2}{(1+\chi(\psi_n))^2}|\nabla^X \psi_n|_{TX}^2\, dm \\
\underset{n\to\infty}{\longrightarrow} &-\int \frac{|\nabla^X f|_{TX}^2}{(1+f)^2}\, dm = -\int \langle \nabla^X(\frac{f}{1+f}), \nabla^X f\rangle_{TX}\, dm \\
= &\int \frac{H_m^X f}{1+f}\, dm - \int H_m^X f\, dm
\end{aligned}$$

(because $\chi(f) = f$, $\chi'(f) = 1$). Therefore,

(4.3.19)
$$\langle \frac{\chi'(\psi_n)}{1+f_n}\Delta^X \psi_n, \cdot\rangle \to \langle \frac{H_m^X f}{1+f}, \cdot\rangle - \rho_{\lambda,m} \cdot \int H_m^X f\, dm \text{ in } L^2(\Gamma_X; \mu_{\lambda,m}) \text{ as } n \to \infty.$$

(4.3.17) – (4.3.19) and the fact that $\xi' \equiv 1$ on $]-\infty, 0]$ imply that

$$\langle \frac{\Delta^X f_n}{1+f_n}, \cdot\rangle\, \xi'(\langle \log(1+f_n), \cdot\rangle) \exp \circ \xi(\langle \log(1+f_n), \cdot\rangle)$$
$$\underset{n\to\infty}{\longrightarrow} \left(\langle \frac{H_m^X f}{1+f}, \cdot\rangle - \rho_{\lambda,m} \cdot \int H_m^X f\, dm\right) \exp(\langle \log(1+f), \cdot\rangle)$$
in $L^1(\Gamma_X; \mu_{\lambda,m})$.

Since by (4.3.14)

$$\frac{|\nabla^X f_n|_{TX}^2}{(1+f_n)^2} = \frac{\chi'(\psi_n)^2\, |\nabla^X \psi_n|_{TX}^2}{(1+f_n)^2} \to \frac{|\nabla^X f|_{TX}^2}{(1+f)^2} \quad \text{in } L^1(X; m) \text{ as } n \to \infty,$$

it follows by (4.3.3), because $\xi' = (\xi')^2 \equiv 1$ on $]-\infty, 0]$ and hence $\xi'' \equiv 0$ on $]-\infty, 0]$, that the remaining terms on the right hand side of (4.3.16) converge to zero in $L^1(\Gamma_X; \mu_{\lambda,m})$. Thus (i) is proved.

Assertion (ii) also follows then easily since $\exp \circ \xi(\langle \log(1+f_n), \cdot\rangle) \in \mathcal{F}C_b^\infty$ and $|F_n| \le \max(1, \exp \|\xi\|_\infty)$ for all $n \in \mathbb{N}$.

□

Proof of Proposition 3.7. (i): By a simple approximation argument it suffices to prove (i), if $f \in \mathcal{D}_1$. Then by Lemmas 4.3.9, 4.3.10

$$\mathbb{R}_+ \ni t \mapsto F(t) := \exp(\langle \log(1+f_t), \cdot\rangle - z \int (f_t - f)\, dm) \in L^2(\Gamma_X; \mu_{\lambda,m})$$

is differentiable with derivative equal to

$$\left(-\langle \frac{H_m^X f_t}{1+f_t}, \cdot\rangle + z \int H_m^X f_t\, dm\right)$$
$$\cdot \exp(\langle \log(1+f_t), \cdot\rangle - z \int (f_t - f)\, dm),$$
$$t \in \mathbb{R}_+ . \tag{4.3.20}$$

Realizing that $\rho_{\lambda,m} \equiv z$ if $\lambda = \varepsilon_z$, we see that by Lemma 4.3.11 this coincides with

$$-H_{\mu_{\lambda,m}}^\Gamma \exp(\langle \log(1+f_t), \cdot\rangle - z \int (f_t - f)\, dm)$$

i.e.,

$$\frac{d}{dt}F(t) = -H^{\Gamma}_{\mu_{\lambda,m}} F(t), \quad F(0) = \exp(\langle \log(1+f), \cdot \rangle). \tag{4.3.21}$$

$\int H^X_m f_t \, dm = 0$ for all $t \in \mathbb{R}_+$, if $(H^X_m, D(H^X_m))$ is conservative. Hence (4.3.21) holds for general probability measures λ on $(\mathbb{R}_+, \mathcal{B}(\mathbb{R}_+))$ satisfying (4.3.1) in this case. But also

$$\mathbb{R}_+ \ni t \mapsto T^{\Gamma}_{\mu_{\lambda,m}}(t) \exp(\langle \log(1+f), \cdot \rangle) \in L^2(\Gamma_X; \mu_{\lambda,m})$$

solves the Cauchy problem (4.3.21). But as is well-known (and can be easily proved using Duhamel's formula) the solution to (4.3.21) is unique. Hence assertion (i) is proved.

(ii): By part (i) and by Remark 4.3.6 (ii)

$$T^{\Gamma}_{\mu_{\lambda,m}}(t)\,(E(\mathcal{D}_1)) \subset E(\mathcal{D}_1)\,. \tag{4.3.22}$$

Since $E(\mathcal{D}_1)$ is obviously dense in $L^2(\Gamma_X; \mu_{\lambda,m})$, hence in $L^1(\Gamma_X; \mu_{\lambda,m})$ and since $E(\mathcal{D}_1) \subset D(H^{\Gamma}_{\mu_{\lambda,m}})$ by Lemma 4.3.11 (i), the assertion follows from a standard theorem on operator semigroups (cf. e.g. [RS75, Theorem X.49]).

(iii): Let $F \in E(\mathcal{D}_1)$ and F_n, $n \in \mathbb{N}$, as in Lemma 4.3.11(ii). Then $F_n \to F$ in $L^2(\Gamma_X; \mu_{\lambda,m})$ as $n \to \infty$ and

$$\int |\nabla^{\Gamma} F_n|^2_{T\Gamma_X} d\mu_{\lambda,m} = -\int F_n \, \Delta^{\Gamma} F_n \, d\mu_{\lambda,m} \underset{n\to\infty}{\longrightarrow} \int F \; H^{\Gamma}_{\mu_{\lambda,m}} F \, d\mu_{\lambda,m}\,.$$

In particular,

$$\sup_n \mathcal{E}^{\Gamma}_{\mu_{\lambda,m}}(F_n, F_n) < \infty.$$

Now by [MR92, Chap. I, Lemma 2.12] there exists a subsequence $(n_k)_{k\in\mathbb{N}}$ such that for $G_N := \frac{1}{N}\sum_{k=1}^{N} F_{n_k} \in E(\mathcal{D}_1 \cap C_0^{\infty}(X))$, $N \in \mathbb{N}$,

$$G_N \longrightarrow F \text{ as } N \to \infty \text{ w.r.t. } \|\cdot\|_{H_0^{1,2}(\Gamma_X;\mu_{\lambda,m})}\,.$$

□

4.3.3 Brownian motion

In this section we shall treat the diffusion process associated with our Dirichlet form $(\mathcal{E}^{\Gamma}_{\mu_{\lambda,m}}, D(\mathcal{E}^{\Gamma}_{\mu_{\lambda,m}}))$ from Subsection 3.1 and identify it (when started with $\mu_{\lambda,m}$) as the well-known independent infinite particle process on X provided condition (A) in Theorem 4.3.3 is fulfilled and $(H^X_m, D(H^X_m))$ is conservative.

The desired process in general will live on the bigger state space $\ddot{\Gamma}$ consisting of all $\mathbb{Z}_+$-valued Radon measures on X (which is Polish, see e.g. [MR97]). Since $\Gamma_X \subset \ddot{\Gamma}_X$ and $\mathcal{B}(\ddot{\Gamma}) \cap \Gamma_X = \mathcal{B}(\Gamma_X)$, we can consider $\mu_{\lambda,m}$ as a measure on $(\ddot{\Gamma}, \mathcal{B}(\ddot{\Gamma}))$ and correspondingly $(\mathcal{E}, D(\mathcal{E}))$ as a Dirichlet form on $L^2(\ddot{\Gamma}; \mu_{\lambda,m})$.

Theorem 4.3.12 *$(\mathcal{E}^{\Gamma}_{\mu_{\lambda,m}}, D(\mathcal{E}^{\Gamma}_{\mu_{\lambda,m}}))$ is quasi–regular. In particular, there exists a conservative diffusion process*

$$\mathbf{M} = (\Omega, \mathbf{F}, (\mathbf{F})_{t\geq 0}, (\Theta_t)_{t\geq 0}, (\mathbf{X}_t)_{t\geq 0}, (\mathbf{P}_\gamma)_{\gamma \in \ddot{\Gamma}})$$

on $\ddot{\Gamma}$ (cf. [Dy65]) which is properly associated with $(\mathcal{E}^{\Gamma}_{\mu_{\lambda,m}}, D(\mathcal{E}^{\Gamma}_{\mu_{\lambda,m}}))$, i.e., for all ($\mu_{\lambda,m}$-versions of) $F \in L^2(\ddot{\Gamma}; \mu_{\lambda,m})$ and all $t > 0$ the function

$$\gamma \mapsto p_t F(\gamma) := \int_\Omega F(X_t) d\mathbf{P}_\gamma, \quad \gamma \in \ddot{\Gamma}, \tag{4.3.23}$$

is an $\mathcal{E}^{\Gamma}_{\mu_{\lambda,m}}$-quasi-continuous version of $\exp(-tH^{\Gamma}_{\mu_{\lambda,m}})F$. $\mathbf{M}$ is up to $\mu_{\lambda,m}$-equivalence unique (cf. [MR92, Ch. IV, Sect. 6]). In particular, $\mathbf{M}$ is $\mu_{\lambda,m}$–symmetric (i.e., $\int G\, p_t F d\mu_{\lambda,m} = \int F\, p_t G\, d\mu_{\lambda,m}$ for all $F, G : \ddot{\Gamma} \to \mathbb{R}_+, \mathcal{B}(\ddot{\Gamma})$-measurable) and has $\mu_{\lambda,m}$ as an invariant measure.

In the above theorem $\mathbf{M}$ is canonical, i.e., $\Omega = C([0, \infty[\to \ddot{\Gamma})$, $\mathbf{X}_t(\omega) := \omega(t)$, $t \geq 0$, $\omega \in \Omega$, $(\mathbf{F}_t)_{t\geq 0}$ together with $\mathbf{F}$ is the corresponding minimum completed admissible family (cf. [F80, Sect.4.1] and also [FOT94]) and Θ_t, $t \geq 0$, are the corresponding natural time shifts.

Theorem 4.3.12 is in fact a special case of Theorems 4.4.12, 4.4.14 in the next section where we shall also recall the definition of quasi–regularity, $\mathcal{E}^{\Gamma}_{\mu_{\lambda,m}}$–quasi–continuity etc. So, we refer to Subsection 4.2 for more details.

Let $\mathbf{M} = (\Omega, \mathbf{F}, (\mathbf{F}_t)_{t\geq 0}, (\Theta_t)_{t\geq 0}, (\mathbf{X}_t)_{t\geq 0}, (\mathbf{P}_\gamma)_{\gamma\in\ddot{\Gamma}})$ be as in Theorem 4.3.12. As usual we define

$$\mathbf{P}_{\mu_{\lambda,m}} := \int \mathbf{P}_\gamma\, \mu_{\lambda,m}(d\gamma). \tag{4.3.24}$$

Let $M = (\Omega, \mathcal{F}, (\mathcal{F}_t)_{t\geq 0}, (X_t)_{t\geq 0}, (P_x)_{x\in X})$ be the diffusion process on X associated with the Dirichlet form $(\mathcal{E}^X_m, D(\mathcal{E}^X_m))$ (cf. e.g. [MR92, Ch.IV, Subsect.4 a) resp. Ch. II, Subsect. 2 a)]). Set

$$P_m := \int_X P_x\, m(dx). \tag{4.3.25}$$

We note that in the non–conservative case $\Omega = C([0,\infty[\to X_\Delta)$ rather than $C([0,\infty[\to X)$. Here X_Δ is the one–point compactification of X.

In order to identify $\mathbf{M}$ as the independent infinite particle process if $(H_m^X, D(H_m^X))$ is conservative, we have to show that $\mathbf{P}_{\mu_{\lambda,m}}$ has the same finite dimensional distributions as $\int_{\mathbb{R}_+} \psi_{i_1\dots i_n zP_m}\,\lambda(dz)$, where $\psi_{i_1\dots i_n zP_m}$ is the Poisson measure on Ω with intensity zP_m, $z \geq 0$ (cf. e.g. [ST74], [M-L76]). For this it suffices to prove the following proposition whose proof is straight-forward by Proposition 4.3.7(i) (see also [ST74, Proposition 1.5]).

Proposition 4.3.13 *Assume that condition* (A) *in Theorem 4.3.3 holds and that* $(H_m^X, D(H_m^X))$ *is conservative. Then for all* $f_1, \dots, f_N \in L^1(X;m)$ *taking values in* $]-1,0]$ *and* $0 \leq t_1 < \cdots < t_N < \infty$

$$\int_\Omega \exp(\langle \log(1+f_1), \mathbf{X}_{t_1}\rangle)\cdots\exp(\langle\log(1+f_N),\mathbf{X}_{t_N}\rangle)d\mathbf{P}_{\mu_{\lambda,m}}$$

$$=\int\exp\left(z\int_\Omega[(1+f_1(X_{t_1}))\cdots(1+f_N(X_{t_N}))-1]dP_m\right)\lambda(dz). \tag{4.3.26}$$

Proof. If $N=1$, then by Proposition 4.3.7(i)

$$\int_\Omega \exp(\langle\log(1+f_1),\mathbf{X}_{t_1}\rangle)d\mathbf{P}_{\mu_{\lambda,m}} = \int_{\Gamma_X}\exp(\langle\log(1+e^{-t_1H_m^X}f_1),\gamma\rangle)\mu_{\lambda,m}(d\gamma)$$

$$=\int\exp\left[z\int_X e^{-t_1H_m^X}f_1\,dm\right]\lambda(dz) = \int\exp\left[z\int_\Omega f_1(X_{t_1})dP_m\right]\lambda(dz).$$

Suppose (4.3.26) holds for $N-1$. Then by Proposition 4.3.7 and the Markov property of $\mathbf{P}_{\mu_{\lambda,m}}$ the left hand side of (4.3.26) is equal to

$$\int_\Omega\exp(\langle\log(1+f_1),\mathbf{X}_{t_1}\rangle)\cdots\exp(\langle\log(1+f_{N-1}),\mathbf{X}_{t_{N-1}}\rangle)$$
$$\times\ \exp(\langle\log(1+e^{-(t_N-t_{N-1})H_m^X}f_N),\mathbf{X}_{t_{N-1}}\rangle)d\mathbf{P}_{\mu_{\lambda,m}}.$$

This in turn by induction hypothesis is equal to

$$\int\exp\Bigg(z\int_\Omega[(1+f_1(X_{t_1}))\cdots(1+f_{N-1}(X_{t_{N-1}}))$$
$$\times\ (1+e^{-(t_N-t_{N-1})H_m^X}f_N(X_{t_{N-1}}))-1]dP_m\Bigg)\lambda(dz)$$

which by the Markov property of P_m is just the right hand side of (4.3.26). □

In the non–conservative case the finite dimensional distributions of $P_{\mu_{\lambda,m}}$ and $\int_{\mathbb{R}_+}\psi_{i_1\dots i_n zP_m}\,\lambda(dz)$ do not coincide in general as the following result shows:

Proposition 4.3.14 *Assume condition (A) in Theorem 4.3.3 holds and let* $z \in [0,\infty[$*. Then for all* $f_1,\dots,f_N \in L^1(X;m)$ *taking values in* $]-1,0]$ *and* $0 \le t_1 < \dots < t_N < \infty$

$$\int_\Omega \exp(\langle \log(1+f_1), \mathbf{X}_{t_1}\rangle)\dots\exp(\langle\log(1+f_N),\mathbf{X}_{t_N}\rangle)\, d\mathbf{P}_{\psi_{i_1\dots i_n}\, z\cdot m}$$

$$= \exp(z\int_\Omega [(1+f_1(X_{t_1}))\dots(1+f_N(X_{t_N}))-1]\,dP_m$$

$$\cdot \exp\Bigg(-z\int (e^{-t_1 H_m^X}-1)f_1 dm - z\sum_{i=2}^N \int (e^{-t_1H_m^X}-1)\Bigg[(1+f_1)e^{-(t_2-t_1)H_m^X}$$

$$(4.3.27) \qquad (1+f_2)e^{-(t_3-t_2)H_m^X}\dots(1+f_{i-1})e^{-(t_i-t_{i-1})H_m^X}f_i\Bigg]\,dm\Bigg)\,.$$

By Proposition 4.3.7 (i) the proof of Proposition 4.3.14 is entirely the same as that of 4.3.13.

Remark 4.3.15 (i) Since for $a_1,\dots,a_N \in \mathbb{R}$,

$$(1+a_1)\dots(1+a_N)-1 = a_1 + \sum_{i=2}^N (1+a_1)\dots(1+a_{i-1})a_i$$

the right hand side of (4.3.27) obviously simplifies to

$$\exp\Bigg(z\int f_1\,dm +$$

$$\sum_{i=1}^N \int (1+f_1)e^{-(t_2-t_1)H_m^X}\Big[(1+f_2)e^{-(t_3-t_2)H_m^X}[\dots(1+f_{i-1})e^{-(t_i-t_{i-1})H_m^X}f_i]\Big]\,dm\Bigg).$$

(ii) Since $\mathbf{M}$ is always conservative, M, however, if and only if so is $(H_m^X, D(H_m^X))$, Proposition 4.3.14 is, of course, not surprising.

(iii) ("σ–case") Let σ be as in the beginning of Remark 4.2.3(i). As mentioned before all the above extends to the case where σ replaces m. The diffusion process M associated to $(\mathcal{E}_\sigma^X, D(\mathcal{E}_\sigma^X))$ is usually called *distorted Brownian motion on* X. Therefore, we call the diffusion associated to $(\mathcal{E}^\Gamma_{\mu_{\lambda,\sigma}}, D(\mathcal{E}^\Gamma_{\mu_{\lambda,\sigma}}))$ *distorted Brownian motion on* Γ_X because of the complete analogy. Likewise, since M is just ordinary Brownian motion on X if $\sigma = m$, it is justified to call $\mathbf{M}$ *Brownian motion* on Γ_X in this case. Therefore, Proposition 4.3.13 shows that, provided M or equivalently $(H_\sigma^X, D(H_\sigma^X))$ is conservative and condition (A) holds, the usual independent particle process known for more than 40 years (cf. [D53], [Do56]) is nothing but (distorted) Brownian motion on Γ_X.

We conclude this section with a result concerning the question whether the bigger state space $\ddot{\Gamma}_X$ rather than just Γ_X is really needed.

Theorem 4.3.16 *Suppose $X = \mathbb{R}^d$, $d \geq 2$. Let $\sigma = \rho \cdot m$ be as specified at the beginning of Remark 4.2.3 (i) and assume, in addition, that $\sigma \in L^2_{loc}(\mathbb{R}^d; m)$. Then $\ddot{\Gamma}_X \setminus \Gamma_X$ is $\mathcal{E}^{\Gamma}_{\mu_{\lambda,\sigma}}$-exceptional.*

This result was obtained in [RS97] to which we refer for its proof. Theorem 4.3.16 just says that for $\mathcal{E}_{\mu_{\lambda,\sigma}}$-q.e. $\gamma \in \Gamma_X$ under $\mathbf{P}_\gamma$, $(\mathbf{X}_t)_{t\geq 0}$ never leaves Γ_X (cf. [MR92, Proposition 5.30 (i)]).

4.3.4 Ergodicity in the free case

Let $\mathbf{M} = (\Omega, \mathbf{F}, (\mathbf{F}_t)_{t\geq 0}, (\Theta_t)_{t\geq 0}, (\mathbf{X}_t)_{t\geq 0}, (\mathbf{P}_\gamma)_{\gamma\in\ddot{\Gamma}_X})$ be as in Theorem 4.3.16. Concerning ergodicity / irreducibility we recall the following well–known result:

Proposition 4.3.17 *The following assertions:*

(i) $\mathbf{P}_{\mu_{\lambda,m}}$ *is (time–)ergodic (i.e., every bounded $\mathbf{F}$–measurable function $G : \Omega \to \mathbb{R}$ which is Θ_t–invariant for all $t \geq 0$, is constant $\mathbf{P}_{\mu_{\lambda,m}}$–a.e.).*

(ii) $(\mathcal{E}^{\Gamma}_{\mu_{\lambda,m}}, D(\mathcal{E}^{\Gamma}_{\mu_{\lambda,m}}))$ *is irreducible (i.e., for $F \in D(\mathcal{E}^{\Gamma}_{\mu_{\lambda,m}})$, $\mathcal{E}^{\Gamma}_{\mu_{\lambda,m}}(F, F) = 0$ implies that $F = \text{const}$).*

(iii) $(T^{\Gamma}_{\mu_{\lambda,m}}(t))_{t>0}$ *is irreducible (i.e., if $G \in L^2(\Gamma_X; \mu_{\lambda,m})$ such that $T^{\Gamma}_{\mu_{\lambda,m}}(t)(GF) = GT^{\Gamma}_{\mu_{\lambda,m}}(t)F$ for all $F \in L^\infty(\Gamma_X; \mu_{\lambda,m}), t > 0$, then $G = \text{const}$).*

(iv) *If $F \in L^2(\Gamma_X; \mu_{\lambda,m})$ such that $T^{\Gamma}_{\mu_{\lambda,m}}(t)F = F$ for all $t > 0$, then $F = \text{const}$.*

(v) $(T^{\Gamma}_{\mu_{\lambda,m}}(t))_{t>0}$ *is ergodic (i.e.,*

$$\int (T^{\Gamma}_{\mu_{\lambda,m}}(t)F - \int F d\mu_{\lambda,m})^2 d\mu_{\lambda,m} \to 0 \text{ as } t \to \infty \text{ for all } F \in L^2(\Gamma_X; \mu_{\lambda,m})).$$

(vi) *If $F \in D(H^{\Gamma}_{\mu_{\lambda,m}})$ with $H^{\Gamma}_{\mu_{\lambda,m}} F = 0$, then $F = \text{const}$.*

Proof. The equivalence of (ii) – (vi) can be shown entirely analogously to [AKR97d, Proposition 2.3]. The equivalence of (i) and (ii) is a direct consequence of the regularization method in [MR92, Chap. VI, Sect. 1] and the main results in [F80], [F83]. □

The main result here is, however, the following:

Theorem 4.3.18 *Assume that $m(X) = \infty$. If condition (A) in Theorem 4.3.3 holds and $(H_m^X, D(H_m^X))$ is conservative, then any of the assertions (i)–(vi) of Proposition 4.3.17 is equivalent to:*

(vii) $\mu_{\lambda,m} = \psi_{i_1 \dots i_n \, z \cdot m}$ *for some* $z \in [0, \infty[$.

This theorem is a special case of Theorem 4.9.9 below. So, we refer to the detailed discussion in Section 9.

Remark 4.3.19 *("σ–case")* There are generalizations of Theorem 4.3.18 and Proposition 4.3.17 in the case where m is replaced by $\sigma = \rho \cdot m$ with $\sigma(X) = \infty$ for a large class of even not weakly differentiable functions $\rho : X \to \mathbb{R}_+$ (cf. [AKRR98]). However, if ρ satisfies all assumptions specified in the beginning of Remark 2.3 (i), then the generalization to the "σ–case is as before straightforward (cf. [AKR97a].)

4.4 Classical Dirichlet forms on configuration spaces w.r.t. general measures

Many results on Dirichlet forms, defined in terms of the gradient ∇^Γ and the "tangent bundle" $(T_\gamma \Gamma_X)_{\gamma \in \Gamma_X} = (L^2(X \to TX; \gamma))_{\gamma \in \Gamma_X}$ introduced before, can be proved for quite arbitrary probability measures μ on $(\Gamma_X, \mathcal{B}(\Gamma_X))$ replacing the mixed Poisson measure $\mu_{\lambda,m}$. The purpose of this section is to summarize some of these results which are of particular importance. In this section we fix a probability measure μ on $(\Gamma_X, \mathcal{B}(\Gamma_X))$ with Radon mean, i.e.,

$$\int \gamma(K) \, \mu(d\gamma) < \infty \text{ for all compact } K \subset X. \tag{4.4.1}$$

We define

$$\mathcal{E}_\mu^\Gamma(F, G) := \int_{\Gamma_X} \langle \nabla^\Gamma F, \nabla^\Gamma G \rangle_{T\Gamma_X} \, d\mu \, ; \quad F, G \in \mathcal{F}C_b^\infty. \tag{4.4.2}$$

Note that by (4.2.8), the Cauchy–Schwarz inequality and (4.4.1) the integral in (4.4.2) exists for all $F, G \in \mathcal{F}C_b^\infty$, hence $\mathcal{E}_\mu^\Gamma$ is well-defined. Furthermore, assume that

$$\nabla^\Gamma F = \nabla^\Gamma G \ \mu\text{–a.e. if } F, G \in \mathcal{F}C_b^\infty \text{ such that } F = G \ \mu\text{-a.e.} \tag{4.4.3}$$

Clearly, (4.4.3) holds if supp $\mu = \Gamma_X$. Let $\mathcal{F}C_b^{\infty,\mu}$ denote the set of μ–equivalence classes determined by $\mathcal{F}C_b^\infty$. (4.4.3) implies that $(\mathcal{E}_\mu^\Gamma, \mathcal{F}C_b^{\infty,\mu})$ is a well–defined symmetric positive defined bilinear form on $L^2(\Gamma_X; \mu)$. It is also densely defined, since by an easy monotone class argument its domain $\mathcal{F}C_b^{\infty,\mu}$ is dense in $L^2(\Gamma_X; \mu)$.

4.4.1 Closability, domains and Sobolev spaces

We first recall the definition of closability: Equip $\mathcal{F}C_b^{\infty,\mu}$ with the norm (cf. (4.3.6))

$$\|\cdot\|_{H_0^{1,2}(\Gamma_X;\mu)} := \left(\mathcal{E}_\mu^\Gamma(\cdot,\cdot) + (\ ,\ )_{L^2(\Gamma_X;\mu)}\right)^{1/2}. \tag{4.4.4}$$

Then the embedding map

$$i : (\mathcal{F}C_b^{\infty,\mu}, \|\cdot\|_{H_0^{1,2}(\Gamma_X;\mu)}) \to (L^2(\Gamma_X;\mu), \|\cdot\|_{L^2(\Gamma_X;\mu)})$$

is continuous, hence i uniquely extends to a continuous (linear) map $\bar{i}$ from the abstract completion $\overline{\mathcal{F}C_b^{\infty,\mu}}$ of $\mathcal{F}C_b^{\infty,\mu}$ w.r.t. $\|\cdot\|_{H_0^{1,2}(\Gamma_X;\mu)}$ to $L^2(\Gamma_X;\mu)$.

Definition 4.4.1 $(\mathcal{E}_\mu^\Gamma, \mathcal{F}C_b^{\infty,\mu})$ *is called closable on* $L^2(\Gamma_X;\mu)$ *if the above map*

$$\bar{i} := \overline{\mathcal{F}C_b^{\infty,\mu}} \longrightarrow L^2(\Gamma_X;\mu)$$

is one-to-one.

Remark 4.4.2 (i) Clearly, $(\mathcal{E}_\mu^\Gamma, \mathcal{F}C_b^{\infty,\mu})$ is by definition closable on $L^2(\Gamma_X;\mu)$ if for all $F_n \in \mathcal{F}C_b^{\infty,\mu}$, $n \in \mathbb{N}$, such that

$$\mathcal{E}_\mu^\Gamma(F_n - F_m, F_n - F_m) \underset{n,m\to\infty}{\longrightarrow} 0, \quad \|F_n\|_{L^2(\Gamma_X;\mu)} \underset{n\to\infty}{\longrightarrow} 0,$$

it follows that $\mathcal{E}_\mu^\Gamma(F_n, F_n) \underset{n\to\infty}{\longrightarrow} 0$.

(ii) If $(\mathcal{E}_\mu^\Gamma, \mathcal{F}C_b^{\infty,\mu})$ is closable on $L^2(\Gamma_X;\mu)$, $\mathcal{E}_\mu^\Gamma$ uniquely extends to a continuous bilinear map on all of $\overline{\mathcal{F}C_b^{\infty,\mu}} =: D(\mathcal{E}_\mu^\Gamma)$. $(\mathcal{E}_\mu^\Gamma, D(\mathcal{E}_\mu^\Gamma))$ is the smallest closed extension of $(\mathcal{E}_\mu^\Gamma, \mathcal{F}C_b^{\infty,\mu})$.

In general it is not true that $(\mathcal{E}_\mu^\Gamma, \mathcal{F}C_b^{\infty,\mu})$ is closable on $L^2(\Gamma_X;\mu)$, but, in fact, quite weak assumptions ensure this. We refer to Subsection 6.3 below and to [MR97], [AKRR98], [dSKR98] for other general methods and other concrete examples. In the rest of this subsection we shall mainly concentrate on measures satisfying an integration by parts formula of type (4.2.14), such as the mixed Poisson measures. Therefore, we introduce the following condition:

(IbP) $\int \gamma(K)^2\, \mu(d\gamma) < \infty$ for all compact $K \subset X$ and for every $v \in V_0(X)$ there exists $B_v^\mu \in L^2(\Gamma_X;\mu)$ such that

$$\int \nabla_v^\Gamma F\, d\mu = -\int F\, B_v^\mu\, d\mu.$$

Since ∇_v^Γ satisfies the product rule, (IbP) implies that

$$\int \nabla_v^\Gamma F \, G \, d\mu = - \int F \, \nabla_v^\Gamma G \, d\mu - \int F \, G B_v^\mu \, d\mu \tag{4.4.5}$$
for all $F, G \in \mathcal{F}C_b^\infty$ and all $v \in V_0(X)$.

Furthermore, (IbP) , obviously, implies (4.4.3).

From now on we assume that (IbP) holds. The mixed Poisson measures of Section 3 and Ruelle measure to be introduced in Subsection 7.1 below are typical examples.

In analogy with (4.2.13), (4.2.24) for $V = \sum_{i=1}^N F_i \, v_i \in \mathcal{VF}C_b^\infty$ we define

$$\operatorname{div}_\mu^\Gamma V := \sum_{i=1}^N (\nabla_{v_i}^\Gamma F_i + B_{v_i}^\mu F_i) \tag{4.4.6}$$

and for $F \in \mathcal{F}C_b^\infty$

$$\Delta_\mu F := \operatorname{div}_\mu^\Gamma \nabla^\Gamma F \, . \tag{4.4.7}$$

The following result is the analogue of Remark 2.3 (ii) and Propositions 4.3.1, 4.3.2. By (4.4.5)–(4.4.7) its proof is the same , and therefore, omitted.

Proposition 4.4.3 *Suppose μ satisfies* (IbP) . *Then:*
(i) The right hand side of (4.4.6) *is μ–a.e. independent of the representation of V.*
(ii) For all $F, G \in \mathcal{F}C_b^\infty$

$$\mathcal{E}_\mu^\Gamma(F, G) = - \int \Delta_\mu^\Gamma F \, G \, d\mu \, . \tag{4.4.8}$$

In particular, $(\mathcal{E}_\mu^\Gamma, \mathcal{F}C_b^{\infty,\mu})$ is closable on $L^2(\Gamma_X; \mu)$. Its closure is a symmetric Dirichlet form on $L^2(\Gamma_X; \mu)$ which is conservative *(i.e., $1 \in D(\mathcal{E}_\mu^\Gamma)$ and $\mathcal{E}_\mu^\Gamma(1,1) = 0$). Its generator, denoted by $(H_\mu^\Gamma, D(H_\mu^\Gamma))$ is the Friedrichs' extension of $(-\Delta_\mu, \mathcal{F}C_b^{\infty,\mu})$ on $L^2(\Gamma_X; \mu)$.*

Clearly, ∇^Γ extends to all of $D(\mathcal{E}_\mu^\Gamma)$. We shall denote its extension by ∇_μ^Γ. As for $\mu = \mu_{\lambda,m}$ we set

$$H_0^{1,2}(\Gamma_X; \mu) := D(\mathcal{E}_\mu^\Gamma) \tag{4.4.9}$$

i.e., the $(1,2)$–*Sobolev space on* Γ_X *(w.r.t. μ)*.

We shall now construct an extension of $(\mathcal{E}_\mu^\Gamma, H_0^{1,2}(\Gamma_X;\mu))$ which is the exact analogue on Γ_X of a weak Sobolev space on a finite dimensional manifold. This is essentially a special case of a construction by A. Eberle in [Eb97].

We first note that, as is easy to check, the μ–equivalence classes $\mathcal{VFC}_b^{\infty,\mu}$ determined by $\mathcal{VFC}_b^\infty$ are dense in $L^2(\Gamma_X \to T\Gamma_X;\mu)$, i.e., the Hilbert space of μ–square integrable sections in $T\Gamma_X$. Let $((\mathrm{div}_\mu^\Gamma)^{*,\mu}, D((\mathrm{div}_\mu^\Gamma)^{*,\mu}))$ be the adjoint of $(\mathrm{div}_\mu^\Gamma, \mathcal{VFC}_b^{\infty,\mu})$ as an operator from $L^2(\Gamma_X \to T\Gamma_X;\mu)$ to $L^2(\Gamma_X;\mu)$.

Remark 4.4.4 As an adjoint the operator $((\mathrm{div}_\mu^\Gamma)^{*,\mu}, D((\mathrm{div}_\mu^\Gamma)^{*,\mu}))$ is automatically closed, and by definition it is an operator from $L^2(\Gamma_X;\mu)$ to $L^2(\Gamma \to T\Gamma_X;\mu)$. Furthermore, again by definition, $G \in L^2(\Gamma_X;\mu)$ belongs to $D((\mathrm{div}_\mu^\Gamma)^{*,\mu})$ if and only if there exists $V_G \in L^2(\Gamma_X \to T\Gamma_X;\mu)$ such that

$$\int G \,\mathrm{div}_\mu^\Gamma\, V\, d\mu = -\int \langle V_G, V\rangle_{T\Gamma_X}\, d\mu \text{ for all } V \in \mathcal{VFC}_b^\infty. \tag{4.4.10}$$

In this case $(\mathrm{div}_\mu^\Gamma)^{*,\mu} G = V_G$.

Because of Remark 4.4.4 we set

$$W^{1,2}(\Gamma_X;\mu) := D((\mathrm{div}_\mu^\Gamma)^{*,\mu}), \quad d^\mu := (\mathrm{div}_\mu^\Gamma)^{*,\mu} \tag{4.4.11}$$

and think of $W^{1,2}(\Gamma_X;\mu)$ as a *weak* $(1,2)$*–Sobolev space on* Γ_X with norm

(4.4.12)

$$W^{1,2}(\Gamma_X;\mu) \ni G \mapsto \|G\|_{W^{1,2}(\Gamma_X;\mu)} := \left(\int \langle d^\mu G, d^\mu G\rangle_{T\Gamma_X} d\mu + \int G^2\, d\mu\right)^{1/2}.$$

By (4.4.5) it follows from Remark 4.4.4 that

$$\mathcal{FC}_b^{\infty,\mu} \subset W^{1,2}(\Gamma_X;\mu) \text{ and } d^\mu = \nabla^\Gamma \text{ on } \mathcal{FC}_b^{\infty,\mu}, \tag{4.4.13}$$

hence the densely defined positive definite symmetric bilinear form

$$(F,G) \mapsto \int \langle d^\mu F, d^\mu G\rangle_{T\Gamma_X} d\mu,$$

with domain $W^{1,2}(\Gamma_X;\mu)$ extends $(\mathcal{E}_\mu^\Gamma, H_0^{1,2}(\Gamma_X;\mu))$ and is, therefore, denoted by $(\mathcal{E}_\mu^\Gamma, W^{1,2}(\Gamma_X;\mu))$.

Proposition 4.4.5 *Let* $F \in H_0^{1,2}(\Gamma_X;\mu) \cap L^\infty(\Gamma_X;\mu)$ *and* $G \in W^{1,2}(\Gamma_X;\mu) \cap L^\infty(\Gamma_X;\mu)$. *Then* $FG \in W^{1,2}(\Gamma_X;\mu)$ *and*

$$d^\mu(FG) = F\, d^\mu G + G\, d^\mu F. \tag{4.4.14}$$

Proof. Let $V \in \mathcal{VFC}_b^\infty$ and assume first that $F \in \mathcal{F}C_b^\infty$. Then by the product rule for ∇^Γ and (4.4.6) for all $G \in W^{1,2}(\Gamma_X;\mu) \cap L^\infty(\Gamma_X;\mu)$

$$\begin{aligned}\int FG \, \mathrm{div}_\mu V \, d\mu &= \int G \, \mathrm{div}_\mu^\Gamma(FV)\, d\mu - \int G \langle \nabla^\Gamma F, V\rangle_{T\Gamma_X} \, d\mu \\ &= -\int \langle d^\mu G, FV\rangle_{T\Gamma_X} \, d\mu - \int \langle G d^\mu F, V\rangle_{T\Gamma_X} \, d\mu\end{aligned}$$

where we used (4.4.10), (4.4.11) and (4.4.13) in the last step. Since $\mathcal{VFC}_b^{\infty,\mu}$ is dense in $L^2(\Gamma_X \to T\Gamma_X;\mu)$, the assertion follows provided $F \in \mathcal{F}C_b^{\infty,\mu}$. The case where $F \in H_0^{1,2}(\Gamma_X;\mu) \cap L^\infty(\Gamma_X;\mu)$ is then an immediate consequence, since $\mathcal{F}C_b^{\infty,\mu}$ is dense in $H_0^{1,2}(\Gamma_X;\mu)$ w.r.t. $\| \ \|_{H_0^{1,2}(\Gamma_X;\mu)}$ by definition. □

Proposition 4.4.6 *Suppose conditions (A) and (B) in Theorem 4.3.3 hold and let $\mu_{\lambda,m}$ (or even $\mu_{\lambda,\sigma}$) be a mixed Poisson measure satisfying* (4.3.1). *Then*

(4.4.15)
$$\mathcal{E}^\Gamma_{\mu_{\lambda,m}}(F,G) = \int H^\Gamma_{\mu_{\lambda,m}} F \, G \, d\mu_{\lambda,m} \text{ for all } F \in E(\mathcal{D}_1), G \in W^{1,2}(\Gamma_X;\mu)$$

(cf. (4.3.10), (4.3.11)*). In particular,*

$$H_0^{1,2}(\Gamma_X;\mu_{\lambda,m}) = W^{1,2}(\Gamma_X;\mu_{\lambda,m}). \tag{4.4.16}$$

Proof. Let $G \in W^{1,2}(\Gamma_X;\mu_{\lambda,m})$ and $F \in E(\mathcal{D}_1)$. By Proposition 4.3.7 (ii) there exist $F_n \in \mathcal{F}C_b^\infty$, $n \in \mathbb{N}$, such that $F_n \to F$ as $n \to \infty$ w.r.t. $\| \cdot \|_1 + \| \cdot \|_{H_0^{1,2}(\Gamma_X;\mu_{\lambda,m})}$. Hence, because $(d^\mu, W^{1,2}(\Gamma_X;\mu_{\lambda,m}))$ extends $(\nabla^\Gamma_\mu, H_0^{1,2}(\Gamma_X;\mu_{\lambda,m}))$ by definition of d^μ

$$\begin{aligned}\mathcal{E}^\Gamma_\mu(F,G) &= \lim_{n\to\infty} \int \langle \nabla^\Gamma F_n, d^\mu G\rangle_{T\Gamma_X} \, d\mu_{\lambda,m} \\ &= -\lim_{n\to\infty} \int \Delta^\Gamma F_n \, G \, d\mu_{\lambda,m} = \int H^\Gamma_{\mu_{\lambda,m}} F \, G \, d\mu_{\lambda,m}.\end{aligned}$$

This implies that the generator of $(\mathcal{E}^\Gamma_\mu; W^{1,2}(\Gamma_X;\mu_{\lambda,m}))$, which is self–adjoint, extends $(H^\Gamma_{\mu_{\lambda,m}}, E(\mathcal{D}_1))$ (cf. e.g. [MR92, Chap. I, Proposition 2.16]), hence it also extends its closure on $L^2(\Gamma_X;\mu_{\lambda,m})$, which by Proposition 3.7 (ii) is self–adjoint, too. Since self–adjoint operators cannot strictly extend each other, they must coincide. But by Proposition 4.3.7 (ii) the closure of $(H^\Gamma_{\mu_{\lambda,m}}, E(\mathcal{D}_1))$ is equal to its Friedrichs' extension, i.e., equal to the generator of $(\mathcal{E}^\Gamma_\mu, H_0^{1,2}(\Gamma_X;\mu_{\lambda,m}))$, hence the assertion follows. □

In general, it might be that $H_0^{1,2}(\Gamma_X;\mu) \subsetneqq W^{1,2}(\Gamma_X;\mu)$ (though we expect that equality holds for Ruelle measures to be discussed in Subsection 7.1 below under quite weak assumptions) and that $(\mathcal{E}_\mu^\Gamma, W^{1,2}(\Gamma_X;\mu))$ might not be a Dirichlet form. Therefore, in [RSch97] two useful "intermediate" spaces $H_0^{1,2}(\Gamma_X;\mu) \subset \mathcal{F}^{(c)} \subset \mathcal{F} \subset W^{1,2}(\Gamma_X;\mu)$ were introduced so that $(\mathcal{E}_\mu^\Gamma, \mathcal{F}^{(c)})$, $(\mathcal{E}_\mu^\Gamma, \mathcal{F})$ are Dirichlet forms on $L^2(\Gamma_X;\mu)$. We shall briefly recall the corresponding definitions and results. We need the following additional assumption on μ:

(QI)(i) For any $n \in \mathbb{N}$ either $\mu(\{\gamma \in \Gamma_X \mid \gamma(X) = n\}) > 0$ or $\mu(\{\gamma \in \Gamma_X \mid \gamma(X) \geq n\}) > 0$ corresponding to whether X is compact or non-compact respectively and for all $v \in V_0(X)$ and $t \in \mathbb{R}$, μ is quasi-invariant w.r.t. ψ_t^v, i.e., $\mu \circ (\psi_t^v)^{-1} \approx \mu$.

(ii) $\operatorname*{essinf}_{r \leq s \leq t} \Phi_s > 0$ μ-a.e. for all $r, t \in \mathbb{R}$, $r < t$, where

$$\Phi_s := \frac{d(\mu \circ (\psi_s^v)^{-1}) \otimes ds}{d\mu \otimes ds}.$$

By [AKR97a, Proposition 2.2] it follows that every mixed Poisson measure satisfies (QI). The same is true for a class of Ruelle measures to be studied in Section 7 below (cf. Remark 4.7.6 (iii)).

Remark 4.4.7 It can be shown that (QI) (i) implies that $\operatorname{supp}\mu = \Gamma_X$ (cf. [RSch97, Proposition 5.6]). Hence, in particular, (4.4.3) holds in this case without assumption (IbP).

We now assume that μ satisfies both (IbP) and (QI)(i). Let $\mathfrak{F}$ denote the set of all $F \in L^\infty(\Gamma_X;\mu)$ such that there exists $\nabla^\Gamma F \in L^2(\Gamma_X \to T\Gamma_X;\mu)$ such that for all $v \in V_0(X)$, $s \in \mathbb{R}$

$$\frac{F \circ \psi_{s+t}^v - F \circ \psi_s^v}{t} \underset{t \to 0}{\longrightarrow} \langle \nabla^\Gamma F, v \rangle_{T\Gamma_X} \circ \psi_s^v \quad \text{in } L^2(\Gamma_X;\mu).$$

Let $\mathfrak{F}^{(c)}$ denote the subset of $\mathfrak{F}$ having a μ-version in $C(\Gamma_X)$, where $C(\Gamma_X)$ denotes the set of all continuous functions on Γ_X. Clearly, $\mathcal{F}C_b^{\infty,\mu} \subset \mathfrak{F}^{(c)}$ and $\nabla_\mu^\Gamma = \nabla^\Gamma$ on $\mathcal{F}C_b^{\infty,\mu}$. But we also have:

Lemma 4.4.8 *Let μ satisfy (IbP) and (QI). Then $\mathfrak{F} \subset W^{1,2}(\Gamma_X;\mu)$ and $\nabla^\Gamma = d^\mu$ on $\mathfrak{F}$.*

Proof. [RSch97, Lemma 6.3]. □

Proposition 4.4.9 *(i) Let μ satisfy (IbP) and (QI)(i). Then $(\mathcal{E}_\mu^\Gamma, \mathfrak{F})$, $(\mathcal{E}_\mu^\Gamma, \mathfrak{F}^{(c)})$ are closable on $L^2(\Gamma_X;\mu)$. Their respective closures $(\mathcal{E}_\mu^\Gamma, \mathcal{F})$, $(\mathcal{E}_\mu^\Gamma, \mathcal{F}^{(c)})$ are symmetric Dirichlet forms on $L^2(\Gamma_X;\mu)$. Furthermore, $D(\mathcal{E}_\mu^\Gamma) \subset \mathcal{F}^{(c)} \subset \mathcal{F} \subset W^{1,2}(\Gamma_X;\mu)$.*

(ii) If $\mu = \mu_{\lambda,m}$ (or $\mu_{\lambda,\sigma}$) is a mixed Poisson measure satisfying (4.3.1) *and if conditions (A) and (B) in Theorem 3.3 hold, then*

$$D(\mathcal{E}_\mu^\Gamma) = \mathcal{F}^{(c)} = \mathcal{F} = W^{1,2}(\Gamma_X;\mu_{\lambda,m}).$$

Proof. (i): The closability of both forms and the last assertion are immediate consequences of Lemma 4.4.8. The proof that both $(\mathcal{E}_\mu^\Gamma, \mathcal{F})$ and $(\mathcal{E}_\mu^\Gamma, \mathcal{F}^{(c)})$ are Dirichlet forms is quite straightforward. We refer to [RSch97, Proof of Proposition 1.4(i)].

(ii): This follows by Proposition 4.4.6. □

We would like to emphasize that even only trying to prove directly that $D(\mathcal{E}_{\mu_{\lambda,m}}^\Gamma) = \mathcal{F}^{(c)}$ (i.e., that $\mathcal{F}C_b^\infty$ is dense in $\mathcal{F}^{(c)}$ w.r.t. $\|\cdot\|_{W^{1,2}(\Gamma_X;\mu_{\lambda,m})}$) appears extremely hard. The elaborate machinery about uniqueness of semigroup generators when restricted to subsets of their full domain, on which the proof of Proposition 4.4.9 is based, seems unavoidable even in this simple case where μ is a mixed Poisson measure.

There are many reasons to study $(\mathcal{E}_\mu^\Gamma, \mathcal{F})$ and $(\mathcal{E}_\mu^\Gamma, \mathcal{F}^{(c)})$. One is that one can prove a Rademacher theorem on Γ_X (cf. Section 5 below). Others are presented in [AKRR98].

4.4.2 Quasi–regularity and corresponding diffusions

Suppose that μ is a probability measure on $(\Gamma_X, \mathcal{B}(\Gamma_X))$ satisfying (4.4.1) and (4.4.3) such that

$$(4.4.17) \qquad (\mathcal{E}_\mu^\Gamma, \mathcal{F}C_b^{\infty,\mu}) \text{ is closable on } L^2(\Gamma_X;\mu).$$

As before the closure is denoted by $(\mathcal{E}_\mu^\Gamma, D(\mathcal{E}_\mu^\Gamma))$. The generators of $(\mathcal{E}_\mu^\Gamma, D(\mathcal{E}_\mu^\Gamma))$ we denote again by $(H_\mu^\Gamma, D(H_\mu^\Gamma))$ but emphasize that because we do not assume that (IbP) holds, it is in general not equal to the Friedrichs' extension of an operator such as Δ_μ^Γ explicitly given on $\mathcal{F}C_b^{\infty,\mu}$. It might be in general not even true that $\mathcal{F}C_b^{\infty,\mu} \subset D(H_\mu^\Gamma)$.

As in Subsection 3.3 the desired processes in general will live on the bigger state space $\ddot{\Gamma}_X$. To state the corresponding existence result precisely we need

Definitions 4.4.10, 4.4.11 below. As in Subsection 3.3 we can consider μ as a measure on $(\ddot{\Gamma}_X, \mathcal{B}(\ddot{\Gamma}_X))$ and correspondingly $(\mathcal{E}_\mu^\Gamma, D(\mathcal{E}_\mu^\Gamma))$ as a Dirichlet form on $L^2(\ddot{\Gamma}_X; \mu)$.

Definition 4.4.10 (*cf.* [MR92, Ch.III, Definitions 2.1 and 3.2]). *Let $(\mathcal{E}, D(\mathcal{E}))$ be a Dirichlet form on $L^2(\ddot{\Gamma}_X; \mu)$.*

(i) *A sequence $(F_n)_{n\in\mathbb{N}}$ of closed subsets of $\ddot{\Gamma}_X$ is called an $\mathcal{E}$-nest, if*

$$\bigcup_{n\in\mathbb{N}} \{F \in D(\mathcal{E}) | F = 0 \ \mu\text{-a.e. on } \ddot{\Gamma}_X \setminus F_n\}$$

is dense in $D(\mathcal{E})$ w.r.t. $\mathcal{E}_1^{1/2}(:= [\mathcal{E} + (\ ,\ )_{L^2(\mu)}]^{1/2})$.

(ii) *A set $N \subset \ddot{\Gamma}_X$ is called $\mathcal{E}$-exceptional, if $N \subset \ddot{\Gamma}_X \setminus \bigcup_{n\in N} F_n$ for some $\mathcal{E}$-nest $(F_n)_{n\in\mathbb{N}}$. We say that a property of points in $\ddot{\Gamma}_X$ holds $\mathcal{E}$-quasi-everywhere (abbeviated $\mathcal{E}$-q.e.) if it holds outside some $\mathcal{E}$-exceptional set.*

(iii) *A function $f : \ddot{\Gamma}_X \to \mathbb{R}$ is called $\mathcal{E}$-quasi-continuous if there exists an $\mathcal{E}$-nest $(F_n)_{n\in\mathbb{N}}$ such that the restriction $f_{|F_n}$ of f to F_n is continuous for all $n \in \mathbb{N}$.*

We recall that the notion "$\mathcal{E}$-q.e." is much finer than "μ-a.e." (cf. [MR92, Exercise 2.6 (ii)]).

Definition 4.4.11 (cf. [MR92, Chap.IV, Definition 3.1]) A Dirichlet form $(\mathcal{E}, D(\mathcal{E}))$ on $L^2(\ddot{\Gamma}_X; \mu)$ is called *quasi–regular* if:

(i) There exists an $\mathcal{E}$–nest $(F_k)_{k\in\mathbb{N}}$ consisting of compact subsets of $\ddot{\Gamma}_X$.

(ii There exists a $\mathcal{E}_1^{1/2}$–dense subset of $D(\mathcal{E})$ whose elements have $\mathcal{E}$–quasi–continuous μ–versions.

(iii) There exist $u_n \in D(\mathcal{E})$, $n \in \mathbb{N}$, having $\mathcal{E}$–quasi–continuous μ–versions $\tilde{u}_n$, $n \in \mathbb{N}$, and an $\mathcal{E}$–exceptional set $N \subset \ddot{\Gamma}_X$ such that $\{\tilde{u}_n \mid n \in \mathbb{N}\}$ separates the points of $\ddot{\Gamma}_X \setminus N$.

The following is a special case of the main result in [MR97].

Theorem 4.4.12 *$(\mathcal{E}_\mu^\Gamma, D(\mathcal{E}_\mu^\Gamma))$ is a quasi–regular Dirichlet form on $L^2(\ddot{\Gamma}_X; \mu)$.*

We do not include a proof of Theorem 4.4.12 here, but refer to [MR97].

Corollary 4.4.13 $(\mathcal{E}_\mu^\Gamma, D(\mathcal{E}_\mu^\Gamma))$ *is* local *(i.e.,* $\mathcal{E}_\mu^\Gamma(F,G) = 0$ *provided* $F, G \in D(\mathcal{E}_\mu^\Gamma)$ *with* $\operatorname{supp}(|F|\mu) \cap \operatorname{supp}(|G|\mu) = \emptyset$*).*

Proof. Since $(\mathcal{E}_\mu^\Gamma, D(\mathcal{E}_\mu^\Gamma))$ is quasi–regular by Theorem 4.4.12 and since ∇_μ^Γ satisfies the product rule on bounded functions in $D(\mathcal{E}_\mu^\Gamma)$, the proof for locality is entirely analogous to that in [MR92, Chap. V, Examples 1.12 (ii)]. (We also refer to the beautiful work [S95] for a complete discussion of locality of Dirichlet forms.) □

As a consequence of Theorem 4.4.12, Corollary 4.4.13 and [MR92, Chap. IV, Theorem 3.5 and Chap. V, Theorem 1.11] we immediately obtain

Theorem 4.4.14 *There exists a conservative diffusion process*

$$\mathbf{M} = (\Omega, \mathbf{F}, (\mathbf{F}_t)_{t\geq 0}, (\Theta_t)_{t\geq 0}, (\mathbf{X}_t)_{t\geq 0}, (\mathbf{P}_\gamma)_{\gamma \in \ddot{\Gamma}_X})$$

on $\ddot{\Gamma}_X$ *(cf. [Dy65]) which is properly associated with* $(\mathcal{E}_\mu^\Gamma, D(\mathcal{E}_\mu^\Gamma))$*, i.e., for all ($\mu$-versions of)* $F \in L^2(\ddot{\Gamma}_X; \mu)$ *and all* $t > 0$ *the function*

$$\gamma \mapsto p_t F(\gamma) := \int_\Omega F(\mathbf{X}_t)\, d\mathbf{P}_\gamma, \quad \gamma \in \ddot{\Gamma}_X, \tag{4.4.18}$$

is an $\mathcal{E}_\mu^\Gamma$*-quasi-continuous version of* $\exp(-tH_\mu^\Gamma)F$*.* $\mathbf{M}$ *is up to* μ*-equivalence unique* (*cf.* [MR92, Chap. IV, Sect. 6]). *In particular,* $\mathbf{M}$ *is* μ*–symmetric (i.e.,* $\int G\, p_t F d\mu = \int F\, p_t G\, d\mu$ *for all* $F, G : \ddot{\Gamma}_X \to \mathbb{R}_+, \mathcal{B}(\ddot{\Gamma}_X)$*-measurable) and has* μ *as an invariant measure.*

Proof. By Theorem 4.4.12 and Corollary 4.4.13 the proof follows directly from [MR92, Chap. V, Theorem 1.11]. □

In the above theorem $\mathbf{M}$ is canonical, i.e., $\Omega = C([0,\infty) \to \ddot{\Gamma}_X)$, $\mathbf{X}_t(\omega) := \omega(t)$, $t \geq 0$, $\omega \in \Omega$, $(\mathbf{F}_t)_{t\geq 0}$ together with $\mathbf{F}$ is the corresponding minimum completed admissible family (cf. [F80, Sect.4.1]) and Θ_t, $t \geq 0$, are the corresponding natural time shifts.

Remark 4.4.15 (i) Theorems 4.4.12, 4.4.14 and Corollary 4.4.13 generalize corresponding results in [Os96], [Y96].

(ii) $\mathbf{M}$ is also called the corresponding *stochastic dynamics*, or *stochastic quantization corresponding to* μ.

Theorem 4.4.16 **M** *from Theorem 4.4.14 is the (up to μ-equivalence, cf. [MR92, Chap. IV, Definition 6.3]) unique diffusion process having μ as an invariant measure and solving the martingale problem for $(-H_\mu^\Gamma, D(H_\mu^\Gamma))$, i.e., for all $G \in D(H_\mu^\Gamma)$ $(\supset \mathcal{F}C_b^\infty(\mathcal{D},\Gamma))$*

$$\tilde{G}(\mathbf{X}_t) - \tilde{G}(\mathbf{X}_0) + \int_0^t H_\mu^\Gamma G(\mathbf{X}_s)\, ds, \ t \geq 0,$$

is an $(\mathbf{F}_t)$-martingale under $\mathbf{P}_\gamma$ (hence starting at γ) for $\mathcal{E}_\mu^\Gamma$-q.e. $\gamma \in \ddot{\Gamma}_X$. (Here $\tilde{G}$ denotes a quasi-continuous version of G, cf. [MR92, Chap. IV, Proposition 3.3]).

Proof. This follows immediately from [AR95, Theorem 3.5]. □

Remark 4.4.17 In fact, the uniqueness statement in Theorem 4.4.16 can be strengthened, as follows:
M is (up to μ-equivalence) unique among all right processes $\mathbf{M}' = (\Omega', \mathbf{F}', (\mathbf{F}'_t)_{t\geq 0}, (\Theta'_t)_{t\geq 0}, (\mathbf{X}')_{t\geq 0}, (\mathbf{P}'_\gamma)_{\gamma \in \ddot{\Gamma}_X})$ on $\ddot{\Gamma}_X$ having μ as a sub-invariant measure and being such that for all $G \in D(H_\mu^\Gamma)$

$$\tilde{G}(\mathbf{X}'_t) - \tilde{G}(\mathbf{X}'_0) + \int_0^t H_\mu^\Gamma G(\mathbf{X}'_s)\, ds, \ t \geq 0,$$

is an $(\mathbf{F}'_t)$-martingale under $\mathbf{P}'_\mu := \int \mathbf{P}'_\gamma \, \mu(d\gamma)$.

The proof is analogous to that of [AR95, Theorem 3.5] except for its ending where one has to use the recent result of P. Fitzsimmons in [Fi96] in order to be able to apply the arguments in [MR92, Chap. IV, Section 6] to deduce the result.

4.5 Intrinsic metric on configuration spaces

In this section we give a summary of the main results in the recent work [RSch97]. Below we fix a probability measure μ on $(\Gamma_X, \mathcal{B}(\Gamma_X))$ satisfying the conditions (IbP) and (QI) specified in Subsection 4.1, to which we also refer for examples. Let also $(\mathcal{E}_\mu^\Gamma, D(\mathcal{E}_\mu^\Gamma))$, $(\mathcal{E}_\mu^\Gamma, \mathcal{F}^{(c)})$, $(\mathcal{E}_\mu^\Gamma, \mathcal{F})$ be as in Subsection 4.1, considered as Dirichlet forms on $L^2(\ddot{\Gamma}_X; \mu)$.

4.5.1 Identification of the intrinsic metric for a class of measures

The following L^2–*Wasserstein type distance* ρ on $\ddot{\Gamma}_X$ was introduced in [RSch97]:

$$\rho(\gamma,\omega) := \inf\left\{ \left(\int d(x,y)^2\, \eta(dxdy) \right)^{1/2} \mid \eta \in \ddot{\Gamma}_{X\times X} \text{ with marginals } \omega, \gamma \text{ respectively} \right\};\ \gamma,\omega \in \ddot{\Gamma}_X \tag{4.5.1}$$

where d is the Riemannian distance on X. Note that ρ is a pseudo–metric, in particular, $\rho(\omega,\gamma) = 0$ implies $\omega = \gamma$. But $\rho(x,y)$ will be infinite if $\omega(X) \neq \gamma(X)$, since there is no $\eta \in \ddot{\Gamma}_{X\times X}$ with marginals ω, γ respectively. But also if both ω and γ are infinite configurations one will find that possibly $\rho(\omega,\gamma) = \infty$ (e.g. take $X = \mathbb{R}$, $\omega = \sum_{z\in\mathbb{Z}} \varepsilon_z$, and $\gamma = \omega - \delta_0$). Obviously, the topology induced on Γ_X by ρ is finer than the vague topology.

Theorem 4.5.1 *The intrinsic metrics of the Dirichlet forms* $(\mathcal{E}_\mu^\Gamma, \mathcal{F})$ *and* $(\mathcal{E}_\mu^\Gamma, \mathcal{F}^{(c)})$ *are both equal to the* L^2*–Wasserstein type distance* ρ *defined in* (4.5.1), *more precisely:*

$$\begin{aligned}
&\sup\{F(\gamma) - F(\omega) \mid F \in \mathcal{F} \cap C(\Gamma_X) \text{ and } |\nabla^\Gamma F|_{T\Gamma_X} \le 1\ \mu\text{–a.e. on } \ddot{\Gamma}_X\} \\
= {} &\sup\{F(\gamma) - F(\omega) \mid F \in \mathcal{F}^{(c)} \cap C(\Gamma_X) \text{ and } |\nabla^\Gamma F|_{T\Gamma_X} \le 1\ \mu\text{–a.e. on } \ddot{\Gamma}_X\} \\
= {} &\sup\{F(\gamma) - F(\omega) \mid F \in \mathfrak{F}^{(c)} \text{ and } |\nabla^\Gamma F|_{T\Gamma_X} \le 1\ \mu\text{–a.e. on } \ddot{\Gamma}_X\} \\
= {} &\rho(\gamma,\omega).
\end{aligned}$$

Proof. See [RSch97, Sect. 8]. □

4.5.2 A Rademacher theorem

We recall that a function $F : \ddot{\Gamma}_X \to \mathbb{R}$ is called ρ–Lipschitz continuous if

$$\operatorname{Lip}(F) := \sup\left\{ \frac{|F(\gamma) - F(\omega)|}{\rho(\gamma,\omega)} \mid \gamma,\omega \in \ddot{\Gamma}_X,\ \gamma \neq \omega \right\} < \infty. \tag{4.5.2}$$

Concerning generalizations, different from the following theorem, of the classical Rademacher theorem (cf. [Ra19]) to infinite dimensional spaces (though in contrast to ours only "flat" ones), we refer to the introduction of [RSch97] and the references therein.

Theorem 4.5.2 *(i) Suppose $F \in L^2(\ddot{\Gamma}_X;\mu)$ is ρ–Lipschitz continuous. Then $F \in \mathcal{F}$ and*

$$|\nabla^\Gamma F|_{T\Gamma_X} \leq \operatorname{Lip}(F) \quad \mu - a.e.$$

(ii) Suppose $F \in \mathcal{F}$ such that $|\nabla^\Gamma F|_{T\Gamma_X} \leq c$ μ–a.e. for some $c \in]0,\infty[$ and such that F has a ρ–continuous μ–version. Then there exists a ρ–Lipschitz continuous μ–version $\widetilde{F}$ of F satisfying

$$\operatorname{Lip}(\widetilde{F}) \leq c\,.$$

Proof. [RSch97, Sections 7 and 8]. □

4.5.3 Potential-theoretic consequences

We now state some of the results on the potential theory on $\ddot{\Gamma}_X$, induced by the Dirichlet forms $(\mathcal{E}^\Gamma_\mu, D(\mathcal{E}^\Gamma_\mu))$ and $(\mathcal{E}^\Gamma_\mu, \mathcal{F}^{(c)})$ on $L^2(\ddot{\Gamma}_X;\mu)$, obtained as consequences of Theorems 4.5.1, 4.5.2 respectively their proofs in [RSch97]. For proofs we refer to [RSch97, Sect. 3].

Define for $A \subset \ddot{\Gamma}_X$

$$\rho_A(\gamma) := \inf\{\rho(\omega,\gamma) \mid \omega \in A\}. \tag{4.5.3}$$

Let $\bar{A}^\rho$ denote the closure of A w.r.t. the topology induced by ρ.

Proposition 4.5.3 *If $K \subset \ddot{\Gamma}_X$ is compact and $c \geq 0$, then $c \wedge \rho_K$ is an $(\mathcal{E}^\Gamma_\mu, \mathcal{F}^{(c)})$–quasi–continuous function in $\mathcal{F}^{(c)}$.*

Corollary 4.5.4 *If $A \in \mathcal{B}(\ddot{\Gamma}_X)$ such that $\mu(A) = 1$, then $\ddot{\Gamma}_X \setminus \bar{A}^\rho = \{\rho_A > 0\}$ is $(\mathcal{E}^\Gamma_\mu, \mathcal{F}^{(c)})$–exceptional.*

Example 4.5.5 If $A = \{\gamma \in \ddot{\Gamma}_X \mid \gamma(X) = \infty\}$ one sees immediately that $A = \{\rho_A < \infty\}$. Hence, as follows from Corollary 4.5.4, $\mu(A) = 1$ implies that $\ddot{\Gamma}_X \setminus A$ is $(\mathcal{E}^\Gamma_\mu, \mathcal{F}^{(c)})$–exceptional, whereas $\mu(A) = 0$ implies that A is $(\mathcal{E}^\Gamma_\mu, \mathcal{F}^{(c)})$–exceptional. For certain cases this result has first been proven by Byron Schmuland (private communication).

We note that since $\mathcal{F}C_b^{\infty,\mu} \subset D(\mathcal{E}^\Gamma_\mu) \subset \mathcal{F}^{(c)} \subset \mathcal{F}$, both $(\mathcal{E}^\Gamma_\mu, \mathcal{F}^{(c)})$ and $(\mathcal{E}^\Gamma_\mu, \mathcal{F})$ obviously fulfill property (iii) in the definition of quasi–regularity (cf. Definition 4.4.11) with $N = \emptyset$. (Just pick $u_n \in \mathcal{F}C_b^\infty$, $n \in \mathbb{N}$, separating the points of $\ddot{\Gamma}_X$). Furthermore, by [MR92, Chap. III, Theorem 2.11(i) and Excercise 2.10] it follows that any sequence $(F_k)_{k\in\mathbb{N}}$ which is a nest w.r.t. $(\mathcal{E}^\Gamma_\mu, D(\mathcal{E}^\Gamma_\mu))$ is also a nest w.r.t. $(\mathcal{E}^\Gamma_\mu, \mathcal{F}^{(c)})$ and $(\mathcal{E}^\Gamma_\mu, \mathcal{F})$. In particular, then by Theorem 4.4.12 both $(\mathcal{E}^\Gamma_\mu, \mathcal{F}^{(c)})$ and $(\mathcal{E}^\Gamma_\mu, \mathcal{F})$ also have property (i) in Definition 4.4.11. Since $(\mathcal{E}^\Gamma_\mu, \mathcal{F}^{(c)})$ by definition also satisfies 4.4.11 (ii). We have:

Proposition 4.5.6 *$(\mathcal{E}_\mu^\Gamma, \mathcal{F}^{(c)})$ is quasi–regular on $L^2(\ddot{\Gamma}_X; \mu)$. Moreover, $(\mathcal{E}_\mu^\Gamma, \mathcal{F}^{(c)})$ is local.*

Proof. We have just seen that $(\mathcal{E}_\mu^\Gamma, \mathcal{F}^{(c)})$ is quasi–regular. Clearly, ∇^Γ satisfies the product rule on $\mathfrak{F}^{(c)}$, hence on $\mathcal{F}^{(c)}$. Consequently, the proof of locality is entirely analogous to that in [MR92, Chap. V, Examples 1.12 (ii)]. □

But in the present situation one can even give an "explicit" nest consisting of compacts:

Proposition 4.5.7 *Let $K_n \subset \ddot{\Gamma}_X$, $K_n \subset K_{n+1}$, $n \in \mathbb{N}$, be compact such that $\mu(K_n) \uparrow 1$. Then*

$$F_n := \{\rho_{K_n} \leq \frac{1}{2}\}, \ n \in \mathbb{N},$$

are compact and form a $(\mathcal{E}_\mu^\Gamma, \mathcal{F}^{(c)})$–nest.

Proof. [RSch97, Corollary 3.4]. □

Theorem 4.5.8 *The analogoes of Theorems 4.4.14 and 4.4.16 hold with $(\mathcal{E}_\mu^\Gamma, \mathcal{F}^{(c)})$ replacing $(\mathcal{E}_\mu^\Gamma, D(\mathcal{E}_\mu^\Gamma))$ resp. with the generator of $(\mathcal{E}_\mu^\Gamma, \mathcal{F}^{(c)})$ replacing $(H_\mu^\Gamma, D(H_\mu^\Gamma))$.*

Proof. Analogous to the proofs of Theorems 4.4.14 and 4.4.16. □

4.6 Gibbs measures on configuration spaces

So far, the only concrete examples of probability measures μ on $(\Gamma_X, \mathcal{B}(\Gamma_X))$ have been the mixed Poisson measures. In this section we will describe another class of measures on $(\Gamma_X, \mathcal{B}(\Gamma_X))$, so-called *Gibbs measures.* Our presentation is (apart from slight notational modifications) very much oriented along the beautiful work by C. Preston (see [P79], [P80], but also [P76] and [Ge79]).

In this section we formulate all results right away for the more general intensity measures $\sigma = \rho \cdot m$ specified at the beginning of Remark 4.2.3 (i). So, the underlying (reference) Poisson measure is $\psi_{i_1 \dots i_n \sigma}$ rather than $\psi_{i_1 \dots i_n m}$.

4.6.1 Grand canonical Gibbs measures

Let $\mathcal{B}_b(\Gamma_X)$ denote the set of all bounded $\mathcal{B}(\Gamma_X)$ - measurable functions $G : \Gamma_X \to \mathbb{R}$. If for $B \in \mathcal{B}(X)$ we define $N_B : \Gamma_X \to \mathbb{Z}_+ \cup \{+\infty\}$ by

$$N_B(\gamma) := \gamma(B), \tag{4.6.1}$$

then

$$\mathcal{B}(\Gamma_X) = \sigma(\{N_\Lambda | \Lambda \in \mathcal{O}_c(X)\}). \tag{4.6.2}$$

For $A \in \mathcal{B}(X)$ we define

$$\mathcal{B}_A(\Gamma_X) := \sigma(\{N_B | B \in \mathcal{B}_c(X), B \subset A\}). \tag{4.6.3}$$

Let Φ be a *potential*, i.e., a function $\Phi : \Gamma_X \to \mathbb{R} \cup \{+\infty\}$ such that

$$\begin{aligned} \Phi &= 1_{\{N_X < \infty\}} \Phi, \ \Phi(\emptyset) = 0, \ \text{and} \\ \gamma &\mapsto \Phi(\gamma_\Lambda) \ \text{is} \ \mathcal{B}_\Lambda(\Gamma_X) - \text{measurable} \\ & \text{for all} \ \Lambda \in \mathcal{B}_c(X). \end{aligned} \tag{4.6.4}$$

For $\Lambda \in \mathcal{B}_c(X)$ the *conditional energy* $E_\Lambda^\Phi : \Gamma_X \to \mathbb{R} \cup \{+\infty\}$ is defined by

$$E_\Lambda^\Phi(\gamma) := \begin{cases} \sum_{\gamma' \subset \gamma, \gamma'(\Lambda) > 0} \Phi(\gamma') & \text{if } \sum_{\gamma' \subset \gamma, \gamma'(\Lambda) > 0} |\Phi(\gamma')| < \infty, \\ +\infty & \text{otherwise,} \end{cases} \tag{4.6.5}$$

(where the sum over the empty set is defined to be zero).

Now we can define grand canonical Gibbs measures:

Definition 4.6.1 *For $\Lambda \in \mathcal{O}_c(X)$ define for $\gamma \in \Gamma_X, \Delta \in \mathcal{B}(\Gamma_X)$*

$$\begin{aligned} \Pi_\Lambda^{\sigma,\Phi}(\gamma, \Delta) &:= 1_{\{Z_\Lambda^{\sigma,\Phi} < \infty\}}(\gamma) [Z_\Lambda^{\sigma,\Phi}(\gamma)]^{-1} \int_{\Gamma_X} 1_\Delta(\gamma_{X\setminus\Lambda} + \gamma'_\Lambda) \\ &\times \exp[-E_\Lambda^\Phi(\gamma_{X\setminus\Lambda} + \gamma'_\Lambda)] \ \psi_{i_1 \dots i_n \sigma}(d\gamma'), \end{aligned}$$

where

$$Z_\Lambda^{\sigma,\Phi}(\gamma) := \int_{\Gamma_X} \exp[-E_\Lambda^\Phi(\gamma_{X\setminus\Lambda} + \gamma'_\Lambda)] \ \psi_{i_1 \dots i_n \sigma}(d\gamma').$$

A probability measure μ on $(\Gamma_X, \mathcal{B}(\Gamma_X))$ is called a grand canonical Gibbs measure with interaction potential Φ *if*

$$\mu\Pi_\Lambda^{\sigma,\Phi} \;=\; \mu \text{ for all } \Lambda \in \mathcal{O}_c(X). \tag{4.6.6}$$

Let $\mathcal{G}_{gc}(\sigma, \Phi)$ denote the set of all such probability measures μ.

Remark 4.6.2 (i) It is well-known that $(\Pi_\Lambda^{\sigma,\Phi})_{\Lambda\in\mathcal{O}_c(X)}$ is a $(\mathcal{B}_{X\setminus\Lambda}(\Gamma_X))_{\Lambda\in\mathcal{O}_c(X)}$ - specification in the sense of [Fö75], [P76] (cf. [P76, Section 6] or [P79]), i.e., for all $\Lambda, \Lambda' \in \mathcal{O}_c(X)$

(S 1) $\Pi_\Lambda^{\sigma,\Phi}(\gamma, \Gamma_X) \in \{0, 1\}$ for all $\gamma \in \Gamma_X$.

(S 2) $\Pi_\Lambda^{\sigma,\Phi}(\cdot, \Delta)$ is $\mathcal{B}_{X\setminus\Lambda}(\Gamma_X)$-measurable for all $\Delta \in \mathcal{B}(\Gamma_X)$.

(S 3) $\Pi_\Lambda^{\sigma,\Phi}(\cdot, \Delta' \cap \Delta) = 1_{\Delta'}\Pi_\Lambda(\cdot, \Delta)$ for all $\Delta \in \mathcal{B}(\Gamma_X), \Delta' \in \mathcal{B}_{X\setminus\Lambda}(\Gamma_X)$.

(S 4) $\Pi_{\Lambda'}^{\sigma,\Phi} = \Pi_{\Lambda'}^{\sigma,\Phi}\Pi_\Lambda^{\sigma,\Phi}$ if $\Lambda \subset \Lambda'$.

Here for $\gamma \in \Gamma_X, \Delta \in \mathcal{B}(\Gamma_X)$

$$(\Pi_{\Lambda'}^{\sigma,\Phi}\Pi_\Lambda^{\sigma,\Phi})(\gamma, \Delta) \;:=\; \int \Pi_\Lambda^{\sigma,\Phi}(\gamma', \Delta)\; \Pi_{\Lambda'}^{\sigma,\Phi}(\gamma, d\gamma')$$

(and $\mu\Pi_\Lambda^{\sigma,\Phi}$ in (4.6.6) is defined correspondingly). (4.6.6) above are called *Dobrushin-Lanford-Ruelle equations.* We also note that as a direct consequence of (S 3) we have for all $\Lambda \in \mathcal{O}_c(X)$ that

$$\mu(\{Z_\Lambda^{\sigma,\Phi} < \infty\}) \;=\; 1$$

for all $\mu \in \mathcal{G}_{gc}(\sigma, \Phi)$.

(ii) It is an easy exercise (see e.g. [P80, Proposition 10.3]) to show that because of (4.2.17) for all $\Lambda \in \mathcal{O}_c(X)$, $\gamma \in \Gamma_X$, $\Delta \in \mathcal{B}(\Gamma_X)$

$$\Pi_\Lambda^{\sigma,\Phi}(\gamma, \Delta) = 1_{\{\tilde{Z}_\Lambda^{\sigma,\Phi}<\infty\}}(\gamma)[\tilde{Z}_\Lambda^{\sigma,\Phi}(\gamma)]^{-1}[1_\Delta(\gamma_{X\setminus\Lambda})+$$

$$\sum_{n=1}^{\infty}\frac{1}{n!}\int_\Lambda\cdots\int_\Lambda 1_\Delta(\gamma_{X\setminus\Lambda}+\sum_{k=1}^{n}\varepsilon_{x_k})\exp[-E_\Lambda^\Phi(\gamma_{X\setminus\Lambda}+\sum_{k=1}^{n}\varepsilon_{x_k})]\;\sigma(dx_1)\ldots\sigma(dx_n)],$$

where

$$\tilde{Z}_\Lambda^{\sigma,\Phi}(\gamma) := 1 + \sum_{n=1}^{\infty} \frac{1}{n!} \int_\Lambda \cdots \int_\Lambda \exp[-E_\Lambda^\Phi(\gamma_{X\setminus\Lambda} + \sum_{k=1}^{n} \varepsilon_{x_k})]\ \sigma(dx_1)\ldots\sigma(dx_n)$$

$$= \exp(\sigma(\Lambda)) Z_\Lambda^{\sigma,\Phi}(\gamma).$$

(iii) Clearly, by properties (S 2), (S 3) a probability measure μ on $(\Gamma_X, \mathcal{B}(\Gamma_X))$ is a grand canonical Gibbs measure if and only if for all $\Lambda \in \mathcal{O}_c(X)$ and all $G \in \mathcal{B}_b(\Gamma_X)$

$$E_\mu[G \mid \mathcal{B}_{X\setminus\Lambda}(\Gamma_X)] \ = \ \Pi_\Lambda^{\sigma,\Phi} G \ \ \mu - \text{a.e.},$$

where for a sub-σ-algebra $\Sigma \subset \mathcal{B}(\Gamma_X)$, $E_\mu[\cdot\ |\Sigma]$ denotes the conditional expectation w.r.t. μ given Σ and for $G \in \mathcal{B}_b(\Gamma_X)$ we set

$$\Pi_\Lambda^{\sigma,\Phi}(\gamma, G) \ := \ (\Pi_\Lambda^{\sigma,\Phi} G)(\gamma) \ = \ \int G(\gamma')\ \Pi_\Lambda^{\sigma,\Phi}(\gamma, d\gamma'). \tag{4.6.7}$$

(iv) Above as well as in the next section we may always replace $\mathcal{O}_c(X)$ by $\mathcal{B}_c(X)$ without any changes (in particular, $\mathcal{G}_{gc}(\sigma,\Phi)$ remains unchanged). We chose $\Lambda \in \mathcal{O}_c(X)$ only for simplicity.

4.6.2 Canonical Gibbs measures

Consider the situation of Subsection 6.1.

Definition 4.6.3 *For $\Lambda \in \mathcal{O}_c(X)$ define for $\gamma \in \Gamma_X, \Delta \in \mathcal{B}(\Gamma_X)$*

$$\hat{\Pi}_\Lambda^{\sigma,\Phi}(\gamma, \Delta) \ := \ \begin{cases} \frac{\Pi_\Lambda^{\sigma,\Phi}(\gamma, \Delta\cap\{N_\Lambda = \gamma(\Lambda)\})}{\Pi_\Lambda^{\sigma,\Phi}(\gamma, \{N_\Lambda=\gamma(\Lambda)\})} & \text{if } \Pi_\Lambda^{\sigma,\Phi}(\gamma, \{N_\Lambda = \gamma(\Lambda)\}) > 0 \\ 0 & \text{otherwise} \end{cases} \tag{4.6.8}$$

A probability measure μ on $(\Gamma_X, \mathcal{B}(\Gamma_X))$ is called a canonical Gibbs measure with interaction potential Φ *if*

$$\mu\hat{\Pi}_\Lambda^{\sigma,\Phi} \ = \ \mu \ \text{ for all } \Lambda \in \mathcal{O}_c(X). \tag{4.6.9}$$

Let $\mathcal{G}_c(\sigma,\Phi)$ denote the set of all such probability measures μ.

Remark 4.6.4 For $\Lambda \in \mathcal{O}_c(X)$ let $\hat{\mathcal{B}}_{X\setminus\Lambda}(\Gamma_X)$ denote the σ - algebra generated by $\mathcal{B}_{X\setminus\Lambda}(\Gamma)$ and $\sigma(N_\Lambda)$. It has been shown in [P79] that $(\hat{\Pi}_\Lambda^{\sigma,\Phi})_{\Lambda\in\mathcal{O}_c(X)}$ is indeed a $(\hat{\mathcal{B}}_{X\setminus\Lambda}(\Gamma_X))_{\Lambda\in\mathcal{O}_c(X)}$ - specification (i.e., satisfies (S 1) - (S 4) in Remark 6.2 (i) with $\hat{\Pi}_\Lambda^{\sigma,\Phi}$ replacing $\Pi_\Lambda^{\sigma,\Phi}$ and $\hat{\mathcal{B}}_{X\setminus\Lambda}(\Gamma_X)$ replacing $\mathcal{B}_{X\setminus\Lambda}(\Gamma_X)$ for $\Lambda \in \mathcal{O}_c(X)$). So, the above definition makes sense. In particular, for all $\Lambda \in \mathcal{O}_c(X)$ we have that

$$\mu(\{\gamma \in \Gamma_X \mid \Pi_\Lambda^{\sigma,\Phi}(\gamma, \{N_\Lambda = \gamma(\Lambda)\}) > 0\}) = 1 \tag{4.6.10}$$

for all $\mu \in \mathcal{G}_c(\sigma, \Phi)$. Clearly, also the analogue of Remark 6.2 (iii) holds.

It can be checked that

$$\mathcal{G}_{gc}(\sigma, \Phi) \subset \mathcal{G}_c(\sigma, \Phi) \tag{4.6.11}$$

(cf. [P79, Proposition 2.1]). The precise relation between the convex sets $\mathcal{G}_c(\sigma, \Phi)$ and $\mathcal{G}_{gc}(\sigma, \Phi)$ resp. between their extreme points $ex\mathcal{G}_c(\sigma, \Phi)$ and $ex\mathcal{G}_{gc}(\sigma, \Phi)$ is one of the major problems in Statistical Mechanics, known under the key word *equivalence of ensembles* (cf. [P79], [Ge79]).

Remark 4.6.5 By a general result due to E.B. Dynkin and H. Föllmer (cf. [Dy78], [Fö75], and for a beautiful exposition [P76, Theorem 2.2]) we have $\mathrm{ex}\mathcal{G}_c(\sigma, \Phi) \neq \emptyset$, $\mathrm{ex}\mathcal{G}_{gc}(\sigma, \Phi) \neq \emptyset$ provided $\mathcal{G}_c(\sigma, \Phi) \neq \emptyset$, $\mathcal{G}_{gc}(\sigma, \Phi) \neq \emptyset$ respectively, and that any μ in $\mathcal{G}_c(\sigma, \Phi)$ resp. in $\mathcal{G}_{gc}(\sigma, \Phi)$ has an integral representation in terms of the respective extreme points (see Theorem 6.6 below for the special case $\Phi \equiv 0$).

In the sequel we shall mainly analyze $\mathcal{G}_c(\sigma, \Phi)$, but using known results on the above problem we shall deduce new results also for $\mathcal{G}_{gc}(\sigma, \Phi)$ (cf. Subsection 9.2). Concrete examples for the potential Φ will be treated in Subsection 7.1 below.

In fact the mixed Poisson measures from the previous sections are special canonical Gibbs measures with potential $\Phi \equiv 0$, i.e., in the *free case.* Let us set $\mathcal{G}_c(\sigma) := \mathcal{G}_c(\sigma, 0)$, $\mathcal{G}_{gc}(\sigma) := \mathcal{G}_{gc}(\sigma, 0)$. Then we have the following well known result (cf. [NZ77, Sect. 2] or [Ge79, Sect. 4.2], resp. [MKM78]).

Theorem 4.6.6 *(i) Suppose $\sigma(X) = \infty$ and let μ be a probability measure on $(\Gamma_X, \mathcal{B}(\Gamma_X))$. Then $\mu \in \mathcal{G}_c(\sigma)$ if and only if there exists a probability measure λ on $(\mathbb{R}_+, \mathcal{B}(\mathbb{R}_+))$ such that*

$$\mu = \int_{\mathbb{R}_+} \psi_{i_1 \ldots i_n\, z\sigma}\ \lambda(dz)$$

(i.e., if and only if μ is a mixed Poisson measure with mean proportional to σ).

(ii) $ex\mathcal{G}_c(\sigma) = \{\psi_{i_1\dots i_n z\sigma} \mid z \in \mathbb{R}_+\}$.

(iii) $ex\mathcal{G}_c(\sigma) = \cup_{z\in\mathbb{R}_+} ex\mathcal{G}_{gc}(z\sigma)$.

We note that since σ is diffuse, the set $\mathfrak{M}^{\cdot\cdot}$ in [NZ77, Sect. 2] (resp. M in [Ge79, Sect. 4.2]) can really be replaced by Γ_X as done in Theorem 6.6 above.

4.6.3 Closability of the corresponding Dirichlet forms

In this subsection we want to prove the closability of $(\mathcal{E}_\mu^\Gamma, \mathcal{F}C_b^\infty)$ defined in (4.4.2) on $L^2(\Gamma_X;\mu)$ for all $\mu \in \mathcal{G}_{gc}^1(\sigma,\Phi)$ where

$$\mathcal{G}_{gc}^1(\sigma,\Phi) := \{\mu \in \mathcal{G}_{gc}(\sigma,\Phi) \mid \mu \text{ satisfies } (4.4.1)\} \tag{4.6.12}$$

(with $\mathcal{G}_{gc}(\sigma,\Phi)$ as defined in Subsection 6.1) under quite general assumptions on the potential Φ. The presentation below is a self–contained and (w.r.t. technical details) completed version of [dSKR98, Sections 5, 6].

We start with giving a simple direct proof for the following identity for $\mu \in \mathcal{G}_{gc}^1(\sigma,\Phi)$ which was derived in [NZ79], [MMW79] from the well–known Mecke identity for $\psi_{i_1\dots i_n\sigma}$ (cf. [Mec67, Satz 3.1]).

Lemma 4.6.7 *Let $\mu \in \mathcal{G}_{gc}^1(\sigma,\Phi)$ and let $h : \Gamma_X \times X \to \mathbb{R}_+$ be $\mathcal{B}(\Gamma_X)\otimes\mathcal{B}(X)$–measurable. Then*

$$\int_{\Gamma_X}\int_X h(\gamma,x)\ \gamma(dx)\mu(d\gamma) = \int_{\Gamma_X}\int_X h(\gamma+\varepsilon_x,x)e^{-E^\Phi_{\{x\}}(\gamma+\varepsilon_x)}\sigma(dx)\mu(d\gamma), \tag{4.6.13}$$

where

$$E^\Phi_{\{x\}}(\gamma+\varepsilon_x) := \begin{cases} \sum\limits_{\{x\}\subset\gamma'\subset\gamma\cup\{x\}} \Phi(\gamma') & \text{if } \sum\limits_{\{x\}\subset\gamma'\subset\gamma\cup\{x\}} |\Phi(\gamma')| < \infty \\ +\infty & \text{else} \end{cases} \tag{4.6.14}$$

Proof. Because of standard monotone class arguments it suffices to prove (4.6.13) for

$$h(\gamma,x) := e^{\langle g,\gamma\rangle}\ f(x), \quad \gamma\in\Gamma_X, x\in X,$$

where $f, g \in C_0^\infty(X)$; $(-g), f \geq 0$. Let $\Lambda \in \mathcal{O}_c(X)$ such that $\operatorname{supp} f, \operatorname{supp} g \subset \Lambda$. By Remark 4.6.2 (ii) the left hand side of (4.6.13) is equal to

$$\begin{aligned}
&\int e^{\langle g,\gamma\rangle} \langle f,\gamma\rangle \, \mu(d\gamma) \\
= &\int \Pi_\Lambda^{\sigma,\Phi}(e^{\langle g,\cdot\rangle}\langle f,\cdot\rangle)\, d\mu \\
= &\int_{\Gamma_X} (\widetilde{Z}_\Lambda^{\sigma,\Phi}(\gamma))^{-1} \sum_{n=1}^\infty \frac{1}{n!} \int_{\Lambda^n} e^{\sum_{k=1}^n g(x_k)} \sum_{i=1}^n f(x_i) \exp[-E_\Lambda^\Phi(\gamma_{X\setminus\Lambda} + \sum_{k=1}^n \varepsilon_{x_k})] \\
&\qquad \sigma(dx_1)\dots\sigma(dx_n)\, \mu(d\gamma) \\
= &\int_\Lambda e^{g(x)} f(x) \int_{\Gamma_X} (\widetilde{Z}_\Lambda^{\sigma,\Phi}(\gamma))^{-1} \sum_{n=1}^\infty \frac{1}{(n-1)!} \int_{\Lambda^{n-1}} e^{\sum_{k=1}^{n-1} g(x_k)} \cdot \\
&\qquad \cdot \exp[-E_\Lambda^\Phi(\gamma_{X\setminus\Lambda} + \sum_{k=1}^{n-1} \varepsilon_{x_k} + \varepsilon_x)]\, \sigma(dx_1)\dots\sigma(dx_{n-1})\, \mu(d\gamma)\, \sigma(dx).
\end{aligned}$$

It is easy to see that

(4.6.15)
$$E_\Lambda^\Phi(\gamma + \varepsilon_x) = E_\Lambda^\Phi(\gamma) + E_{\{x\}}^\Phi(\gamma + \varepsilon_x) \text{ for all } \gamma \in \Gamma_X, x \in \Lambda \text{ such that } x \notin \gamma,$$

and that

(4.6.16)
$$(\gamma, x) \mapsto E_{\{x\}}^\Phi(\gamma + \varepsilon_x) \text{ is } \mathcal{B}(\Gamma_X) \times \mathcal{B}(X)\text{–measurable.}$$

Consequently, the left hand side of (4.6.13) is equal to

$$\begin{aligned}
&\int_\Lambda e^{g(x)} f(x) \int_{\Gamma_X} (\widetilde{Z}_\Lambda^{\sigma,\Phi}(\gamma))^{-1} \sum_{n=0}^\infty \frac{1}{n!} \int_{\Lambda^n} e^{\sum_{k=1}^n g(x_k)} \\
&\cdot \exp[-E_{\{x\}}^\Phi(\gamma_{X\setminus\Lambda} + \varepsilon_x + \sum_{k=1}^n \varepsilon_x)] \cdot \exp[-E_\Lambda^\Phi(\gamma_{X\setminus\Lambda} + \sum_{k=1}^n \varepsilon_{x_k})] \\
&\sigma(dx_1)\dots\sigma(dx_n)\, \mu(d\gamma)\, \sigma(dx) \\
= &\int_\Lambda e^{g(x)} f(x) \int_{\Gamma_X} \Pi_\Lambda^{\sigma,\Phi}\left(e^{\langle g,\cdot\rangle} e^{-E_{\{x\}}^\Phi(\cdot+\varepsilon_x)}\right)(\gamma)\, \mu(d\gamma)\sigma(dx) \\
= &\int_\Lambda e^{g(x)} f(x) \int_{\Gamma_X} e^{\langle g,\gamma\rangle} e^{-E_{\{x\}}^\Phi(\gamma+\varepsilon_x)} \mu(d\gamma)\, \sigma(dx)
\end{aligned}$$

which by (4.6.16) and Fubini's theorem is equal to the right hand side of (4.6.13). □

Remark 4.6.8 (i) We note that if $x \notin \gamma$ then the definition of $E^{\Phi}_{\{x\}}(\gamma+\varepsilon_x)$ in (4.6.14) coincides with (4.6.5) for $\Lambda = \{x\}$. On the other hand if

$$M_x := \{\gamma \in \Gamma_X \mid x \in \gamma\},$$

then for $\Lambda \in \mathcal{O}_c(X)$ such that $x \in \Lambda$ and every $\mu \in \mathcal{G}^1_{gc}(\sigma, \Phi)$.

$$\begin{aligned}
&\mu(M_x)\\
&= \int \Pi^{\sigma,\Phi}_{\Lambda}(\gamma, M_x)\ \mu(d\gamma)\\
&= \int_{\Gamma_X} (\widetilde{Z}^{\sigma,\Phi}_{\Lambda}(\gamma))^{-1} \sum_{n=1}^{\infty} \frac{1}{n!} \int_{\Lambda^n} 1_{M_x}(\gamma_{X\setminus\Lambda} + \sum_{k=1}^{n} \varepsilon_{x_k}) \exp[-E^{\Phi}_{\Lambda}(\gamma_{X\setminus\Lambda} + \sum_{k=1}^{n} \varepsilon_{x_k})]\\
&\qquad\qquad \sigma(dx_1)\ldots\sigma(dx_n)\ \mu(d\gamma)\\
&= \int_{\Gamma_X} (\widetilde{Z}^{\sigma,\Phi}_{\Lambda}(\gamma))^{-1} \sum_{n=1}^{\infty} \frac{1}{n!} \int_{\tilde{\Lambda}^n} \sum_{i=1}^{n} 1_{\{x\}}(x_i) \exp[-E^{\Phi}_{\Lambda}(\gamma_{X\setminus\Lambda} + \sum_{k=1}^{n} \varepsilon_{x_k})]\\
&\qquad\qquad \sigma(dx_1)\ldots\sigma(dx_n)\ \mu(d\gamma)
\end{aligned}$$

(4.6.17)

$$= 0\,,$$

where $\tilde{\Lambda}^n$ is as defined in (4.2.2). Hence (4.6.14) and (4.6.5) coincide for μ–a.e. $\gamma \in \Gamma_X$.

(ii) We emphasize that Lemma 4.6.7, of course, holds for arbitrary positive Radon measures σ on $(X, \mathcal{B}(X))$. We never used our specific form $\sigma = \rho \cdot m$ in the above proof.

We define new intensity measures on X by $\sigma_\gamma := \rho_\gamma \cdot m$, where

$$\rho_\gamma(x) := e^{-E^{\Phi}_{\{x\}}(\gamma+\varepsilon_x)}\rho(x),\ \ x \in X, \gamma \in \Gamma_X\ . \tag{4.6.18}$$

Now we can prove the following representation of $(\mathcal{E}^{\Gamma}_{\mu}, \mathcal{F}C^{\infty,\mu}_b)$ (cf. [dSKR98, Theorem 5.1]).

Theorem 4.6.9 *Let $\mu \in \mathcal{G}^1_{gc}(\sigma, \Phi)$ and let $F, G \in \mathcal{F}C^{\infty}_b$. Then*

$$\begin{aligned}
&\int_{\Gamma_X} \langle \nabla^{\Gamma} F, \nabla^{\Gamma} G\rangle_{T\Gamma_X}\ d\mu\\
&= \int_{\Gamma_X} \int_X \langle \nabla^X (F(\gamma+\varepsilon_\cdot) - F(\gamma)), \nabla^X (G(\gamma+\varepsilon_\cdot) - G(\gamma))\rangle_{T\Gamma_X}(x)\\
&\qquad\qquad \sigma_\gamma(dx)\ \mu(d\gamma).
\end{aligned} \tag{4.6.19}$$

In particular, (4.4.3) holds, so $(\mathcal{E}^{\Gamma}_{\mu}, \mathcal{F}C^{\infty,\mu}_b)$ is a well-defined positive definite symmetric bilinear form on $L^2(\Gamma_X; \mu)$.

Proof. By (4.2.8) and Lemma 4.6.7, if $F = g_F(\langle f_1, \cdot\rangle, \dots, \langle f_N, \cdot\rangle)$, $G = g_G(\langle g_1, \cdot\rangle, \dots, \langle g_M, \cdot\rangle)$, where without loss of generality we may assume $N = M$, we have

$$\begin{aligned}
&\int_{\Gamma_X} \langle \nabla^\Gamma F, \nabla^\Gamma G\rangle_{T\Gamma_X} d\mu \\
&= \int_{\Gamma_X}\int_X \sum_{i,j=1}^N \partial_i g_F(\langle f_1, \gamma\rangle, \dots, \langle f_N, \gamma\rangle)\, \partial_j g_G(\langle(g_1, \gamma\rangle, \dots, \langle g_N, \gamma\rangle) \\
&\qquad \langle \nabla^X f_i, \nabla^X g_j\rangle_{TX}(x)\ \gamma(dx)\ \mu(d\gamma) \\
&= \int_{\Gamma_X}\int_X \sum_{i,j=1}^N \partial_i g_F(\langle f_1, \gamma\rangle + f_1(x), \dots, \langle f_N, \gamma\rangle + f_N(x)) \\
&\qquad \partial_j g_G(\langle g_1, \gamma\rangle + g_1(x), \dots, \langle g_N, \gamma\rangle + g_N(x)) \\
&\qquad \langle \nabla^X f_i, \nabla^X g_j\rangle_{TX}(x)\ \sigma_\gamma(dx)\ \mu(d\gamma).
\end{aligned}$$

The latter clearly is equal to the right hand side of (4.6.19).
If now $F = G$ μ–a.e., then by Lemma 4.6.7 for all $\Lambda \in \mathcal{O}_c(X)$ and $F_0 := F - G$

$$\begin{aligned}
0 &= \int_{\Gamma_X} \gamma(\Lambda)\ F_0(\gamma)\mu(d\gamma) = \int_{\Gamma_X}\int_X 1_\Lambda(x) F_0(\gamma)\ \gamma(dx)\mu(d\gamma) \\
&= \int_{\Gamma_X}\int_X 1_\Lambda(x) F_0(\gamma + \varepsilon_x)\sigma_\gamma(dx)\ \mu(d\gamma).
\end{aligned}$$

Since $\Lambda \in \mathcal{O}_c(X)$ was arbitrary, it follows that for μ–a.e. $\gamma \in \Gamma_X$

$$F_0(\gamma + \varepsilon_x) = 0 \text{ for } \sigma_\gamma\text{–a.e. } x \in X.$$

This implies that the right hand side of (4.6.19) applied to $F := G := F_0$ is zero, hence by (4.6.19)

$$\nabla^\Gamma G - \nabla^\Gamma F = \nabla^\Gamma F_0 = 0 \quad \mu\text{–a.e.}$$

and (4.4.3) is proven. □

Remark 4.6.10 The reader might wonder why we took $F(\gamma + \varepsilon_\cdot) - F(\gamma)$ rather than only $F(\gamma + \varepsilon_\cdot)$ in (4.6.19). The reason is that for all $F \in \mathcal{F}C_b^\infty$

$$x \mapsto F(\gamma + \varepsilon_x) - F(\gamma) \in C_0^\infty(X)$$

for all $\gamma \in \Gamma_X$, whereas $x \mapsto F(\gamma + \varepsilon_x)$ is merely in $C_b^\infty(X)$.

Now we are going to prove the main result of this subsection, i.e., the said closability. This was first proved in [dSKR98, Sect. 6]. Because this result is of fundamental importance for the analysis on configurations spaces we repeat the complete proof here:

It was shown in [AR90, Theorem 5.3] (in the case $X = \mathbb{R}^d$) that the components of the Dirichlet form $(\mathcal{E}^X_{\sigma_\gamma}, \mathcal{D}^{\sigma_\gamma})$ corresponding to the measure σ_γ are closable on $L^2(\mathbb{R}^d; \sigma_\gamma)$ if and only if σ_γ is absolutely continuous with respect to Lebesgue measure on $\mathbb{R}^d$ and the Radon-Nikodym derivative satisfies some condition (see (4.6.21) below for details). This result allows us to prove the closability of $(\mathcal{E}^\Gamma_\mu, \mathcal{F}C_b^{\infty,\mu})$ on $L^2(\Gamma_X; \mu)$. Let us first recall the above mentioned result.

Theorem 4.6.11 (cf. [AR90, Theorem 5.3]) *Let ν by a probability measure on $(\mathbb{R}^d, \mathcal{B}(\mathbb{R}^d))$, $d \in \mathbb{N}$, and let $\mathcal{D}^\nu$ denote the ν-classes determined by $\mathcal{D}$. Then the forms $(\mathcal{E}_{\nu,i}, \mathcal{D}^\nu)$ defined by*

$$\mathcal{E}_{\nu,i}(u, v) := \int_{\mathbb{R}^d} \frac{\partial u}{\partial x_i} \frac{\partial v}{\partial x_i} d\nu, \ u, v \in \mathcal{D},$$

are well-defined and closable on $L^2(\mathbb{R}^d; \nu)$ for $1 \leq i \leq d$ if and only if ν is absolutely continuous with respect to Lebesgue measure λ^d on $\mathbb{R}^d$, and the Radon-Nikodym derivative $\rho = d\nu/d\lambda^d$ satisfies the condition:

$$\text{for any } 1 \leq i \leq d \text{ and } \lambda^{d-1}\text{–a.e. } x \in \left\{ y \in \mathbb{R}^{d-1} \,|\, \int_{\mathbb{R}} \rho_y^{(i)}(s) \lambda^1(ds) > 0 \right\},$$

$$\rho_x^{(i)} = 0 \ \lambda^1\text{–a.e. on } \mathbb{R} \backslash R(\rho_x^{(i)}), \text{ where } \rho_x^{(i)}(s) := \rho(x_1, \ldots, x_{i-1}, s, x_i, \ldots, x_d),$$

(4.6.20)

$$s \in \mathbb{R}, \text{ if } x = (x_1, \ldots, x_{d-1}) \in \mathbb{R}^{d-1}$$

and where

$$R(\rho_x^{(i)}) := \left\{ t \in \mathbb{R} \,|\, \int_{t-\varepsilon}^{t+\varepsilon} \frac{1}{\rho_x^{(i)}(s)} \lambda^1(ds) < \infty \textit{ for some } \varepsilon > 0 \right\}. \tag{4.6.21}$$

There is an obvious generalization of Theorem 4.6.11 to the case where a Riemannian manifold X is replacing $\mathbb{R}^d$, to be formulated in terms of local charts. Since here we are only interested in the "if part" of Theorem 4.6.11, we now recall a slightly weaker sufficient condition for closability in the general case where X is a manifold as before.

Theorem 4.6.12 *Suppose $\sigma_1 = \rho_1 \cdot m$, where $\rho_1 : X \to \mathbb{R}_+$ is $\mathcal{B}(X)$-measurable such that*

(4.6.22)

$$\rho_1 = 0 \ m\text{-a.e. on}\, X \setminus \left\{ x \in X \,|\, \int_{\Lambda_x} \frac{1}{\rho_1} dm < \infty \text{ for some open neighbourhood } \Lambda_x \text{ of } x \right\}$$

Then $(\mathcal{E}^X_{\sigma_1}, \mathcal{D}^{\sigma_1})$ defined by

$$\mathcal{E}^X_{\sigma_1}(u,v) := \int_X \langle \nabla^X u, \nabla^X v \rangle_{TX} \, d\sigma_1; \ u, v \in \mathcal{D},$$

is closable on $L^2(X;\sigma_1)$.

The proof is a straightforward adaptation of the line of arguments in [MR92, Chap. II, Subsection 2a] (see also [ABR89, Theorem 4.2] for details). We emphasize that (4.6.22) (hence (4.6.20)) e.g. always holds, if ρ_1 is lower semicontinuous, and that (4.6.20) holds if e.g. ρ is weakly differentiable. Neither ν in Theorem 4.6.11 nor σ_1 in Theorem 4.6.12 is required to have full support, so e.g. ρ_1 is not necessarily strictly positive m-a.e. on X.

Theorem 4.6.13 *Let $\mu \in \mathcal{G}^1_{gc}(\sigma, \Phi)$. Suppose that for μ-a.e. $\gamma \in \Gamma_X$ the function ρ_γ defined in (4.6.18) satisfies (4.6.22) (resp. (4.6.21) in case $X = \mathbb{R}^d$). Then the form $(\mathcal{E}^\Gamma_\mu, \mathcal{F}C_b^{\infty,\mu})$ is closable on $L^2(\Gamma_X;\mu)$.*

Proof. Let $(F_n)_{n\in\mathbb{N}}$ be a sequence in $\mathcal{F}C_b^{\infty,\mu}$ such that $F_n \to 0$ in $L^2(\Gamma_X;\mu)$ as $n \to \infty$ and

$$\mathcal{E}^\Gamma_\mu(F_n - F_m, F_n - F_m) \underset{n,m\to\infty}{\longrightarrow} 0. \tag{4.6.23}$$

We have to show that

$$\mathcal{E}^\Gamma_\mu(F_{n_k}, F_{n_k}) \underset{k\to\infty}{\longrightarrow} 0 \tag{4.6.24}$$

for some subsequence $(n_k)_{k\in\mathbb{N}}$. Let $(n_k)_{k\in\mathbb{N}}$ be a subsequence such that

$$\left(\int_{\Gamma_X} F_{n_k}^2 d\mu \right)^{1/2} + \mathcal{E}^\Gamma_\mu(F_{n_{k+1}} - F_{n_k}, F_{n_{k+1}} - F_{n_k})^{1/2} < \frac{1}{2^k} \text{ for all } k \in \mathbb{N}.$$

Then

$$\begin{aligned}\infty &> \sum_{k=1}^{\infty} \mathcal{E}_{\mu}^{\Gamma}(F_{n_{k+1}} - F_{n_k}, F_{n_{k+1}} - F_{n_k})^{1/2} \\ &\geq \sum_{k=1}^{\infty} \int_{\Gamma_X} \left(\int_X |\nabla^X((F_{n_{k+1}} - F_{n_k})(\gamma + \varepsilon.) - (F_{n_{k+1}} - F_{n_k})(\gamma))|_{TX}^2(x) \right. \\ &\qquad\qquad \left. e^{-E_{\{x\}}^{\Phi}(\gamma+\varepsilon_x)}\sigma(dx)\right)^{1/2} \mu(d\gamma) \\ &- \int_{\Gamma_X} \sum_{k=1}^{\infty} \left(\int_X |\nabla^X((F_{n_{k+1}} - F_{n_k})(\gamma + \varepsilon.) - (F_{n_{k+1}} - F_{n_k})(\gamma))|_{TX}^2(x) \right. \\ &\qquad\qquad \left. \sigma_\gamma(dx)\right)^{1/2} \mu(d\gamma),\end{aligned}$$

where we used Theorem 4.6.9 and (4.6.18). From the last expression we obtain that

$$\begin{aligned}&\sum_{k=1}^{\infty} \mathcal{E}_{\sigma_\gamma}^{X}(u_{n_{k+1}}^{(\gamma)} - u_{n_k}^{(\gamma)}, u_{n_{k+1}}^{(\gamma)} - u_{n_k}^{(\gamma)})^{1/2} \\ &= \sum_{k=1}^{\infty} \left(\int_X |\nabla^X((F_{n_{k+1}} - F_{n_k})(\gamma + \varepsilon.) - (F_{n_{k+1}} - F_{n_k})(\gamma))|_{TX}^2(x) \right. \\ &\qquad\qquad \left. \cdot e^{-E_{\{x\}}^{\Phi}(\gamma+\varepsilon_x)}\sigma(dx)\right)^{1/2} \\ &< \infty \qquad \text{for } \mu\text{-a.e. } \gamma \in \Gamma_X, \end{aligned} \tag{4.6.25}$$

where for $k \in \mathbb{N}$, $\gamma \in \Gamma_X$,

$$u_{n_k}^{(\gamma)}(x) := F_{n_k}(\gamma + \varepsilon_x) - F_{n_k}(\gamma),\ x \in X.$$

We note that $u_{n_k}^{(\gamma)} \in \mathcal{D}$. (4.6.25) implies that for μ-a.e. $\gamma \in \Gamma_X$

$$\mathcal{E}_{\sigma_\gamma}^{X}(u_{n_k}^{(\gamma)} - u_{n_l}^{(\gamma)}, u_{n_k}^{(\gamma)} - u_{n_l}^{(\gamma)}) \underset{k,l\to\infty}{\longrightarrow} 0. \tag{4.6.26}$$

Let $\Lambda \subset \mathcal{O}_c(X)$.

Claim 1: For μ-a.e. $\gamma \in \Gamma_X$

$$\int_X (u_{n_k}^{(\gamma)}(x))^2\, 1_\Lambda(x)\sigma_\gamma(dx) \underset{k\to\infty}{\longrightarrow} 0.$$

To prove Claim 1 we first note that for μ-a.e. $\gamma \in \Gamma_X$

$$\sigma_\gamma(\Lambda) < \infty,$$

as follows immediately from (4.6.13) (taking $h(\gamma, x) := 1_\Lambda(x)$ for x$\in X$, $\gamma \in \Gamma_X$), since $\mu \in \mathcal{G}^1_{gc}(\sigma, \Phi)$. Therefore, for μ-a.e. $\gamma \in \Gamma_X$

$$\int_X F^2_{n_k}(\gamma)\, 1_\Lambda(x)\sigma_\gamma(dx) = F^2_{n_k}(\gamma)\sigma_\gamma(\Lambda) \underset{k\to\infty}{\longrightarrow} 0. \tag{4.6.27}$$

Furthermore, by (4.6.13)

$$\begin{aligned}
&\int_{\Gamma_X}\int_X F^2_{n_k}(\gamma + \varepsilon_x)\, 1_\Lambda(x)\sigma_\gamma(dx)(1+\gamma(\Lambda))^{-1}\mu(d\gamma) \\
= &\int_{\Gamma_X} F^2_{n_k}(\gamma)\int_X \frac{1_\Lambda(x)}{1+\gamma(\Lambda) - 1_\Lambda(x)}\gamma(dx)\mu(d\gamma) \\
\leq &\int_{\Gamma_X} F^2_{n_k}(\gamma)\mu(d\gamma) < \frac{1}{2^k},
\end{aligned}$$

because the integral w.r.t. γ is dominated by 1 for all $\gamma \in \Gamma_X$. Hence

$$\begin{aligned}
\infty > &\sum_{k=1}^{\infty}\left(\int_{\Gamma_X}\int_X F^2_{n_k}(\gamma+\varepsilon_x)\, 1_\Lambda(x)\sigma_\gamma(dx)(1+\gamma(\Lambda))^{-1}\mu(d\gamma)\right)^{1/2} \\
\geq &\int_{\Gamma_X}\sum_{k=1}^{\infty}\left(\int_X F^2_{n_k}(\gamma+\varepsilon_x)\, 1_\Lambda(x)\sigma_\gamma(dx)\right)^{1/2}(1+\gamma(\Lambda))^{-1}\mu(d\gamma).
\end{aligned}$$

Therefore, for μ-a.e. $\gamma \in \Gamma_X$

$$\int_X F^2_{n_k}(\gamma+\varepsilon_x)\, 1_\Lambda(x)\sigma_\gamma(dx) \underset{k\to\infty}{\longrightarrow} 0. \tag{4.6.28}$$

Now Claim 1 follows by (4.6.27) and (4.6.28).

Claim 2: For μ-a.e. $\gamma \in \Gamma_X$

$$|\nabla^X u^{(\gamma)}_{n_k}|_{TX} \underset{k\to\infty}{\longrightarrow} 0 \quad \sigma_\gamma\text{-a.e.}$$

To prove Claim 2 we first note that clearly (4.6.26) implies that for μ-a.e. $\gamma \in \Gamma_X$

$$\mathcal{E}^X_{1_\Lambda\sigma_\gamma}(u^{(\gamma)}_{n_k} - u^{(\gamma)}_{n_l}, u^{(\gamma)}_{n_k} - u^{(\gamma)}_{n_l}) \underset{k,l\to\infty}{\longrightarrow} 0. \tag{4.6.29}$$

Hence we can apply Theorem 4.6.12 (resp. 4.6.11) to $\rho_1 := 1_\Lambda \rho_\gamma$ and conclude by Claim 1 and (4.6.29) that for μ-a.e. $\gamma \in \Gamma_X$

$$\mathcal{E}^X_{1_\Lambda \sigma_\gamma}(u^{(\gamma)}_{n_k}, u^{(\gamma)}_{n_k}) \underset{k\to\infty}{\longrightarrow} 0,$$

hence by (4.6.25)

$$1_\Lambda |\nabla^X u^{(\gamma)}_{n_k}|_{TX} \underset{k\to\infty}{\longrightarrow} 0 \quad \sigma_\gamma\text{-a.e.}$$

Since Λ was arbitrary, Claim 2 is proven.

¿From Claim 2 we now easily deduce (4.6.24) by (4.6.19) and Fatou's Lemma as follows:

$$\begin{aligned} \mathcal{E}^\Gamma_\mu(F_{n_k}, F_{n_k}) &\le \int_{\Gamma_X} \liminf_{l\to\infty} \int_X |\nabla^X (u^{(\gamma)}_{n_k} - u^{(\gamma)}_{n_l})|^2_{TX}\, \sigma_\gamma(dx)\, \mu(d\gamma) \\ &\le \liminf_{l\to\infty} \mathcal{E}^\Gamma_\mu(F_{n_k} - F_{n_l}, F_{n_k} - F_{n_l}), \end{aligned}$$

which by (4.6.23) can be made arbitrarily small for k large enough. □

Remark 4.6.14 (i) The above method to prove closability of pre-Dirichlet forms on configuration spaces Γ_X extends immediately to the case where X is replaced by an infinite dimensional "manifold" such as the loop space (cf. [MR97]).

(ii) We also emphasize that all results in Subsection 4.2 now apply to every $\mu \in \mathcal{G}^1_{gc}(\sigma, \Phi)$ that satisfies the condition of Theorem 4.6.13.

4.7 Integration by parts characterization of canonical Gibbs measures

In this section we shall first recall the definition of a class of grand canonical Gibbs measures, so–called Ruelle measures. They serve as examples for the measures in $\mathcal{G}^1_{gc}(\sigma, \Phi)$ studied in the previous section, with explicit (pair) potentials Φ. Subsequently, we shall present the analogue of Theorem 4.2.2 for canonical Gibbs measures which was proven in [AKR97b, Subsection 4.3]. In this section we consider for simplicity the case $X = \mathbb{R}^d$ with the Euclidean metric and $\sigma = z \cdot m$, $z \in]0, \infty[$ (with $m =$ Lebesgue measure). We also simplify notations by setting $\Gamma := \Gamma_{\mathbb{R}^d}$, $\psi^z_{i_1\ldots i_n} := \psi_{i_1\ldots i_n\, z\cdot m}$, $\nabla := \nabla^{\mathbb{R}^d}$. We follow the presentation of [AKR97b], but omit the technically rather involved proofs and, for simplicity, only consider the "finite range case" in Subsection 7.2.

4.7.1 Ruelle measures

We will now describe (following essentially [Ru70]) a class of grand canonical Gibbs measures which appears in Classical Statistical Mechanics of continuous systems, see also [Do70], [KY93], [Ru69].

A *pair potential* is a $\mathcal{B}(\mathbb{R}^d)$–measurable function $\phi : \mathbb{R}^d \to \mathbb{R} \cup \{+\infty\}$ such that $\phi(-x) = \phi(x)$. Any pair potential ϕ defines a potential $\Phi = \Phi_\phi$ as follows: we set $\Phi(\gamma) := 0$, $|\gamma| \neq 2$ and $\Phi(\gamma) := \phi(x-y)$ for $\gamma = \{x, y\} \subset \mathbb{R}^d$. It is useful to rewrite the conditional energy E_Λ^Φ (see (4.6.5)) in the following form for $\Lambda \in \mathcal{B}_c(\mathbb{R}^d)$:

$$E_\Lambda^\Phi(\gamma) = E_\Lambda^\Phi(\gamma_\Lambda) + W(\gamma_\Lambda|\gamma_{\Lambda^c}), \tag{4.7.1}$$

where the term

$$W(\gamma_\Lambda|\gamma_{\Lambda^c}) := \sum_{x\in\gamma_\Lambda, y\in\gamma_{\Lambda^c}} \phi(x-y) \tag{4.7.2}$$

describes the *interaction energy* between γ_Λ and γ_{Λ^c}. (Here as usual $\Lambda^c := \mathbb{R}^d \setminus \Lambda$). Analogously, we define $W(\gamma'|\gamma'')$ when γ', γ'' are located in disjoint regions.

For every $r = (r^1, \dots, r^d) \in \mathbb{Z}^d$ we define a cube

$$Q_r := \{x \in \mathbb{R}^d \mid r^i - \frac{1}{2} \leq x^i < r^i + \frac{1}{2}\}.$$

These cubes form a partition of $\mathbb{R}^d$. For any $\gamma \in \Gamma$ we set $\gamma_r := \gamma_{Q_r} = \gamma \cap Q_r$, $r \in \mathbb{Z}^d$. For $N \in \mathbb{N}$ let Λ_N be the cube with side length $2N-1$ centered at the origin in $\mathbb{R}^d$; Λ_N is then a union of $(2N-1)^d$ unit cubes of the form Q_r.

Now we are able to formulate conditions on the interaction.

(S) (*Stability*) There exists $B \geq 0$ such that for any $\Lambda \in \mathcal{B}_c(\mathbb{R}^d)$ and for all $\gamma \in \Gamma_\Lambda$

$$E_\Lambda^\Phi(\gamma) \geq -B|\gamma|.$$

A condition stronger than stability is the following.

(SS) (*Superstability*) There exist $A > 0$, $B \geq 0$ such that if $\gamma \in \Gamma_{\Lambda_N}$ for some N, then

$$E_{\Lambda_N}^\Phi(\gamma) \geq \sum_{r\in\mathbb{Z}^d} [A|\gamma_r|^2 - B|\gamma_r|].$$

(LR) (*Lower regularity*) There exists a decreasing positive function $a : \mathbb{N} \to \mathbb{R}_+$ such that

$$\sum_{r\in\mathbb{Z}^d} a(\|r\|) < \infty$$

and for any Λ', Λ'' which are each finite unions of unit cubes of the form Q_r and disjoint, with $\gamma' \in \Gamma_{\Lambda'}$, $\gamma'' \in \Gamma_{\Lambda''}$,

$$W(\gamma'|\gamma'') \geq - \sum_{r',r'' \in \mathbb{Z}^d} a(\|r' - r''\|) \, |\gamma'_{r'}| \cdot |\gamma''_{r''}|.$$

Here $\|\cdot\|$ denotes the maximum norm on $\mathbb{R}^d$

We say that the pair potential ϕ is *stable* (resp. *superstable*, *lower regular*) if (S) (resp (SS), (LR)) is satisfied. In [Ru70] there are given many criteria for stability, superstability and lower regularity of a pair potential. For example, if $\phi_1 \geq 0$ is continuous and $\phi_1(0) > 0$, then ϕ_1 is superstable. If ϕ_2 is stable than $\phi = \phi_1 + \phi_2$ is superstable etc. Let us mention also the following Dobrushin-Fisher-Ruelle criterium.

Proposition 4.7.1 *Let $0 < d_1 < d_2 < +\infty$ and let*

$$s_1 : [0, d_1] \to \mathbb{R} \cup \{+\infty\}, \quad s_2 : [d_2, +\infty[\to \mathbb{R}$$

be positive, decreasing and such that

$$\int_0^{d_1} t^{d-1} s_1(t) dt = +\infty, \quad \int_{d_2}^{\infty} t^{d-1} s_2(t) dt < +\infty.$$

If the pair potential ϕ is bounded below and satisfies

$$\phi(x) \geq s_1(\|x\|) \text{ for } \|x\| \leq d_1,$$

$$|\phi(x)| \leq s_2(\|x\|) \text{ for } \|x\| \geq d_2 ,$$

then ϕ is superstable and lower regular.

Definition 4.7.2 *A probability measure μ on $(\Gamma, \mathcal{B}(\Gamma))$ is called* tempered *if μ is supported by*

$$S_\infty := \cup_{n=1}^{\infty} S_n,$$

where

$$S_n := \{\gamma \in \Gamma \mid \forall N \in \mathbb{N} \sum_{r \in \Lambda_N \cap \mathbb{Z}^d} |\gamma_r|^2 \leq n^2 |\Lambda_N \cap \mathbb{Z}^d| \}.$$

By $\mathcal{G}^t_{gc}(z, \phi) \subset \mathcal{G}_{gc}(z, \phi) := \mathcal{G}_{gc}(zm, \Phi_\phi)$ (cf. Subsection 6.1) we denote the set of all tempered grand canonical Gibbs measures (*Ruelle measures* for short).

Remark 4.7.3 Let $z \in]0, \infty[$. Due to [Ru70, Sect. 5]

$$\mathcal{G}_{gc}^t(z, \phi) \neq \emptyset$$

and for all $\mu \in \mathcal{G}_{gc}^t(z, \phi)$ and all $q \in [1, \infty[$

$$\int \gamma(K)^q \, \mu(d\gamma) < \infty \quad \text{for all compact } K \subset \mathbb{R}^d$$

if ϕ is a super-stable, lower regular potential and satisfies the following condition:

(I) (*Integrability*) $\displaystyle\int_{\mathbb{R}^d} |1 - e^{-\phi(x)}| \, m(dx) \; < \; +\infty \, .$

For a given probability measure μ on $(\Gamma, \mathcal{B}(\Gamma))$ and any $\Lambda \in \mathcal{B}_c(\mathbb{R}^d)$ we can introduce a probability measure μ_Λ on $(\Gamma_\Lambda, \mathcal{B}(\Gamma_\Lambda))$ as $\mu_\Lambda = \mu \circ p_\Lambda^{-1}$, cf. (4.2.17). Since there is a decomposition of Γ_Λ analogous to (4.2.13) for all $\Lambda \in \mathcal{B}_c(\mathbb{R}^d)$ (and not only for $\Lambda \in \mathcal{O}_c(\mathbb{R}^d)$), we have an induced decomposition of μ_Λ: $\mu_\Lambda = \sum_{n=0}^\infty \mu_\Lambda^{(n)}$. In the case of a Ruelle measure μ which corresponds to a potential ϕ with properties (SS), (LR) and (I) we have the representations

$$\mu_\Lambda^{(n)}(d\gamma_n) = \frac{1}{n!} \tau_{\Lambda,\mu}^n(\gamma_n) m_{\Lambda,n}(d\gamma_n), \;\; \gamma_n \in \Gamma_\Lambda^{(n)}, \; n \geq 0,$$

where $\tau_{\Lambda,\mu}^n(\cdot)$ are positive measurable functions on $\Gamma_\Lambda^{(n)}$. They can be considered as positive Lebesgue integrable functions $\tau_{\Lambda,\mu}^n(x_1, \ldots, x_n)$ on Λ^n invariant w.r.t. all permutations from the group S_n. These functions are called the *system of density distributions.* We refer to [Ru70] for details.

The following theorem gives necessary properties of the so-called correlation functions associated to Ruelle measures, for the proof see [Ru70].

Theorem 4.7.4 *Let ϕ be a pair potential satisfying conditions (SS), (LR) and (I). For a given $\mu \in \mathcal{G}_{gc}^t(\phi, z)$ let $\{\tau_{\Lambda,\mu}^n | n \geq 0\}$ be the associated system of density distributions. We define correlation functions by*

$$\varrho_\mu(x_1, \ldots, x_n) = \sum_{k=0}^\infty \frac{1}{k!} \int_{\Lambda^n} \tau_{\Lambda,\mu}^{n+k}(x_1, \ldots, x_{n+k}) m(dx_{n+1}) \ldots m(dx_{n+k})$$

for $x_1, \ldots, x_n \in \Lambda$. The correlation functions satisfy the Mayer equations

$$\varrho_\mu(x_1, \ldots, x_n) = \exp[-\sum_{i<j} \phi(x_j - x_i)] \times$$

$$\sum_{p=0}^{\infty}\frac{1}{p!}K((x_1,\dots,x_n),(y_1,\dots,y_p))\varrho_\mu(y_1,\dots,y_p)m(dy_1)\dots m(dy_p),$$

where

$$K((x_1,\dots,x_n),(y_1,\dots,y_p))=\prod_{j=1}^{p}K((x_1,\dots,x_n),y_j),$$

$$K((x_1,\dots,x_n),y)=\exp[-\sum_{i=1}^{n}\phi(y-x_j)]\;-\;1.$$

Moreover, there exists a constant $\xi>0$ such that for all n and $x_1,\dots,x_n\in\mathbb{R}^d$ we have:

$$\varrho_\mu(x_1,\dots,x_n)\;\le\;\xi^n.$$

Remark 4.7.5 For pair potentials ϕ the condition in Theorem 4.6.13 ensuring closability of $(\mathcal{E}_\mu^\Gamma,\mathcal{F}C_b^{\infty,\mu})$ on $L^2(\Gamma;\mu)$ for $\mu\in\mathcal{G}_{gc}^1(z,\phi)$ can be now easily formulated as follows: for μ–a.e. $\gamma\in\Gamma$ and m–a.e. $x\in\{y\in\mathbb{R}^d\mid\sum_{y'\in\gamma}|\phi(y-y')|<\infty\}$ it holds that

$$\int_{V_x}e^{\sum_{y'\in\gamma}\phi(y-y')}\,m(dy)<\infty$$

for some open neighbourhood V_x of x. This condition trivially holds e.g. if $\operatorname{supp}\phi$ is compact, $\{\phi<\infty\}$ is open, and $\phi^+\in L^\infty_{loc}(\{\phi<\infty\};m)$. If even $\mu\in\mathcal{G}_{gc}^t(z,\phi)$ and ϕ satisfies the assumptions in Proposition 4.7.1, then it suffices to merely assume that $\{\phi<\infty\}$ is open and $\phi^+\in L^\infty_{loc}(\{\phi<\infty\};m)$. This follows by an elementary consideration. This generalizes the closability result in [Os96]. The a priori bigger domain for $\mathcal{E}_\mu^\Gamma$ considered there, can also be treated by our method to prove closability in Subsection 6.3.

4.7.2 Integration by parts characterization

In this subsection we shall prove an integration by parts formula for a certain convex subset of canonical Gibbs measures, to be introduced below, which contains the Ruelle measures. In fact, we shall characterize this subset by integration by parts in a way analogous to the free case in Theorem 4.2.2. For simplicity we assume that

(C) $\quad\operatorname{supp}\phi$ is compact.

The general case is more complicated to formulate. We refer to [AKR97b, Subsection 4.3] for this case including all proofs of the results stated below.

Let us introduce another condition on the potential ϕ:

(D) (*Differentiability*) $e^{-\phi}$ is weakly differentiable on $\mathbb{R}^d$, ϕ is weakly differentiable on $\mathbb{R}^d \setminus \{0\}$ and the weak gradient $\nabla\phi$ (which is a locally m-integrable function on $\mathbb{R}^d \setminus \{0\}$) considered as an m-a.e. defined function on $\mathbb{R}^d$ satisfies

$$\nabla\phi \in L^1(\mathbb{R}^d; e^{-\phi}m) \cap L^2(\mathbb{R}^d; e^{-\phi}m).$$

(We refer to [Y96] for a similar condition). Note that for many typical potentials in Statistical Physics we have $\phi \in C^\infty(\mathbb{R}^d \setminus \{0\})$. For such "regular outside the origin" potentials condition (D) nevertheless does not exclude a singularity at the point $0 \in \mathbb{R}^d$.

Remark 4.7.6 (i) Suppose that (D) holds. Then $e^{-\phi}$ is weakly differentiable, hence, in particular, $e^{-\phi} \in L^1_{loc}(\mathbb{R}^d; m)$. So, (C) and (D) imply (I).

(ii) It is an easy exercise to check that (D) implies that

$$\nabla e^{-\phi} \;=\; -\nabla\phi\, e^{-\phi} \quad m-\text{a.e. on } \mathbb{R}^d.$$

(iii) By [RSch97, Proposition 2.1] any $\mu \in \mathcal{G}^t_{gc}(z,\phi)$ satisfies condition (QI) in Subsection 4.1, provided ϕ satisfies (SS), (LR), (C) and (D).

Example 4.7.7 A concrete example of a pair potential which is especially important in atomic and molecular physics is given for $d = 3$ by the so-called *Lennard-Jones potential*

$$\phi_{a,b}(x) := \frac{a}{|x|^{12}} - \frac{b}{|x|^6}, \quad x \in \mathbb{R}^d \setminus \{0\},$$

where $a, b > 0$. This potential obviously satisfies all conditions of Proposition 4.7.1 and is, therefore, super-stable and lower regular. Furthermore, we have

$$\nabla\phi_{a,b}(x) \;=\; \frac{6x}{r^8(x)} - \frac{12x}{r^{14}(x)}$$

and it is easy to see that (D) is also true. To also satisfy (C) we e.g. just have to multiply $\phi_{a,b}$ by some $\chi \in C_0^\infty(\mathbb{R}^d)$, $\chi \geq 0$.

Lemma 4.7.8 *Let ϕ be a pair potential satisfying conditions (SS), (LR), (C) and (D). For any vector field $v \in V_0(\mathbb{R}^d)$ we consider the function*

$$\Gamma \ni \gamma \mapsto L^\phi_v(\gamma) \;:=\; - \sum_{\{x,y\}\subset\gamma} \langle \nabla\phi(x-y), v(x)-v(y)\rangle_{\mathbb{R}^d}. \tag{4.7.3}$$

Then for any Ruelle measure μ and all $v \in V_0(\mathbb{R}^d)$ we have that

$$L^\phi_v \in L^2(\Gamma; \mu).$$

Remark 4.7.9 An analysis of the proof of Lemma 4.7.8 (cf. [AKR97b, Lemma 4.1]) shows that condition (D) can be weakened by replacing $L^1(\mathbb{R}^d; e^{-\phi}m)$, $L^2(\mathbb{R}^d; e^{-\phi}m)$ by the weak L^p–spaces $L^1_w(\mathbb{R}^d; e^{-\phi}m)$, $L^2_w(\mathbb{R}^d; e^{-\phi}m)$ respectively.

In order to prove the main result of this section (i.e., Theorem 4.7.11 below) we need to introduce the convex set $\mathcal{P}_q, q \in [1, \infty[$, of all probability measures μ on $(\Gamma, \mathcal{B}(\Gamma))$ satisfying the following properties:

($\mathcal{P}$ 1) $\mu(S_\infty) = 1$, i.e., μ is tempered.

($\mathcal{P}$ 2) $\int \gamma(K)^q \, \mu(d\gamma) < \infty$ for all compact $K \subset \mathbb{R}^d$.

($\mathcal{P}$ 3) For all $v \in V_0(\mathbb{R}^d)$, $L_v^\phi \in L^q(\Gamma; \mu)$, where L_v^ϕ is as defined in (4.7.3).

For $\mu \in \mathcal{P}_q$ and $v \in V_0(\mathbb{R}^d)$ we define

$$B_v^\phi := L_v^\phi + \langle \operatorname{div} v, \cdot \rangle (\in L^q(\Gamma; \mu)). \tag{4.7.4}$$

and for $V = \sum_{i=1}^N F_i \, v_i \in \mathcal{VFC}_b^\infty$

$$\operatorname{div}_\phi^\Gamma V := \sum_{i=1}^N (\nabla_{v_i}^\Gamma F_i + B_{v_i}^\phi F_i). \tag{4.7.5}$$

Lemma 4.7.10 *$\mathcal{G}^t_{gc}(z, \phi) \subset \mathcal{G}_c(m, \Phi) \cap \mathcal{P}_2$ for all $z > 0$ (where $\Phi = \Phi_\phi$ as above).*

Proof. Let $z > 0$, $\mu \in \mathcal{G}_{gc}(z, \phi)$. We already know that $\mu \in \mathcal{G}_c(m, \Phi)$ by (4.6.11) and by Remark 4.7.3 that μ satisfies ($\mathcal{P}$ 2). By Lemma 4.7.8 also ($\mathcal{P}$ 3) with $q = 2$ holds.

□.

Now we are prepared to prove the main result of this section.

Theorem 4.7.11 *Let ϕ be a pair potential satisfying properties (SS), (LR), (C), (D). Then for $\mu \in \mathcal{P}_1$ the following assertions are equivalent:*

(i) $\mu \in \mathcal{G}_c(m, \Phi)$ (where $\Phi = \Phi_\phi$ as above).

(ii) For all $v \in V_0(\mathbb{R}^d)$ and all $F, G \in \mathcal{FC}_b^\infty$ the following integration by parts formula holds:

$$\int \nabla_v^\Gamma F \; G \, d\mu \; = \; -\int F \; \nabla_v^\Gamma G \, d\mu \; - \; \int F \; G \; B_v^\phi \, d\mu.$$

In particular, any Ruelle measure (with interaction given by ϕ) satisfies (ii).

Remark 4.7.12 (i) We note that if even $\mu \in \mathcal{P}_2$, then (ii) in Theorem 4.7.11 is equivalent to

$$(4.7.6) \qquad (\nabla_v^\Gamma)^* = -\nabla_v^\Gamma - B_v^\phi \quad \text{for all } v \in V_0(\mathbb{R}^d)$$

as an operator equality on the domain $\mathcal{F}C_b^\infty$ in $L^2(\Gamma;\mu)$, where $(\nabla_v^\Gamma)^*$denotes the adjoint of ∇_v^Γ on $L^2(\Gamma;\mu)$.

(ii) If $\mu \in \mathcal{P}_1$, then Theorem 4.7.11 (ii) is obviously equivalent to:

$$(4.7.7) \qquad \int \langle \nabla^\Gamma F, V\rangle_{T\Gamma}\, d\mu = -\int F \operatorname{div}_\phi^\Gamma V\, d\mu \text{ for all } F \in \mathcal{F}C_b^\infty,\ V \in \mathcal{V}\mathcal{F}C_b^\infty.$$

This shows the analogy of Theorem 4.7.11 with Theorem 4.2.2.

(iii) As mentioned before, condition (C) is not really necessary for Theorem 4.7.11 to hold. The more general analogue of this result, i.e., Theorem 4.3 in [AKR97b] applies to the Lennard–Jones potential $\phi_{a,b}$ in Example 4.7.7.

4.8 Infinite interacting particle systems

The purpose of this section is to briefly summarize all results on the stochastic dynamics associated with Gibbs measures μ, i.e., on the corresponding *infinite interacting particle systems* (cf. below). By all the previous preparations in Sections 6 and 7, these results are direct applications of the general results obtained in Subsection 4.2.

4.8.1 Stochastic dynamics corresponding to Gibbs states

Let X be a Riemannian manifold as before. Let $\mathcal{G}_{gc}^1(\sigma,\Phi)$ (cf. (4.6.12)) for a general potential Φ and $\sigma = \rho\cdot m$ as specified at the beginning of Remark 4.2.3 (i).

Theorem 4.8.1 *Suppose that μ, Φ satisfy the conditions of Theorem 4.6.13. Then $(\mathcal{E}_\mu^\Gamma, \mathcal{F}C_b^{\infty,\mu})$ is (well-defined and) closable on $L^2(\Gamma_X;\mu)$. The closure $(\mathcal{E}_\mu^\Gamma, D(\mathcal{E}_\mu^\Gamma))$ is a symmetric quasi-regular Dirichlet form on $L^2(\ddot{\Gamma}_X;\mu)$ and, thus has an associated conservative diffusion process* **M** *satisfying all properties in Theorems 4.4.14, 4.4.16. In particular, all this holds if $X = \mathbb{R}^d$ with the Euclidean metric and if $\mu \in \mathcal{G}_{gc}^t(z,\phi)$ for some pair potential ϕ satisfying (SS), (LR), (I) and the condition specified in Remark 4.7.5.*

Proof. Theorems 4.6.13, 4.4.12, 4.4.14, 4.4.16. □

4.8.2 Solutions of the martingale problem resp. of the corresponding "heuristic" SDE

Let $X = \mathbb{R}^d$ with the Euclidean metric, $\Gamma := \Gamma_{\mathbb{R}^d}$, $\ddot{\Gamma} := \ddot{\Gamma}_{\mathbb{R}^d}$. Let $\mu \in \mathcal{G}_c(m, \Phi) \cap \mathcal{P}_2$ (cf. Subsection 7.2), (e.g. $\mu \in \mathcal{G}^t_{gc}(z \cdot m, \phi)$, $z \in]0, \infty[$ where $\Phi = \Phi_\phi$ is associated with a pair potential ϕ).

Theorem 4.8.2 *Suppose ϕ satisfies (SS), (LR), (C) and (D). Then:*

(i) μ satisfies (IbP) in Subsection 4.1 and for all $F, G \in \mathcal{F}C_b^\infty$

$$\mathcal{E}^\Gamma_\mu(F, G) = -\int \Delta^\Gamma_\phi F\, G\, d\mu$$

where

$$\Delta^\Gamma_\phi = div^\Gamma_\phi \nabla^\Gamma$$

and div^Γ_ϕ is given by (4.7.5) (see also (4.7.4), (4.7.3)).

(ii) $(\mathcal{E}^\Gamma_\mu, \mathcal{F}C_b^{\infty,\mu})$ is (well-defined and) closable on $L^2(\Gamma; \mu)$. The closure $(\mathcal{E}^\Gamma_\mu, D(\mathcal{E}^\Gamma_\mu))$ is a symmetric quasi-regular Dirichlet form on $L^2(\ddot{\Gamma}; \mu)$ and is thus associated with a conservative diffusion process **M** *satisfying all properties in Theorems 4.4.14, 4.4.16.*

(iii) If $\mu \in \mathcal{G}^t_{gc}(z \cdot m, \phi)$, then μ satisfies (QI), hence $(\mathcal{E}^\Gamma_\mu, \mathcal{F}^{(c)})$ (as defined in Proposition 4.4.9) also is a symmetric quasi-regular Dirichlet form on $L^2(\ddot{\Gamma}; \mu)$ and has thus an associated conservative diffusion process **M** *satisfying all properties in the analogues of Theorems 4.4.14, 4.4.16 with $(\mathcal{E}^\Gamma_\mu, \mathcal{F}^{(c)})$ replacing $(\mathcal{E}^\Gamma_\mu, D(\mathcal{E}^\Gamma_\mu))$ resp. with the generator of $(\mathcal{E}^\Gamma_\mu, \mathcal{F}^{(c)})$ replacing $(H^\Gamma_\mu, D(H^\Gamma_\mu))$.*

(iv) If $\mu \in \mathcal{G}^t_{gc}(z \cdot m, \phi)$, then $\ddot{\Gamma} \setminus \Gamma$ is $(\mathcal{E}^\Gamma_\mu, D(\mathcal{E}^\Gamma_\mu))$-exceptional, hence also $(\mathcal{E}^\Gamma_\mu, \mathcal{F}^{(c)})$-exceptional.

Proof. (i), (ii): Theorem 4.7.11, Proposition 4.4.3 (ii), Theorems 4.4.12, 4.4.14, 4.4.16.
(iii): By [RSch97, Proposition 2.1] μ satisfies (QI). Hence, the assertions follow from Propositions 4.4.9, 4.5.6, and Theorem 4.5.8.
(iv): The first part is just [RS97, Corollary 1] the second part then follows from the first (cf. the discussion in Subsection 5.3 which led to the proof of Proposition 4.5.6. □

Theorem 4.4.16 states that both in the situation of Theorem 4.8.2 and 4.8.1 the process $\mathbf{M} = (\Omega, \mathbf{F}, (\mathbf{F}_t)_{t\geq 0}, (\Theta_t)_{t\geq 0}, (\mathbf{X}_t)_{t\geq 0}, (\mathbf{P}_\gamma)_{\gamma\in\ddot{\Gamma}_X})$ solves the martingale problem for $(-H_\mu^\Gamma, D(H_\mu^\Gamma))$, in particular, implies that for all $F \in \mathcal{F}C_b^\infty$

$$F(\mathbf{X}_t) - F(\mathbf{X}_0) + \int_0^t H_\mu^\Gamma F(\mathbf{X}_s)\, ds\ ,\ t \geq 0, \tag{4.8.1}$$

is an $(\mathbf{F}_t)$–martingale under $\mathbf{P}_\gamma$ for $\mathcal{E}_\mu^\Gamma$–q.e. $\gamma \in \ddot{\Gamma}_X$. (The reader should note that if this only holds for all $F \in \mathcal{F}C_b^\infty$, rather than all $F \in D(H_\mu^\Gamma)$, it is not clear whether this uniquely determines the diffusion $\mathbf{M}$. This is in fact an open question which so far has only been answered positively if μ is a pure Poisson measure resp. a mixed Poisson measure and $(H_\sigma^X, D(H_\sigma^X))$ is conservative. The latter follows immediately by Theorem 4.3.3). In the case of Theorem 4.8.2 we can, however, write an explicit formula for $H_\mu^\Gamma F$ for all $F \in \mathcal{F}C_b^\infty$ (which is not possible in the situation of Theorem 4.8.1). It namely follows from Theorem 4.8.2(i), (4.2.13) and (4.2.23) that

$$H_\mu^\Gamma F(\gamma) = \Delta^\Gamma F(\gamma) - 2\sum_{y\in\gamma} \langle \nabla\phi(\cdot - y), \nabla^\Gamma F(\gamma)\rangle_{T\Gamma}, \quad \gamma \in \Gamma.$$

Since the closure of $(\Delta^\Gamma, \mathcal{F}C_b^\infty)$ on $L^2(\ddot{\Gamma}; \psi_{i_1\ldots i_n}{}^z)$ generates the Brownian motion on $\mathbb{R}^d$ (cf. Theorems 4.3.3, 4.3.12 and Remark 4.3.15(iii)), (4.8.1) can be interpreted that $\mathbf{M}$ solves the following "heuristic" stochastic differential equation

$$dX_t^i = dW_t^i + \sum_{j:j\neq i} \nabla\phi(X_t^i - X_t^j)dt, \quad i \in \mathbb{N}, \tag{4.8.2}$$

where $(W_t^i)_{t\geq 0}$, $i \in \mathbb{N}$, are independent Brownian motions on $\mathbb{R}^d$ and $(X_t^i)_{t\geq 0}$ are such that $\mathbf{X}_t = \sum_{i=1}^\infty \varepsilon_{X_t^i}$, $t \geq 0$ (cf. [La77], [Fr87] for the "smooth case", i.e., with $\phi \in C_0^3(\mathbb{R}^d)$). The solution to (4.8.2) is interpreted as an infinite particle system undergoing interactions given by the drift determined by $\nabla\phi$. But, of course, (4.8.2) is *purely heuristic*.

4.9 Ergodicity

We now want to prove the analogue of Theorem 4.3.18 for $\mu \in \mathcal{G}_c(m, \Phi) \cap \mathcal{P}_2$ first proven in [AKR97b, Sect. 6]. We start, however, with a general result on ergodicity which can be formulated for large classes of probability measures μ on $(\Gamma_X, \mathcal{B}(\Gamma_X))$. In this general situation also the proofs become more transparent, since they are quite universal. Therefore, in this section we include all proofs from [AKR97b, Sect. 6].

4.9.1 A general result on irreducibility resp. ergodicity

Consider the situation studied in Subsection 4.1 after Remark 4.4.2, i.e., μ is a probability measure on $(\Gamma_X, \mathcal{B}(\Gamma_X))$ (X a Riemannian manifold) satisfying (IbP), in particular, hence (4.4.3) holds and $(\mathcal{E}_\mu^\Gamma, \mathcal{F}C_b^{\infty,\mu})$ (as defined in (4.4.2)) is closable on $L^2(\Gamma_X;\mu)$ and its closure $(\mathcal{E}_\mu^\Gamma, D(\mathcal{E}_\mu^\Gamma))$ is a quasi–regular local symmetric Dirichlet form on $L^2(\ddot{\Gamma}_X;\mu)$ (cf. Proposition 4.4.3, Theorem 4.4.12, and Corollary 4.4.13). We recall that according to (4.4.9) we set $H_0^{1,2}(\Gamma_X;\mu) := D(\mathcal{E}_\mu^\Gamma)$.

We start with the following simple lemma.

Lemma 4.9.1 *Let $(\mathcal{E}, D(\mathcal{E}))$ be a symmetric Dirichlet form on $L^2(\Gamma_X;\mu)$. Then $(\mathcal{E}, D(\mathcal{E}))$ is irreducible (i.e., F = const. provided $F \in D(\mathcal{E})$ such that $\mathcal{E}(F,F) = 0$) if and only if for all* bounded *$F \in D(\mathcal{E})$ with $\mathcal{E}(F,F) = 0$ it follows that* $F = \text{const}$.

Proof. Suppose $F \in D(\mathcal{E})$ with $\mathcal{E}(F,F) = 0$. Then (e.g. by [MR92, Chap. I, Proposition 4.17]) we have that for all $n \in \mathbb{N}$

$$F_n := (F \wedge n) \vee (-n) \in D(\mathcal{E})$$

and $F_n \underset{n\to\infty}{\longrightarrow} F$ w.r.t. $(\mathcal{E}(\cdot,\cdot) + (\cdot,\cdot)_{L^2(\Gamma_X;\mu)})^{1/2}$. Furthermore, $\mathcal{E}(F_n, F_n) \le \mathcal{E}(F,F)$ for all $n \in \mathbb{N}$ (cf. e.g. [MR 92, Chap. I, Theorem 4.12]). Now the assertion follows easily. □

We recall the following known result characterizing the irreducibility of $(\mathcal{E}_\mu^\Gamma, H_0^{1,2}(\Gamma_X;\mu))$. Let $\mathbf{M} = (\Omega, \mathbf{F}, (\mathbf{F}_t)_{t\ge0}, (\Theta_t)_{t\ge0}, (\mathbf{X}_t)_{t\ge0}, (\mathbf{P}_\gamma)_{\gamma\in\ddot{\Gamma}_X})$ be the conservative diffusion process associated with $(\mathcal{E}_\mu^\Gamma, H_0^{1,2}(\Gamma_X;\mu))$ (cf. Theorem 4.4.14). As usual we set

$$\mathbf{P}_\mu := \int \mathbf{P}_\gamma \, \mu(d\gamma) .$$

We have the analogue of Proposition 4.3.17 whose proof also works in this case.

Proposition 4.9.2 *The following assertions are equivalent:*

(i) $\mathbf{P}_\mu$ is (time) ergodic (i.e., every bounded $\mathbf{F}$–measurable function $G : \Omega \to \mathbb{R}$, which is Θ_t–invariant for all $t \ge 0$, is constant $\mathbf{P}_\mu$–a.e.

(ii) $(\mathcal{E}_\mu^\Gamma, H_0^{1,2}(\Gamma_X;\mu))$ is irreducible.

(iii) $(e^{-H_\mu^\Gamma t})_{t>0}$ is irreducible (i.e., if $G \in L^2(\Gamma_X;\mu)$ such that $e^{-H_\mu^\Gamma t}(GF) = G\, e^{-H_\mu^\Gamma t} F$ for all $F \in L^\infty(\Gamma_X;\mu)$, $t > 0$, then $G = const.$).

(iv) If $F \in L^2(\Gamma_X;\mu)$ such that $e^{-H^\Gamma_\mu t}F = F$ for all $t>0$, then $F = const.$.

(v) $(e^{-H^\Gamma_\mu t})_{t>0}$ is ergodic (i.e.,

$$\int \left(e^{-H^\Gamma_\mu t}F - \int F\, d\mu \right)^2 d\mu \to 0 \text{ as } t \to \infty \text{ for all } F \in L^2(\Gamma_X;\mu))$$

(vi) If $F \in D(H^\Gamma_\mu)$ with $H^\Gamma_\mu F = 0$, then $F = const.$ ("uniqueness of ground state").

Now we consider the weak (1,2)–Sobolev space $W^{1,2}(\Gamma_X;\mu)$ introduced in (4.4.11) and the corresponding bilinear form $(\mathcal{E}^\Gamma_\mu, W^{1,2}(\Gamma_X;\mu))$.

In view of Lemma 4.9.1 we define $(\mathcal{E}^\Gamma_\mu, W^{1,2}(\Gamma_X;\mu))$ to be *irreducible* if for all bounded $F \in W^{1,2}(\Gamma_X;\mu)$ with $\mathcal{E}^\Gamma_\mu(F,F) = 0$, it follows that $F = const.$ However, unlike as for $(\mathcal{E}^\Gamma_\mu, H_0^{1,2}(\Gamma_X;\mu))$ we *cannot* drop the term "bounded" in this definition, since it is not clear whether the larger bilinear form $(\mathcal{E}^\Gamma_\mu, W^{1,2}(\Gamma_X;\mu))$ is also a Dirichlet form. We emphasize here that the Dirichlet property was heavily used in the proof of Lemma 4.9.1. In particular, we do not know whether an analogue of Proposition 4.9.2 holds for $(\mathcal{E}^\Gamma_\mu, W^{1,2}(\Gamma_X;\mu))$

Remark 4.9.3 Obviously, the irreducibility of $(\mathcal{E}^\Gamma_\mu, W^{1,2}(\Gamma_X;\mu))$ implies that of $(\mathcal{E}^\Gamma_\mu, H_0^{1,2}(\Gamma_X;\mu))$.

Suppose for every $v \in V_0(X)$ we are given a sequence $(B_{v,n})_{n\in\mathbb{N}}$ of $\mathcal{B}(\Gamma_X)$–measurable $\mathbb{R}$–valued functions on Γ_X. Let $(\mathrm{IbP})^B$ denote the set of all measures μ on $(\Gamma_X, \mathcal{B}(\Gamma_X))$ such that

(IbP 1) $B^\mu_v := L^1(\Gamma_X;\mu) - \lim_{n\to\infty} B_{v,n}$ exists and $B^\mu_v \in L^2(\Gamma_X;\mu)$ for all $v \in V_0(X)$.

(IbP 2) $\int \gamma(K)^2\, \mu(d\gamma) < \infty$ for all compact $K \subset X$ and

$$\int \nabla^\Gamma_v F\, d\mu = -\int F\, B^\mu_v\, d\mu \text{ for all } F \in \mathcal{F}C_b^\infty \text{ and all } v \in V_0(X).$$

Obviously, $(\mathrm{IbP})^B$ is a convex set. Let $\mathrm{ex}(\mathrm{IbP})^B$ denote the set of its extreme points. Clearly, if $\mu_1, \mu_2 \in (\mathrm{IbP})^B$ and $M \in]0,\infty[$ such that $M\mu_1 - \mu_2$ is a non–zero positive measure on $(\Gamma_X, \mathcal{B}(\Gamma_X))$, then $[(M\mu_1 - \mu_2)(\Gamma_X)]^{-1}(M\mu_1 - \mu_2) \in (\mathrm{IbP})^B$. For every $\mu \in (\mathrm{IbP})^B$ both $(\mathcal{E}^\Gamma_\mu, H_0^{1,2}(\Gamma_X;\mu))$ and $(\mathcal{E}, W^{1,2}(\Gamma_X;\mu))$ are defined according to Subsection 4.1.

Theorem 4.9.4 *Let $\mu \in (\mathrm{IbP})^B$. Then the following assertions are equivalent:*

(i) $\mu \in ex(\mathrm{IbP})^B$.

(ii) $\{\nu \in (\mathrm{IbP})^B \mid \nu = \rho \cdot \mu$ *for some bounded,* $\mathcal{B}(\Gamma_X)$*–measurable function* $\rho : \Gamma_X \to \mathbb{R}_+\} = \{\mu\}$.

(iii) $(\mathcal{E}_\mu^\Gamma, W^{1,2}(\Gamma_X;\mu))$ *is irreducible.*

Proof. (i) $\Rightarrow$ (ii): Assume (i) holds. Let $\rho : \Gamma_X \to \mathbb{R}_+$, bounded, $\mathcal{B}(\Gamma_X)$–measurable such that $\nu := \rho \cdot \mu \in (\mathrm{IbP})^B$, and let $M := \sup_{\gamma\in\Gamma_X} \rho(\gamma)$. Define

$$\mu_1 := \frac{M-\rho}{M-1} \cdot \mu \, .$$

Then $\mu_1 \in (\mathrm{IbP})^B$ and $\mu = \frac{M-1}{M}\mu_1 + \frac{1}{M}\nu$. By assumption (i) it follows that $\mu_1 = \nu$ which implies $\rho = 1$, and (ii) is proved.

(ii) $\Rightarrow$ (i): Assume (ii) holds. Let $\mu_1, \mu_2 \in (\mathrm{IbP})^B$ and $t \in]0,1[$ such that $\mu = t\mu_1 + (1-t)\mu_2$. Then they are both absolutely continuous w.r.t. μ with bounded densities. By assumption (ii) it follows that $\mu_1 = \mu = \mu_2$. Consequently, $\mu \in \mathrm{ex}(\mathrm{IbP})^B$.

(ii) $\Rightarrow$ (iii): Assume (ii) holds. Let $G \in W^{1,2}(\Gamma_X;\mu) \cap L^\infty(\Gamma_X;\mu)$ such that $\mathcal{E}_\mu^\Gamma(G,G) = 0$. We have to show that $G = const$. Since $1 \in \mathcal{F}C_b^\infty \subset W^{1,2}(\Gamma_X;\mu)$ and $d^\mu 1 = \nabla_\mu^\Gamma 1 = 0$, replacing G by $G - \operatorname{essinf} G$ we may assume that $G \geq 0$, and, in addition, that $\int G \, d\mu = 1$. Define $\nu := G \cdot \mu$. Then since $d^\mu G = 0$, Proposition 4.4.6 and (IbP2) imply that for all $v \in V_0(X)$ and all $F \in \mathcal{F}C_b^\infty$

$$\begin{aligned} \int \nabla_v^\Gamma F \, d\nu &= \int \langle d^\mu(FG), v\rangle_{T\Gamma_X} \, d\mu \\ &= -\int F \, B_v^\mu \, d\nu \end{aligned}$$

where we used (4.4.10) in the last step. Hence (IbP2) holds. Since G is bounded, clearly $B_{v,n} \to B_v^\mu$ as $n \to \infty$ in $L^1(\Gamma_X;\nu)$ and $B_v^\mu \in L^2(\Gamma_X;\nu)$. So, $\nu \in (\mathrm{IbP})^B$ and assumption (ii) implies that $G = 1$, hence (iii) is proved.

(iii) $\Rightarrow$ (ii): Assume (iii) holds. Let $\rho : \Gamma_X \to \mathbb{R}_+$, bounded, $\mathcal{B}(\Gamma_X)$–measurable such that $\nu := \rho \cdot \mu \in (\mathrm{IbP})^B$. Then $B_{v,n} \to B_v^\mu$ as $n \to \infty$ in $L^1(\Gamma_X;\nu)$ and $B_v^\mu \in L^2(\Gamma_X;\nu)$ for all $v \in V_0(X)$, hence by (IbP2) applied to ν

$$\int \mathrm{div}_\mu V \; \rho \, d\mu = 0 \quad \text{for all } V \in \mathcal{V}\mathcal{F}C_b^\infty.$$

Hence by Remark 4.4.4 and (4.4.11) it follows that

$$\rho \in W^{1,2}(\Gamma_X;\mu) \text{ and } d^\mu \rho = 0,$$

i.e., $\mathcal{E}_\mu^\Gamma(\rho,\rho)=0$. By assumption (iii) it follows that $\rho\equiv 1$ and (ii) is proved. □

Corollary 4.9.5 *Let $\mu\in \mathrm{ex}(\mathrm{IbP})^B$. Then $(\mathcal{E}_\mu^\Gamma, H_0^{1,2}(\Gamma_X;\mu))$ is irreducible and all equivalent assertions in Proposition 4.9.2 hold.*

Proof. Theorem 4.9.4 and Remark 4.9.3. □

4.9.2 Applications to canonical Gibbs and Ruelle measures

We start with the free case.

(a) Mixed Poisson measures

We consider the situation described at the end of Subsection 6.2, i.e., the general "manifold–case" with $\Phi\equiv 0$. So, by Theorem 4.6.6 our canonical Gibbs measures $\mathcal{G}_c(\sigma)$ are exactly the mixed Poisson measures $\mu_{\lambda,\sigma}$.

So let $\sigma=\rho\cdot m$ be as in Remark 4.2.3 (i), i.e., we assume

(a.1) $\rho\in L^1_{loc}(X;m)$ such that $\rho^{1/2}\in H^{1,2}_{loc}(X;m)$.

Furthermore, consider the following conditions

(a.2) $\sigma(X)=\infty$ and $\beta^\sigma\in L^1_{loc}(X;m)$ where $\beta^\sigma := \dfrac{\nabla^X\rho}{\rho}$ (as in (4.2.21)).

(a.3) $(H_\sigma^X,\mathcal{D}^\sigma)$ is essentially self–adjoint on $L^2(X;\sigma)$ (which is e.g. the case if X is complete and $|\beta|_{TX}\in L^p_{loc}(X;m)$ for some $p>\dim X$; cf. Remark 4.3.4).

(a.4) $(H_\sigma^X, D(H_\sigma^X))$ is conservative (cf. condition (B) in Theorem 4.3.3).

In conditions (a.3), (a.4) the operator $(H_\sigma^X, D(H_\sigma^X))$ is the generator of the Dirichlet form $(\mathcal{E}_\sigma^X, D(\mathcal{E}_\sigma^X))$ defined as the closure of $(\mathcal{E}_\sigma^X,\mathcal{D}^\sigma)$ on $L^2(X;m)$ (cf. Theorem 4.6.12).

Let $(\mathrm{IbP})^B$ be defined as in Subsection 9.1 with $B_{v,n}:=\langle \mathrm{div}_\sigma^X v,\cdot\rangle$, $n\in\mathbb{N}$, $v\in V_0(X)$ (cf. (4.2.22)). Let $\mathcal{M}_2$ denote the set of all probability measures μ on $(\Gamma_X,\mathcal{B}(\Gamma_X))$ such that $\int\gamma(K)^2\,\mu(d\gamma)<\infty$ for all compact $K\subset X$.

Lemma 4.9.6 *Assume that (a.1) and (a.2) hold. Then*

$$\begin{aligned}(\mathrm{IbP})^B &= \{\mu_{\lambda,\sigma} \mid \lambda \textit{ satisfies } (4.3.1)\} \\ &= \mathcal{G}_c(\sigma) \cap \mathcal{M}_2(\Gamma_X)\ .\end{aligned}$$

Proof. The first equality immediately follows from Remark 2.3 (i), the second from Theorem 4.6.6. □

As before, for a convex set $\mathcal{K}$ of measures on $(\Gamma_X, \mathcal{B}(\Gamma_X))$ we denote the set of its extreme points by $\mathrm{ex}\mathcal{K}$.

Lemma 4.9.7 *Assume (a.1) and (a.2) hold. Then*

$$\begin{aligned}\mathrm{ex}(\mathrm{IbP})^B &= \mathrm{ex}\{\mu_{\lambda,\sigma} \mid \lambda \textit{ satisfies } (4.3.1)\} \\ &= \{\psi_{i_1 \dots i_n z\sigma} \mid z \in [0, \infty[\}.\end{aligned}$$

Proof. The first equality follows from Lemma 4.9.6. To prove the second we note that since

$$\begin{aligned}&\{\mu_{\lambda,\sigma} \mid \lambda \text{ satisfies } (4.3.1)\} \\ = \ &\{\mu_{\lambda,\sigma} \mid \lambda \text{ any probability measure on } \ (\mathbb{R}_+, \mathcal{B}(\mathbb{R}_+))\} \cap \mathcal{M}_2(\Gamma_X),\end{aligned}$$

it is straightforward to check that

$$\begin{aligned}&\mathrm{ex}\{\mu_{\lambda,\sigma} \mid \lambda \text{ satisfies } (4.3.1)\} \\ \subset \ &\mathrm{ex}\{\mu_{\lambda,\sigma} \mid \lambda \text{ any probability measure on } (\mathbb{R}_+, \mathcal{B}(\mathbb{R}_+))\}.\end{aligned}$$

Since $\psi_{i_1 \dots i_n z\sigma} \in \mathcal{M}_2(\Gamma_X)$ for all $z \in \mathbb{R}_+$, the second equality in the assertion follows by Theorem 4.6.6 (i) and (ii). □

Theorem 4.9.8 *Assume (a.1) and (a.2) hold and let $\mu \in \mathcal{G}_c(\sigma) \cap \mathcal{M}_2(\Gamma_X)$ $(= \{\mu_{\lambda,\sigma} \mid \lambda$ satisfies $(4.3.1)\}$, cf. Lemma 4.9.6). Then the following assertions are equivalent:*

(i) $\mu = \psi_{i_1 \dots i_n z\sigma}$ for some $z \in [0, \infty[$.

(ii) $(\mathcal{E}_\mu^\Gamma, W^{1,2}(\Gamma_X; \mu))$ is irreducible.

Furthermore, the assertions in Corollary 4.9.5 hold for $\mu := \psi_{i_1 \dots i_n z\sigma}$ and any $z \in \mathbb{R}_+$.

Proof. Lemmas 4.9.6, 4.9.7 and Theorem 4.9.4. □

Theorem 4.3.18, left unproved in Subsection 3.4, is now a special case of the following result:

Theorem 4.9.9 *Assume that conditions (a.1) – (a.4) hold and let μ be as in Theorem 4.9.8. Then $H_0^{1,2}(\Gamma_X;\mu) = W^{1,2}(\Gamma_X;\mu)$ and any of the assertions (i) – (vi) of Proposition 4.9.2 is equivalent to:*

(vii) $\mu = \psi_{i_1 \dots i_n z\sigma}$ for some $z \in [0,\infty[$.

Proof. Proposition 4.4.6 and Theorem 4.9.8. □

(b) Canonical Gibbs measures

We now consider the situation described in Section 7. In particular, $X := \mathbb{R}^d$, $\Gamma := \Gamma_{\mathbb{R}^d}$. We assume that the pair potential ϕ satisfies (SS), (LR), (C), and (D). (For cases where (C) does not necessarily hold see [AKR97b, Subsection 6.3 (b) and (c)]).

Obviously, if $(\mathrm{IbP})^B$ is defined as in Subsection 9.1 with $B_{v,n} := B_v^\phi = L_v^\phi + \langle \operatorname{div} v, \cdot \rangle$, $n \in \mathbb{N}$, $v \in V_0(\mathbb{R}^d)$ (and L_v^ϕ as in (4.7.5)), then by Theorem 4.7.11

$$(\mathrm{IbP})_2^B = \mathcal{G}_c(m,\Phi)_2 \;, \tag{4.9.1}$$

where for a set $\mathcal{K}$ of probability measures on $(\Gamma, \mathcal{B}(\Gamma))$ we set

$$\mathcal{K}_2 := \mathcal{K} \cap \mathcal{P}_2$$

with $\mathcal{P}_2$ as in Subsection 7.2.

Lemma 4.9.10 *For any convex set $\mathcal{K}$ of probability measures on $(\Gamma, \mathcal{B}(\Gamma))$*

$$\mathrm{ex}(\mathcal{K} \cap \mathcal{P}_2) = \mathrm{ex}\mathcal{K} \cap \mathcal{P}_2 \;.$$

The proof of Lemma 4.9.10 is straightforward, hence omitted.

Theorem 4.9.11 *Let $\mu \in \mathcal{G}_c(m,\Phi)_2$. Then the following assertions are equivalent:*

(i) $\mu \in \mathrm{ex}\mathcal{G}_c(m,\Phi)$.

(ii) $(\mathcal{E}_\mu^\Gamma, W^{1,2}(\Gamma;\mu))$ is irreducible.

Furthermore, the assertion in Corollary 4.9.5 holds for $\mu \in \mathrm{ex}\mathcal{G}_c(m,\Phi)_2$.

Proof. (4.9.1), Lemma 4.9.10 and Theorem 4.9.4. □.

Remark 4.9.12 Note that since $\emptyset \neq \mathcal{G}_{gc}^t(z,\phi) \subset \mathcal{G}_c(m,\Phi)_2$ for all $z > 0$, it immediately follows from Remark 4.6.5 that $\mathrm{ex}\mathcal{G}_c(m,\Phi)_2 \neq \emptyset$.

(c) Ruelle measures

We are still in the situation of Subsection 9.2 (b).

Theorem 4.9.13 *Let $z \in]0,\infty[$ and $\mu \in \mathcal{G}_{gc}^t(z,\phi)$, i.e., μ is a Ruelle measure. Then the following assertions are equivalent:*

(i) $\mu \in \mathrm{ex}\mathcal{G}_{gc}^t(z,\phi)$.

(ii) $(\mathcal{E}_\mu^\Gamma, W^{1,2}(\Gamma;\mu))$ *is irreducible.*

Furthermore, the assertion in Corollary 4.9.5 holds for $\mu \in \mathrm{ex}\mathcal{G}_{gc}^t(z,\phi)$.

Proof. Since $\mathcal{G}_{gc}^t(z,\phi) \subset \mathcal{G}_c(m,\Phi)_2$ by Lemma 4.7.10 and since $\mathrm{ex}\mathcal{G}_{gc}^t(z,\phi) \subset \mathrm{ex}\mathcal{G}_c(m,\phi)$ for $z > 0$ by [Ge79, Theorem 6.14], all assertions follow from Theorem 4.9.11. □

By [Ru70] there exists $z_0 \in]0,\infty[$ such that

$$\#\mathcal{G}_{gc}^t(z,\phi) = 1 \text{ for all } z \in]0,z_0[.$$

As an immediate consequence of Theorem 4.9.13 we thus obtain:

Corollary 4.9.14 *Let $z \in]0,z_0[$ and $\mu \in \mathcal{G}_{gc}^t(z,\phi)$. Then $(\mathcal{E}_\mu^\Gamma, W^{1,2}(\Gamma;\mu))$ is irreducible and the assertion in Corollary 4.9.5 holds for μ. In particular, the corresponding stochastic dynamics started with μ is (time) ergodic.*

Acknowledgment It is a pleasure to thank the organizers for very pleasant stays and very fruitful conferences in Anogia and at the MSRI in Berkeley. We also thank the EU and the MSRI for financial support. Financial support from DFG through Project Ro1195/2–1 and the SFB–343–Bielefeld is also gratefully acknowledged.

Bibliography

[ABR89] S. Albeverio, J. Brasche, M. Röckner: Dirichlet forms and generalized Schrödinger operators. In: Schrödinger Operators, Editors: H. Holden, A. Jensen. Lecture Notes in Physics 345, 1–42 Berlin: Springer 1989.

[AKRR98] S. Albeverio, Y.G. Kondratiev, J. Ren, M. Röckner: Dirichlet forms for mixed Poisson measures with general intensities. Preprint 1998.

[AKR96a] S. Albeverio, Y.G. Kondratiev, M. Röckner: Differential geometry of Poisson spaces. C.R. Acad. Sci. Paris, t. 323, Série I, 1129–1134 (1996).

[AKR96b] S. Albeverio, Y.G. Kondratiev, M. Röckner: Canonical Dirichlet operator and distorted Brownian motion on Poisson spaces. C.R. Acad. Sci. Paris, t. 323, Série I, 1179–1184 (1996).

[AKR97a] S. Albeverio, Y.G. Kondratiev, M. Röckner: Geometry and Analysis on configuration spaces . SFB–343–Preprint 1997. To appear in: J. Funct. Anal.

[AKR97b] S. Albeverio, Y.G. Kondratiev, M. Röckner: Geometry and analysis on configuration spaces. The Gibbsian case. SFB–343–Preprint 1997. To appear in: J. Funct. Anal.

[AKR97c] S. Albeverio, Y.G. Kondratiev, M. Röckner: Diffeomorphism groups, current algebras and all that: configuration space analysis in quantum theory. SFB–343–Preprint 1997. To appear in: Reviews in Math. Phys.

[AKR97d] S. Albeverio, Y.G. Kondratiev, M. Röckner: Ergodicity of L^2–semigroups and extremality of Gibbs states. J. Funct. Anal. 144, 364–423 (1997).

[AR90] S. Albeverio, M. Röckner: Classical Dirichlet forms on topological vector spaces – closability and a Cameron–Martin formula. J. Funct. Anal. 88, 395–436 (1990).

[AR95] S.Albeverio, M.Röckner: Dirichlet form methods for uniqueness of martingale problems and applications, In: Stochastic Analysis. Proceedings of Symposia in Pure Mathematics Vol. 57, 513–528. Editors: M.C. Cranston, M.A. Pinsky. Am. Math. Soc.: Providence, Rhode Island 1995.

[Ba87] D. Bakry: Étude des transformations de Riesz dans les variétés a courbure minorée. Lect. Notes Math. 1247, 137–172, Berlin: Springer 1987.

[BiGJ87] K. Bichteler, J.–J. Gravereaux, J. Jacod: Malliavin calculus for processes with jumps. Stochastic Monographs, Vol. 2, London: Gordon and Breach 1987.

[BKR97] V. I. Bogachev, N. V. Krylov, M. Röckner: Elliptic regularity and essential self-adjointness of Dirichlet operators on $\mathbb{R}^d$. To appear in: Ann. Scuola Norm. di Pisa, 1997.

[Ch84] I. Chavel: Eigenvalues in Riemannian geometry. New York: Academic Press 1984.

[dSKR98] J.L. da Silva, Y.G. Kondratiev, M. Röckner: On a relation between intrinsic and extrinsic Dirichlet forms for interacting particle systems. Preprint 1998.

[dSKSt97] J.L. da Silva, Y.G. Kondratiev, L. Streit: Differential geometry on compound Poisson space. Preprint 1997.

[Da89] E.B. Davies: Heat kernels and spectral theory. Cambridge: Cambridge University Press 1989.

[Do56] R. L. Dobrushin:, On Poisson distribution of particles in space. Ukr. Math. J., 8, 127–134, 1956.

[D53] J. L. Doob: Stochastic processes. New York – London: Wiley & Sons 1953.

[Dy65] E. B. Dynkin: Markov processes. Berlin: Springer 1965.

[Dy78] E.B.Dynkin: Sufficient statistics and extreme points. Ann. Probab. 6, 705-730 (1978).

[Eb97] A. Eberle: Uniqueness and Non–uniqueness of singular diffusion operators. Doktorarbeit, Universität Bielefeld (1997).

[Fi96] P.J.Fitzsimmons: On the quasi-regularity of semi-Dirichlet forms. Preprint 1996.

[Fö75] H. Föllmer: Phase transition and Martin boundary. Séminaire de Probabilités IX, Lecture Note in Math. 465, Berlin: Springer 1975.

[Fr87] J.Fritz: Gradient dynamics of infinite point systems. Ann. Probab. 15, 478-514 (1987).

[F80] M. Fukushima: Dirichlet forms and Markov processes. Amsterdam–Oxford–New York: North Holland 1980.

[F82] M. Fukushima: A note on irreducibility and ergodicity of symmetric Markov processes. In Lect. Notes Phys. 173, 200–207, Berlin: Springer 1982.

[F83] M.Fukushima: Capacitary maximal inequalities and an ergodic theorem. In Lect. Notes Math. 1021, Berlin: Springer 1983.

[FOT94] M. Fukushima, Y. Oshima, M. Takeda: Dirichlet forms and symmetric Markov processes. Berlin: de Gruyter 1994.

[Ge79] H.O.Georgii: Canonical Gibbs measures. LNM 760, Berlin: Springer 1979.

[KLRRSh97] Yu. G. Kondratiev, E.W. Lytvynov, A.L. Rebenko, M. Röckner, G.V. Shchepan'uk: Euclidean Gibbs states for quantum systems with Boltzmann statistics via cluster expansions. Methods of Funct. Anal. and Top. 3, 62–81 , (1997).

[KY93] D.Klein, W.-Sh. Yang: A characterization of first-order phase transition for superstable interactions in classical statistical mechanics. J. Stat. Phys. 71, 1043-1062, (1993).

[La77] R. Lang: Unendlichdimensionale Wienerprozesse mit Wechselwirkung I., II., Z. Wahrsch. verw. Gebiete 38, 55-72 (1977) and 39, 277-299 (1977).

[MR92] Z.M. Ma, M. Röckner: Introduction to the theory of (non–symmetric) Dirichlet forms. Berlin: Springer 1992.

[MR97] Z.M. Ma, M. Röckner: Construction of diffusions on configuration spaces. Preprint 1997.

[M-L76] A. Martin-Löf: Limit theorems for the motion of a Poisson system of independent Markovian particles with high density. Z. Wahrsch. verw. Gebiete 34, 205-223 (1976).

[MKM78] K.Matthes, J.Kerstan, J.Mecke: Infinite divisible point processes. Berlin: Akademie-Verlag 1978.

[MMW79] K. Matthes, J. Mecke, W. Warmuth: Bemerkungen zu einer Arbeit von Nguyen Xuan Xuan und Hans Zessin. Math. Nachr. 88, 117–127 (1979).

[Mec67] J. Mecke: Stationäre zufällige Mae auf lokalkompakten abelschen Gruppen. Z. Wahrsch. verw. Geb. 9, 36–58 (1967).

[NZ77] X.X.Nguyen, H.Zessin: Martin-Dynkin boundary of mixed Poisson processes. Z. Wahrsch. verw. Gebiete. 37, 191-200 (1977).

[NZ79] X.X. Nguyen, H. Zessin: Integral and differential characterization of the Gibbs process. Math. Nachr. 88, 105–115 (1979).

[Os96] H.Osada: Dirichlet form approach to infinite-dimensional Wiener processes with singular interactions. Commun. Math. Phys. 176, 117-131 (1996).

[Pa67] K.R.Parthasarathy: Probability measures on metric spaces. New York: Academic Press 1967.

[P76] C.J.Preston: Random fields. LNM 534, Berlin: Springer 1976.

[P79] C.J.Preston: Canonical and microcanonical Gibbs states. Z. Wahrsch. verw. Gebiete 46, 125-158 (1979).

[P80] C.J.Preston: Specifications and their Gibbs states. Manuscript (unpublished), 1980.

[Ra19] H. Rademacher: Über partielle und totale Differenzierbarkeit von Funktionen mehrerer Variablen und über die Transformation der Doppelintegrale. Math. Ann. 79, 340–359 (1919).

[RS75] M. Reed, S. Simon: Methods of modern mathematical physics: Vol. II, Fourier analysis, self–adjointness. New York: Academic Press 1975.

[RSch97] M. Röckner, A. Schied: Rademacher's theorem on configuration spaces and applications. SFB-343–Preprint 1998.

[RS97] M. Röckner, B. Schmuland: A support property for infinite dimensional interacting diffusion processes. SFB–343–Preprint 1997. To appear in: C.R. Acad. Sci. Paris, Série I.

[Ru69] D.Ruelle: Statistical mechanics. Rigorous results. New York–Amsterdam: Benjamin 1969.

[Ru70] D.Ruelle: Superstable interactions in classical statistical mechanics. Commun. Math. Phys. 18, 127-159 (1970).

[S95] B. Schmuland: On the local property for positivity preserving coercive forms. In: Dirichlet forms and Stochastic Processes, 345–354, Editors: Z.M. Ma et al., Berlin: de Gruyter 1995.

[ST74] T. Shiga, Y. Takahashi: Ergodic properties of the equilibrium process associated with infinitely many Markovian particles. Publ. RIMS, Kyoto Univ. 9, 505–516 (1974).

[Sm88] N.V. Smorodina: Differential calculus on configuration spaces and stable measures I. Theory Probab. Appl. 33, 488–499 (1988).

[Y96] M.W.Yoshida: Construction of infinite-dimensional interacting diffusion processes through Dirichlet forms, Probab. Th. Rel. Fields 106, 265-297 (1996).

AMS/IP Studies in Advanced Mathematics
Volume 8, 1998

Karl-Theodor Sturm: The Geometric Aspect of Dirichlet Forms

5.1 Introduction

Dirichlet forms play a prominent role in various fields of mathematics like potential theory, differential geometry, calculus of variations, stochastic processes and partial differential equations. This is mainly due to the fact that they allow to develop higly nontrivial extensions of classical theories under minimal regularity hypotheses. The other reason is that they constitute an effective link between different mathematical theories and towards applications.

The theory of Dirichlet forms originates in the pioneering work of Beurling and Deny [BD1], [BD2] on a functional analytic approach to potential theory. Fukushima [F1], [F2] and Silverstein [Si1], [Si2] developed the beautiful and fruitful correspondence between Dirichlet forms and symmetric Markov processes. Since that time, Dirichlet forms have turned out to be a most useful tool in the interplay of *stochastics* and *analysis*.

Recently, a new aspect of Dirichlet forms attracted attention. The crucial observation of Biroli and Mosco [BM1], [BM2] was that every Dirichlet form $(\mathcal{E}, \mathcal{F})$ on a topological space X defines in an intrinsic way a (pseudo) metric

$$\rho(x,y) = \sup\left\{\psi(x) - \psi(y) : \ \psi \in \mathcal{F} \cap \mathsf{C}, \|\Gamma_{\langle\psi\rangle}\| \leq 1\right\}$$

on the state space X. Here C denotes a suitable subspace of $\mathsf{C}(X)$ and $\|\Gamma_{\langle\psi\rangle}\|$ some kind of L^∞-norm of the "gradient" of ψ. Imposing some quantitative conditions on the Dirichlet form in terms of the intrinsic metric (doubling condition, scale invariant Poincaré inequality on intrinsic balls), Biroli and Mosco [BM1], [BM2] could prove the elliptic Harnack inequality and Hölder continuity of the harmonic functions. The parabolic case was treated in [S2], [S3]. Both contributions unify (and generalize) several of the important classical results from the theory of second order partial differential equations.

In this article, we want to illustrate that even *without any quantitative assumption* the intrinsic metric (or some generalization of it) plays an important role and that the geometric point of view leads to new insights. It allows to describe various properties of harmonic functions, heat kernels or stochastic processes in terms of geometric quantities derived from the associated Dirichlet form. This *geometric aspect of Dirichlet forms* suggests to regard the theory of Dirichlet forms as an extension of Riemannian geometry applicable to non-differentiable structures.

The first aim of this contribution is to give a brief survey about properties of the intrinsic metric and results in terms of it. The second aim is to indicate how to deal with non-regular Dirichlet forms (on non-locally compact spaces) where it is not clear how to define and handle the intrinsic metric. Concerning this last point, our solution is that in the non-regular set-up we do not use the intrinsic metric explicitely (whatever it might be). Instead of it, we work directly with the functions which might be used to define the metric.

We start with the second topic and present two new results in this respect. The first is an integrated upper estimate for the semigroup of the type

$$(T_t 1_A, 1_B) \leq \sqrt{m(A)m(B)} \cdot \exp\left(-\frac{\rho_0^2(A,B)}{2t}\right).$$

This holds true for any strongly local Dirichlet form $(\mathcal{E}, \mathcal{F})$ on any measurable state space X. We do not assume that the Dirichlet form is regular or quasi-regular. In particular, we do not assume that X is a topological space. The crucial geometric quantity here is the intrinsic distance of measurable sets

$$\rho_0(A,B) = \sup\{\psi_0(A,B) : \ \psi \in \mathcal{F} \cap L^\infty(X,m), \ \|\Gamma_{\langle\psi\rangle}\| \leq 1\}$$

which replaces the intrinsic metric $\rho(x,y)$.

The second result is an upper estimate for the 0-order capacity of the type

$$\mathrm{Cap}_0(A) \leq \frac{1}{2}\left(\int_0^\infty \frac{dr}{w'(r)}\right)^{-1}$$

where $w(r) = w_\psi(r) = \mu_{\langle\psi\rangle}(\{0 < \psi < r\})$ denotes the energy growth of an exhaustion function $\psi \geq 0$ with $\psi = 0$ on A. This estimate improves similar previous results in several respects: (i) it applies to *quasi-regular* Dirichlet forms, not only to regular ones, (ii) it works for *any exhaustion function* ψ, not only for the distance function $\psi = \rho(.,A)$ of A, (iii) for suitable choice of exhaustion functions it *always yields equality*. We also give various applications of this estimate to questions of recurrence, transience and polarity of sets.

In Chapter 3, we then restrict to locally compact state spaces and present some of the basic properties of the intrinsic metric. We also give several applications of it which relate the longtime behaviour (of the semigroup or the process) with the volume growth at infinity. More precisely, we present criteria for recurrence, conservativeness and positivity of the spectrum.

We do not discuss questions of local (elliptic) regularity, like Poincaré inequality or Harnack inequality. For these questions, we refer to [BM1], [BM2], [S2] and [S3]. See also [CG] for further developments in these directions.

Chapter 4 is devoted to some of the basic and most easiest examples of regular Dirichlet forms and associated intrinsic metrics. Here again we restrict ourselves to regular Dirichlet forms on locally compact spaces. However, we emphasize that there are many important examples of quasi-regular Dirichlet forms on infinite dimensional spaces with very interesting intrinsic metrics. See, for instance, [Be] for Dirichlet forms on the infinite dimensional torus, [OR] as well as [Sc] for Dirichlet forms on the space of probability measures and [AKR] as well as the contribution of Röckner in this volume for Dirichlet forms on configuration spaces.

In this article, we will concentrate on *some* geometric aspects of Dirichlet forms. Of course, there are other important interrelations between the theory of Dirichlet forms and geometry which will be not treated in this article. One of them is the fact, that on every measured metric space, one can define in a canonical way a local Dirichlet form which fits to the given metric. See [S8] and [S9]. Another is the theory of nonlinear Dirichlet forms and its application to harmonic maps. For this we refer to the article by Jost in this volume as well as to [Jo], [S6] and [S7].

For basic properties and other applications of Dirichlet forms we recommend the monographs [Si1,2], [F2], [BH], [MR] and [FOT]. We mainly follow the exposition in [FOT] and freely use results from it.

5.2 Estimates for Semigroups and Resolvents

$\longrightarrow$ In this chapter, X will be an arbitrary measurable space, m a σ-finite measure on X and $(\mathcal{E},\mathcal{F})$ a strongly local Dirichlet form on $L^2(X,m)$.

We do *not* assume that X is a topological space. In particular, we neither assume that $(\mathcal{E}, \mathcal{F})$ is regular not that it is quasi-regular.

5.2.1 The Dirichlet Form

Throughout this article a measure space (X, Ξ, m) with a σ-finite measure m will be fixed and $L^2(X, m)$ will denote the real Hilbert space of square integrable functions with inner product $(u, v) = \int_X uv\, dm$ and norm $\|u\| = (u, u)^{1/2}$. X is called *state space*, and m *reference measure*. We also fix a *strongly local Dirichlet form* $(\mathcal{E}, \mathcal{F})$ on $L^2(X, m)$.

A *Dirichlet form* $(\mathcal{E}, \mathcal{F})$ is a closed symmetric form $\mathcal{E}$ with domain $\mathcal{F} \subset L^2(X, m)$ which is Markovian. The latter means that $u \in \mathcal{F}$ implies $u^\sharp := (u \vee 0) \wedge 1 \in \mathcal{F}$ and

$$\mathcal{E}(u^\sharp, u^\sharp) \leq \mathcal{E}(u, u).$$

It is called *strongly local* iff $(u + \alpha)v = 0$ implies $\mathcal{E}(u, v) = 0$ (for all $u, v \in \mathcal{F}$ and $\alpha \in \mathbb{R}$). In other words, if $\mathcal{E}(u, v) = 0$ whenever u is constant on $\mathrm{supp}(v)$.

Occasionally we use the notation $\mathcal{E}(u) := \mathcal{E}(u, u)$. The generator of the Dirichlet form $(\mathcal{E}, \mathcal{F})$ will be denoted by $(A, \mathcal{D}(A))$. It is the (uniquely determined) positive self-adjoint operator with

$$(\sqrt{A}u, \sqrt{A}v) = \mathcal{E}(u, v) \quad \text{and} \quad \mathcal{D}(\sqrt{A}) = \mathcal{F}.$$

In terms of the generator we define the (sub-Markovian, strongly continuous, contraction) semigroup

$$T_t = e^{-tA} \quad \text{and} \quad \mathcal{D}(T_t) = L^2(X, m), \quad t > 0$$

and the resolvent

$$G_\alpha = (\alpha + A)^{-1} \quad \text{and} \quad \mathcal{D}(G_\alpha) = L^2(X, m), \quad \alpha > 0.$$

We denote by

$$\lambda = \inf \mathrm{spec}(A) = \inf \left\{ \frac{\mathcal{E}(u, u)}{\|u\|^2} : u \in \mathcal{F}, u \neq 0 \right\}$$

the bottom of the spectrum of the self-adjoint operator A. Since $A \geq 0$ we always have $\lambda \geq 0$.

In order to define a distance on X, in this general context, an appropriate class of test functions ("exhaustion functions") is the space $\mathcal{F} \cap L^\infty(X, m)$. We freely use the fact that $\varphi, \psi \in \mathcal{F} \cap L^\infty(X, m)$ implies $\varphi\psi \in \mathcal{F} \cap L^\infty(X, m)$ and

$\varphi e^{\psi} \in \mathcal{F} \cap L^{\infty}(X,m)$ (see e.g. [BH], I.5). The crucial quantity is the square field operator $\Gamma_{\langle\psi\rangle}$ defined for $\psi, \varphi \in \mathcal{F} \cap L^{\infty}(X,m)$ by

$$\Gamma_{\langle\psi\rangle}(\varphi^2) = 2\mathcal{E}(\varphi,\varphi) - 2\mathcal{E}(\varphi e^{\psi}, \varphi e^{-\psi})$$

and its norm

$$\|\Gamma_{\langle\psi\rangle}\| = \sup\left\{\Gamma_{\langle\psi\rangle}(\varphi^2) : \ \varphi \in \mathcal{F} \cap L^{\infty}(X,m), \|\varphi\| \leq 1\right\}$$

(with $\|\varphi\|^2 = \int \varphi^2 dm$). This can be interpreted as the L^{∞}-norm of the "gradient square" of ψ. It can be utilized to define a kind of distance on the state space X. Let us mention that if the Dirichlet form $(\mathcal{E}, \mathcal{F})$ is quasi-regular then

$$\Gamma_{\psi}(\varphi^2) = \int \varphi^2 d\mu_{\langle\psi\rangle}$$

for all $\psi, \varphi \in \mathcal{F} \cap L^{\infty}(X,m)$ and $\|\Gamma_{\psi}\| \leq 1$ if and only if $\mu_{\langle\psi\rangle} \leq m$.

For subsets A and B of X we define their *distance*

$$\rho_0(A,B) = \sup\{\psi_0(A,B) : \ \psi \in \mathcal{F} \cap L^{\infty}(X,m), \ \|\Gamma_{\langle\psi\rangle}\| \leq 1\}$$

where

$$\begin{aligned} \psi_0(A,B) &:= \operatorname{ess\,inf}\{\psi(x) - \psi(y) : \ x \in A, y \in B\} \\ &:= \sup\{C : \psi(x) - \psi(y) \geq C \text{ for } m\text{-}a.e. x \in A, y \in B\}. \end{aligned}$$

Note that $\rho_0(A,B) = 0$ whenever $m(A) = 0$ or $m(B) = 0$.

5.2.2 Estimates for the Semigroup

The basic step in order to obtain Gaussian estimates for $T_t = e^{-At}$ consists in an estimate for the norm of the operator $f \to e^{-\psi} T_t(e^{\psi} f)$ on $L^2(X,m)$. This type of estimate is called (weighted) *integrated maximum principle.*

1.1 LEMMA. *Let $\psi \in \mathcal{F} \cap L^{\infty}(X,m)$ with $\|\Gamma_{\langle\psi\rangle}\| \leq \gamma^2$. Then for all $f \in L^2(X,m)$ and all $t \geq 0$*

$$\|e^{\psi} T_t f\| \leq e^{(\gamma^2/2 - \lambda)t} \|e^{\psi} f\|.$$

Proof. Write $u_t = T_t f$. Then

$$\begin{aligned} \|e^{\psi} u_t\|^2 - \|e^{\psi} u_s\|^2 &= 2\int_s^t (\frac{\partial}{\partial r} u_r, e^{2\psi} u_r) dr = -2\int_s^t \mathcal{E}(u_r, e^{2\psi} u_r) dr = \\ &= \int_s^t \Gamma_{\langle\psi\rangle}(e^{2\psi} u_r^2) - 2\int_s^t \mathcal{E}(e^{\psi} u_r, e^{\psi} u_r) dr \leq \\ &\leq (\gamma^2 - 2\lambda) \int_s^t \|e^{\psi} u_r\|^2 dr. \end{aligned}$$

From this inequality, the claim follows by Gronwall's Lemma. □

1.2 THEOREM. *For any measurable subsets A, $B \subset X$ and any $t > 0$*

$$(T_t 1_A, 1_B) \leq \sqrt{m(A)m(B)} \cdot \exp\left(-\frac{\rho_0^2(A,B)}{2t}\right) \cdot \exp(-\lambda t).$$

This type of estimate is called *integrated Gaussian estimate.* The idea of the proof is the so-called "Method of Davies".

Proof. Without restriction we may assume that A and B have finite measure. For $\gamma > 0$ and $\psi \in \mathcal{F} \cap L^\infty(X,m)$ with $\mu_{\langle\psi\rangle} \leq m$ the preceding integrated maximum principle yields

$$\begin{aligned}(T_t 1_A, 1_B) &= (e^{\gamma\psi} T_t 1_A, e^{-\gamma\psi} 1_B) \leq \\ &\leq e^{(\gamma^2/2-\lambda)t} \cdot \|e^{\gamma\psi} 1_A\| \cdot \|e^{-\gamma\psi} 1_B\| \leq \\ &\leq e^{(\gamma^2/2-\lambda)t} \cdot e^{-\gamma\psi_0(A,B)} \cdot \sqrt{m(A)} \cdot \sqrt{m(B)}.\end{aligned}$$

Taking the supremum over all such ψ and choosing $\gamma = \frac{\rho_0(A,B)}{t}$ yields the claim. □

Whenever X is a topological space, the quantity $\rho_0(A,B)$ in the previous integrated Gaussian estimate, could be replaced by the smaller quantity

$$\tilde{\rho}(A,B) = \sup\{\psi_0(A,B) : \ \psi \in \mathcal{F} \cap C_0(X), \ \|\Gamma_{\langle\psi\rangle}\| \leq 1\}.$$

Similarly, in the definition of $\rho_0(A,B)$ and $\tilde{\rho}(A,B)$ the quantity $\psi_0(A,B)$ can be replaced by $\psi(A,B) = \inf\{\psi(x) - \psi(y) : \ x \in A, y \in B\}$. These modifications lead to weaker results.

1.3 PROPOSITION. *Assume that the state space X is locally compact, that the Dirichlet form $(\mathcal{E}, \mathcal{F})$ is regular and that*

$$\rho(x,y) = \sup\{\psi(x) - \psi(y) : \ \psi \in \mathcal{F} \cap C_0(X), \ \mu_{\langle\psi\rangle} \leq m\}$$

defines a complete metric on X which is compatible with the original topology. Then in the previous integrated Gaussian estimate, the quantity $\rho_0(A,B)$ could be replaced by

$$\rho(A,B) := \operatorname{ess\,inf}\{\rho(x,y) : \ x \in A, y \in B\}.$$

Proof. There exists measurable sets A_0 and B_0 which differ from A or B, resp., at most by m-zero sets such that

$$\rho(A,B) = \rho(A_0,B_0) = \inf\{\rho(x,y) : \ x \in A_0, y \in B_0\}$$

(*not* ess inf $\{\dots\}$!). Moreover, A_0, B_0 can be chosen in such a way that

$$\rho_0(A_0, B_0) = \sup\{\psi(A_0, B_0) : \ \psi \in \mathcal{F} \cap L^\infty(X, m), \ \|\Gamma_{\langle\psi\rangle}\| \leq 1\}$$

where $\psi(A_0, B_0) = \inf\{\psi(x) - \psi(y) : \ x \in A_0, y \in B_0\}$ (not ess inf $\{\dots\}$!). Of course, $(T_t 1_A, 1_B) = (T_t 1_{A_0}, 1_{B_0})$. Hence, without restriction $A = A_0$ and $B = B_0$. Then one easily sees that always $\rho(A, B) \geq \rho_0(A, B)$.

Under the given assumptions and if the sets $A, B \subset X$ are compact, the function

$$\psi(.) = (\rho(A, B) - \rho(A, .))_+$$

lies in $\mathcal{F} \cap \mathsf{C}_0(X)$, [S2]. Hence, $\rho_0(A, B) \geq \rho(A, B)$. That is, $\rho_0(A, B) = \rho(A, B)$ for all compact sets $A, B \subset X$. The rest follows by appoximation. □

REMARKS. (i) *Heat Kernel.* If there exists a fundamental solution $p_t(x, y)$ for $A + \frac{\partial}{\partial t}$ then for any $t > 0$ and any measurable subsets A, B of X

$$\int_A \int_B p_t(x, y)\, m(dx) m(dy) \leq \sqrt{m(A) m(B)} \cdot \exp\left(-\frac{\rho_0^2(A, B)}{2t}\right) \cdot \exp(-\lambda t).$$

The existence of a fundamental solution is a rather weak assumption. It is equivalent to the *absolute continuity hypothesis* in [FOT]. See also [F3] for sufficient conditions.

(ii) The *integrated* Gaussian estimate can be combined with subsolution estimates of the form

$$p_t(x, y) \leq C \cdot \frac{1}{m(A)} \cdot \frac{1}{m(B)} \cdot (T_t 1_A, 1_B) \tag{$*$}$$

in order to obtain *pointwise* Gaussian estimates of the form

$$p_t(x, y) \leq C' \frac{1}{\sqrt{m(A) m(B)}} \exp\left(-\frac{\rho^2(x, y)}{2t}\right) \cdot \exp(-\lambda t)$$

where typically $A = B_{\sqrt{t}}(x)$ and $B = B_{\sqrt{t}}(y)$. Subsolution estimates of the type $(*)$ hold true if $(\mathcal{E}, \mathcal{F})$ satisfies the doubling condition and a scale invariant Poincare (or Sobolev inequality) inequality. See [S2] for details.

(iii) For various former versions of Lemma 1.1 and Theorem 1.2 we refer to [Ga], [D], [CKS], [G3] and [S2].

5.2.3 Estimates for the Resolvent

Using Laplace transformation, we always get upper estimates for the resolvent involving the quantity ρ_0 from the previous section.

1.4 THEOREM. *For any measurable subsets A, $B \subset X$ and any $\alpha > -\lambda$*

$$(G_\alpha 1_A, 1_B) \le \sqrt{m(A)m(B)} \cdot \frac{1 + \sqrt{2(\alpha+\lambda)}\rho_0(A,B)}{\alpha+\lambda} \cdot \exp\left(-\sqrt{2(\alpha+\lambda)}\rho_0(A,B)\right).$$

Proof. The estimate follows immediately from the integrated Gaussian estimate since

$$\int_0^\infty \exp\left(-\frac{r^2}{2t} - \beta t\right) dt = \int_0^{2\frac{r}{\sqrt{2\beta}}} \exp\left(-\frac{r^2}{2t} - \beta t\right) dt + \int_{2\frac{r}{\sqrt{2\beta}}}^\infty \exp\left(-\frac{r^2}{2t} - \beta t\right) dt$$
$$\le 2\frac{r}{\sqrt{2\beta}} \cdot \exp\left(-\sqrt{2\beta} r\right) + \frac{1}{\beta} \cdot \exp\left(-\sqrt{2\beta} r\right)$$

for all $r, \beta > 0$. □

REMARK. The exponential decay is of the correct order. For instance, if $(\mathcal{E}, \mathcal{F})$ is the classical Dirichlet form $(\mathbf{D}, H^1(\mathbb{R}))$ on $\mathbb{R}$ then

$$G_\alpha(x,y) = \frac{1}{\sqrt{2\alpha}} \cdot \exp\left(-\sqrt{2\alpha} \cdot |x-y|\right).$$

For the classical Dirichlet form $(\mathbf{D}, H^1(\mathbb{R}))$ on $\mathbb{R}^N$, $N \ge 1$, we still have

$$G_\alpha(x,y) \sim \exp\left(-\sqrt{2\alpha} \cdot |x-y|\right)$$

provided $|x-y|$ is large.

5.3 Capacity Estimates and Applications

⟶ Throughout this chapter, let X be a Lusin topological space and m a σ-finite measure with full support on X. Moreover, let $(\mathcal{E}, \mathcal{F})$ be a quasi-regular, strongly local Dirichlet form.

5.3.1 Basic Properties of Quasi-regular Dirichlet Forms

QUASI-REGULAR DIRICHLET FORMS. We say that an increasing sequence $(K_n)_n$ of closed sets $K_n \subset X$ is a *nest* iff $\bigcup_n \mathcal{F}_n$ is dense in $\mathcal{F}$ (w.r.t. the norm $\sqrt{\mathcal{E}(.) + \|.\|^2}$) where

$$\mathcal{F}_n = \{u \in \mathcal{F} : \; u = 0 \text{ on } X \setminus K_n\}.$$

A subset $N \subset X$ is called *m-polar* or *exceptional* iff $N \subset \bigcap_n (X \setminus K_n)$ for some nest $(K_n)_n$. A statement depending on $x \in X$ which holds for all $x \in A \setminus N$ with an exceptional set N is said to hold *quasi everywhere* on the set $A \subset X$ (q.e. on A in abbreviation). Note that each exceptional set has measure zero. The aim is to specify assertions (e.g. on pointwise values of function $u \in \mathcal{F}$) not only up to a set of measure zero but up to an exceptional set.

A function $u : X \to [-\infty, \infty]$ is called *quasi-continuous* iff there exists a nest $(K_n)_n$ such that $u|_{K_n}$ is continuous (as a map into $[-\infty, \infty]$). This definition extends without changes to functions which are defined only q.e. on X. Note that our definition slightly differs from the usual one which implies that any quasi-continuous function is necessarily finite q.e. on X. Given a function u another function v is called *quasi-continuous modification of* u iff v is a quasi-continuous function with $v = u$ m-a.e. on X. Similarly, we can define the notions of *quasi-open* or *quasi-closed* sets $A \subset X$ as well as the notion of *quasi-connectedness* of X.

The Dirichlet form $(\mathcal{E}, \mathcal{F})$ is called *quasi-regular* iff

- there exists a nest $(K_n)_n$ consisting of compact sets $K_n \subset X$;
- for every $u \in \mathcal{F}$ there exists a quasi-continuous modification $\tilde{u} \in \mathcal{F}$;
- there exists a sequence $(u_n)_n \subset \mathcal{F}$ and an exceptional set $N \subset X$ such that $\{\tilde{u}_n\}_n$ separates the point of $X \setminus N$.

Note that two quasi-continuous functions which are equal (or $\leq$) m-a.e. on a quasi-open set $U \subset X$ are automatically equal (or $\leq$) q.e. on this set U. In particular, the function $\tilde{u}$ is determined q.e. on X. In other words, the pointwise value $\tilde{u}(x)$ is defined for q.e. $x \in X$ (whereas the pointwise value $u(x)$ of an arbitrary representative of u is defined only for m-a.e. $x \in X$). In the sequel, we freely use the fact that for any quasi-regular Dirichlet form $(\mathcal{E}, \mathcal{F})$ on a Lusin space X can be transferred into (or represented by) a regular Dirichlet form $(\mathcal{E}', \mathcal{F}')$ on a locally compact space X' in such a way that all relevant measure theoretic and potential theoretic properties (like quasi-continuity, measurability etc.) remain unchanged, – but of course not the topological properties (like local compactness, continuity etc.). See [MR] and [FOT]. Hence, all results for regular Dirichlet forms are at our disposal.

CAPACITIES. Associated with any quasi-regular Dirichlet form $(\mathcal{E}, \mathcal{F})$ and any $\alpha > 0$ there is the *α-capacity* $\mathrm{Cap}_\alpha(.)$ defined for *arbitrary* $A \subset X$ as follows

$$\mathrm{Cap}_\alpha(A) = \inf\{\mathcal{E}(u) + \alpha\|u\|^2 : u \in \mathcal{F}, \tilde{u} \geq 1 \text{ q.e. on } A\}$$

(with $\inf \emptyset := \infty$) where $\tilde{u}$ denotes a quasi-continuous modification of u. A set $A \subset X$ is exceptional if and only if $\mathrm{Cap}_\alpha(A) = 0$ for some (hence all) $\alpha > 0$.

An exceptional but important role is played by the 0-order capacitiy. It is defined by

$$\mathrm{Cap}_0(A) = \inf\{\mathcal{E}(u) : u \in \mathcal{F}_e, \tilde{u} \geq 1 \text{ q.e. on } A\}$$

for any set $F \subset X$. The space $\mathcal{F}_e$ consists of those m-measurable functions $u : X \to \mathbb{R}$ for which there exists an $\mathcal{E}$-Cauchy sequence $\{u_n\}_n \subset \mathcal{F}$ such that $u_n \to u$ m-a.e. on X. Of course, $\mathcal{F} \subset \mathcal{F}_e$ and $\mathcal{F}_e \cap L^2(X, m) = \mathcal{F}$.

REMARK. If $(\mathcal{E}, \mathcal{F})$ is *regular*, then for any compact set $K \subset X$

$$\mathrm{Cap}_\alpha(K) = \inf\{\mathcal{E}(u) + \alpha\|u\|^2 : u \in \mathcal{F} \cap \mathsf{C}_0(X), u \geq 1_K\}$$

and

$$\mathrm{Cap}_0(K) = \inf\{\mathcal{E}(u) : u \in \mathcal{F}_e \cap \mathsf{C}_0(X), u \geq 1_K\}.$$

ENERGY MEASURES. For every $u \in \mathcal{F} \cap L^\infty(X, m)$ a positive finite measure $\mu_{\langle u \rangle}$ on X is uniquely defined by the formula

$$\int_X \varphi d\mu_{\langle u \rangle} = 2\mathcal{E}(u, \varphi u) - \mathcal{E}(u^2, \varphi)$$

for all bounded Borel functions $\varphi \in \mathcal{F}$. Of course, with the square field operator from Chapter 1

$$\int_X \varphi^2 d\mu_{\langle u \rangle} = \Gamma(\varphi^2)$$

for all $\varphi, u \in \mathcal{F} \cap L^\infty(X, m)$. For $u \in \mathcal{F}$ we define $\mu_{\langle u \rangle} = \sup_{n \in N} \mu_{\langle u_n \rangle}$ where $u_n = (u \wedge n) \vee (-n)$. Notice that the sequence $\{u_n\}_n$ is increasing. The measure $\mu_{\langle u \rangle}$ is called the *energy measure* of u.

The map $u \mapsto \mu_{\langle u \rangle}$ is a quadratic form on $\mathcal{F}$ with values in the positive finite measures on X. By polarization one obtains a symmetric bilinear form $(u, v) \mapsto \mu_{\langle u,v \rangle} := \frac{1}{4}(\mu_{\langle u+v \rangle} - \mu_{\langle u-v \rangle})$ on $\mathcal{F}$ with values in the signed finite measures on X.

EXHAUSTION FUNCTIONS AND ENERGY GROWTH. In order to estimate the 0-order capacity, one should look for test functions or exhaustion functions ψ which are in some sense "locally" in the extended Dirichlet space $\mathcal{F}_e$. The main point in the following definition is that we entirely avoid any compactness condition. In particular, we do not introduce the set of functions which are locally in $\mathcal{F}_e$. Our choice of exhaustion functions applies to quasi-regular Dirichlet forms and is independent of the regular representation which might be choosen for the Dirichlet space $(\mathcal{E}, \mathcal{F})$.

Throughout this chapter, let us fix numbers a, b with $-\infty < a < b \leq \infty$. Typically, $a = 0$ and $b = \infty$.

2.1 DEFINITION. We say that a function $\psi : X \to [-\infty, \infty]$ is an $[a, b)$-*exhaustion function* iff ψ is quasi-continuous and $(r - (\psi \vee a))_+ \in \mathcal{F}_e$ for all $r < b$.

For such an exhaustion function ψ we define the *energy growth function* $w_\psi : [a, b) \to I\!R$ by

$$w_\psi(r) = 2\mathcal{E}(\psi_r)$$

where $\psi_r = (r - (\psi \vee a))_+$. Denote by $\mu_{\langle\psi\rangle}$ the measure on $\{a < \psi < b\}$ which restricted to the set $\{a < \psi < r\}$ equals the energy measure $\mu_{\langle\psi_r\rangle}$ of ψ_r (for each $r < b$). It coincides on $\{a < \psi < r\}$ with the usual energy measure of ψ provided the latter exists. Then

$$w_\psi(r) = \int_{\{a<\psi<r\}} d\mu_{\langle\psi\rangle}.$$

Obviously, every $[a, b)$-exhaustion function is an $[a', b')$-exhaustion function for every $a \leq a' < b' \leq b$. The function ψ is an $[a, b)$-exhaustion function if and only if $(\psi \vee a) \wedge b$ is such a function. Therefore, without restriction we may assume $a \leq \psi \leq b$ on X.

Intuitively, an $[a, b)$-exhaustion function is a function $\psi : X \to [a, b]$ which "exhausts" the space X such that $\psi < b$ on X and $\psi(x) \to b$ as $x \to \infty$ in a certain (very weak) sense.

2.2 EXAMPLES. (i) Every quasi-continuous function $\psi \in \mathcal{F}_{loc}$ is an $[0, \infty)$-exhaustion function provided the set $\{\psi < r\}$ is relatively compact for each $r > 0$. (Namely, in this case for any $r > 0$ the function $(r - \psi_+)_+$ is in $\mathcal{F}_{loc}$ and has compact support, hence, it actually is in $\mathcal{F}$.)

(ii) Let $(\mathcal{E}, \mathcal{F})$ be a regular Dirichlet form on a locally compact state space X and assume that the intrinsic metric ρ is complete and compatible with the original topology. Then for each non-empty set $A \subset X$ the distance function $\psi = \rho(., A)$ is an $[0, \infty)$-exhaustion function. (According to [S2], it satisfies the assumptions of the previous example.)

(iii) Let $g : [-\infty, \infty] \to [-\infty, \infty]$ be any smooth, decreasing function with $g(0) \geq b$. Then for any nonnegative $u \in \mathcal{F}_e$ the function $\psi = g \circ \tilde{u}$ is an $[a, b)$-exhaustion function. For instance, if $b < \infty$ we can put $\psi = b - \tilde{u}$ and if $b = \infty$ $\psi = \frac{1}{\tilde{u}}$.

(iv) The function $\psi \equiv b$ is always an $[a, b)$-exhaustion function.

2.3 LEMMA. *The function w_ψ is absolutely continuous on $[a, b)$.*

Proof. Fix $R < b$. On $[a, R)$ the Stieltjes measure $w_\psi(dr)$ is the image measure $\psi_* \mu_{\langle\psi\rangle}$ of $\mu_{\langle\psi\rangle}$ (which coincides on $\{a < \psi < R\}$ with the energy measure $\mu_{\langle\psi_R\rangle}$). Here $\psi_R = (R - \psi \vee a)_+ \in \mathcal{F}_e$, hence $\in \mathcal{F}_e \cap L^2(X, \varphi m) = \mathcal{F}_\varphi$ for some $\varphi > 0$. Then the claim follows from the "energy image density property", see [BH], Theorem I.7.1.1. □

Let us denote the essential infimum or quasi-infimum $\tilde{\sup} u$ of a quasi-continuous funcion $u : X \to \overline{I\!R}$ as follows

$$\begin{aligned} \tilde{\inf} u &= \sup\{C \in \overline{I\!R} : u \geq C \text{ q.e. on } X\} \\ &= \sup\{C \in \overline{I\!R} : u \geq C \; m - \text{a.e. on } X\}. \end{aligned}$$

Analogously, we define the essential or quasi-supremum $\tilde{\sup} u$.

2.4 LEMMA. *Assume that X is quasi-connected (in other words, $\mathcal{E}$ is irreducible) or ψ is continuous and X is connected. Put $a' = a \vee \tilde{\inf} \psi$, $b' = b \wedge \tilde{\sup} \psi$. Then w_ψ is strictly increasing on $[a', b')$ (and of course constant on $[a, a') \cup [b', b)$).*

Proof. Assume that w_ψ is not strictly increasing. Then there exist numbers r, R with $\tilde{\inf} \psi \leq r < R < b \wedge \tilde{\sup} \psi$ such that

$$0 = w_\psi(R) - w_\psi(r) = \int_{\{r < \psi < R\}} d\mu_{\langle\psi\rangle}.$$

Hence, ψ is constant $= C$ q.e. on $\{r < \psi < R\}$. Therefore, the sets $\{\psi > C\}$ and $\{\psi \leq C\} = \{\psi < R\}$ are both quasi-open and non-polar. Therefore, X cannot be quasi-connected. (If ψ is continuous, these sets are both open and non-empty. Therefore, X cannot be connected.) □

VOLUME GROWTH. For an $[a, b)$-exhaustion function, let

$$v_\psi(r) = m(\{a < \psi < r\})$$

denote the *volume growth function*. In many applications one chooses exhaustion functions ψ with

$$\mu_{\langle\psi\rangle} \leq m.$$

In this case, it suggests itself to use the volume growth v_ψ instead of the energy growth w_ψ. In general, however, the volume growth function v_ψ will not be continuous, in particular, it will not be absolutely continuous.

EXAMPLES.

(i) Let $(\mathcal{E}, \mathcal{F})$ be the classical Dirichlet form on $L^2(\mathbb{R}^N, dx)$. Now replace the reference measure dx (=Lebesgue measure) by $m(dx) = dx + \sigma(dx)$ where σ denotes the normalized surface measure of the unit sphere $\partial B_1(0)$ (which is a smooth measure w.r.t. the classical Dirichlet form). Then with $\psi(x) = |x|$ we have $\mu_{\langle\psi\rangle} \leq m$ but

$$v_\psi(r) = \begin{cases} c_N \cdot r^N & \text{if } r \leq 1 \\ 1 + c_N \cdot r^N & \text{if } r > 1 \end{cases}$$

(ii) Let $v : \mathbb{R}_+ \to \mathbb{R}_+$ be an arbitrary left continuous, strictly increasing function. Let $X = \mathbb{R}^1$ and define $m(dx) = \frac{1}{2}v(dx)$ on $]0,\infty[$, $m(0) = 0$ and $m(dx) = v(-dx)$ on $]-\infty, 0[$. Then the form $(\mathcal{E}, C_0^\infty(\mathbb{R}))$ with

$$\mathcal{E}(u,u) = \frac{1}{2}\int |\nabla u|^2\, dm$$

is closable in $L^2(\mathbb{R}, m)$ and its closure $(\mathcal{E}, \mathcal{F})$ is a regular, strongly local Dirichlet form. If we choose $\psi(x) = |x|$ then $\mu_{\langle\psi\rangle} \leq m$ and the volume growth function v_ψ is just the given function v, i.e.

$$m(\{\psi < r\}) = v(r).$$

5.3.2 The Capacity Estimate

2.5 THEOREM. *(i) For all sets $A \subset X$ and all $[a,b)$-exhaustion functions ψ with $\psi \leq a$ q.e. on A:*

$$Cap_0(A) \leq \frac{1}{2}\left(\int_a^b \frac{dr}{w'_\psi(r)}\right)^{-1}.$$

(ii) Conversely, for any set $A \subset X$ there exits an $[a,b)$-exhaustion functions ψ with $\psi = a$ q.e. on A and

$$Cap_0(A) = \frac{1}{2}\left(\int_a^b \frac{dr}{w'_\psi(r)}\right)^{-1}.$$

REMARKS.

(i) *Exceptional Sets.* In the above Theorem, the condition "$\psi = a$ q.e. on A" can always be replaced by "$\psi = a$ on A" (just replacing ψ by another suitable quasi-continuous modification). If A is quasi-open, it also can be replaced by "$\psi = a$ m-a.e. on A" (since the latter implies that the quasi-open set $A \cap \{\psi > a\}$ has m-measure zero, hence, it is m-exceptional).

(ii) *Transformation Invariance.* If $f : \mathbb{R} \to \mathbb{R}$ is increasing and smooth then ψ is an $[a, b)$-exhaustion function if and only if $f \circ \psi$ is an $[f(a), f(b))$-exhaustion function and

$$\int_a^b \frac{dr}{w'_\psi(r)} = \int_{f(a)}^{f(b)} \frac{dr}{w'_{f\circ\psi}(r)}.$$

An anlogous result holds for decreasing f.

Proof. (i) Fix $\varepsilon > 0$, $R < b$ and put

$$C = \int_a^R \frac{ds}{w'_\psi(s) + \varepsilon}, \quad \Phi(t) = \frac{1}{C} \int_{R-t}^R \frac{ds}{w'_\psi(s) + \varepsilon}.$$

Note that $0 < C < \infty$ since $C \leq (R-a)/\varepsilon$ and $C \geq (R-a)^2/(w_\psi(R) + \varepsilon(R-a))$. Moreover, put

$$u = \Phi \circ \psi_R$$

where $\psi_R = (R - \psi \vee a)_+$. Then $u \in \mathcal{F}_e$ (since Φ is Lipschitz, $\Phi(0) = 0$ and $\psi_R \in \mathcal{F}_e$) and $u \geq 1_A$ q.e. (since $\psi_R = R - a$ q.e. on A and $\Phi(R-a) = 1$). Hence,

$$\begin{aligned} \mathrm{Cap}_0(A) &\leq \mathcal{E}(u) = \frac{1}{2} \int \Phi'(\psi_R)^2 d\mu_{\langle \psi_R \rangle} \\ &= \frac{1}{2} \int_a^R \Phi'(R-s)^2 w'_\psi(s) ds = \frac{1}{2C^2} \int_a^R \frac{w'(s) ds}{(w'(s) + \varepsilon)^2} \\ &\leq \frac{1}{2} \left(\int_a^R \frac{ds}{w'(s) + \varepsilon} \right)^{-1} \longrightarrow \frac{1}{2} \left(\int_a^b \frac{ds}{w'(s)} \right)^{-1} \end{aligned}$$

as $\varepsilon \to 0$ and $R \to b$.

(ii) Without restriction we may assume that $\mathcal{E}$ is transient. Otherwise, we use the fact that

$$\mathrm{Cap}_0(A) = \mathrm{Cap}_0(A \cap X_d)$$

is the 0-order capacity of A in the transient (quasi-regular) Dirichlet space $(\mathcal{E}, \mathcal{F}^d)$ on the quasi-open set X_d ("dissipative part" of X) where

$$\mathcal{F}^d = \{u \in \mathcal{F} : \tilde{u} = 0 \text{ q.e. on } X \setminus X_d\}$$

and $X_d = \{x \in X : \tilde{G}f(x) < \infty\}$ for some fixed $f > 0$, $f \in L^1$.

Given an $[a, b)$-exhaustion function ψ w.r.t. the Dirichlet space $(\mathcal{E}, \mathcal{F}^d)$ and any number $c \in I\!R$ we obtain a new $[a, b)$-exhaustion function ψ_c w.r.t. the Dirichlet space $(\mathcal{E}, \mathcal{F})$ by

$$\psi_c = \psi 1_{X_d} + c 1_{X \setminus X_d}.$$

Namely,

$$(r - \psi_c \vee a)_+ = (r - \psi)_+ \cdot 1_{X_d} + (r - c \vee a)_+ \cdot 1_{X \setminus X_d} \in \mathcal{F}_e$$

since $u \in (\mathcal{F}^d)_e$ implies $u \cdot 1_{X_d} \in (\mathcal{F}^d)_e \subset \mathcal{F}_e$ and since $1_{X \setminus X_d} \in \mathcal{F}_e$, see [FOT], Corollary 1.6.2.

For instance, if $\mathcal{E}$ is recurrent we can choose $\psi \equiv c$ (with any $c \in I\!R$) as an $[a, b)$-exhaustion function.

Now we assume that $\mathcal{E}$ is transient. Then for $A \subset X$ there exists the equilibrium potential u_A which we choose to be quasi-continuous. Let $g : [0, 1] \to [a, b]$ be any smooth, decreasing function with $g(1) = a$ and $g(0) = b$ and put $\psi = g \circ u_A$. Then $\psi = a$ q.e. on A, $a \le \psi \le b$ q.e. on X and $(r - \psi)_+ \in \mathcal{F}_e$ for all $r < b$ since $u_A \in \mathcal{F}_e$. Therefore,

$$\begin{aligned} \mathrm{Cap}_0(A) &= \mathcal{E}(u_A) \\ &\ge \inf\left\{\mathcal{E}(f \circ \psi) : \quad f \in \mathcal{C}_0^{Lip}([a, b)), f(a) \ge 1\right\} \\ &= \inf\left\{\frac{1}{2}\int_a^b f'(r)^2 w'_\psi(r)dr : \quad f \in \mathcal{C}_0^{Lip}([a, b)), f(a) \ge 1\right\} \\ &= \frac{1}{2}\left(\int_a^b \frac{dr}{w'_\psi(r)}\right)^{-1} \end{aligned}$$

□

2.6 Corollary. *For all sets $A \subset X$ and all $[a, b)$-exhaustion functions ψ with $\psi \le a$ q.e. on A:*

$$Cap_0(A) \le \left(\int_a^b \frac{r - a}{w_\psi(r)} dr\right)^{-1}$$

and for all $a < r < R < b$

$$Cap_0(A) \le \frac{w_\psi(R) - w_\psi(r)}{2(R - r)^2}.$$

In particular, $Cap_0(A) = 0$ whenever $a = 0$ and

$$w_\psi(r) \leq Cr^2 \qquad \text{for } r \to 0$$

or $b = \infty$ and

$$w_\psi(r) \leq Cr^2 \qquad \text{for } r \to \infty.$$

Proof. The first claim follows from Theorem 2.6 and the following inequality

$$\frac{1}{2}\left(\int_a^b \frac{dr}{w'(r)}\right)^{-1} \leq \left(\int_a^b \frac{r-a}{w(r)}dr\right)^{-1}.$$

The latter is a straightforward consequence of Cauchy-Schwarz inequality and integration by parts:

$$\begin{aligned}\left(\int_a^R \frac{r-a}{w(r)+\varepsilon}dr\right)^2 &\leq \left(\int_a^R \frac{(r-a)^2}{(w(r)+\varepsilon)^2}dr\right)\cdot\left(\int_a^R \frac{1}{w'(r)}dr\right)\\ &\leq 2\left(\int_a^R \frac{r-a}{w(r)+\varepsilon}dr\right)\cdot\left(\int_a^R \frac{1}{w'(r)}dr\right).\end{aligned}$$

Hence,

$$\int_a^R \frac{r-a}{w(r)+\varepsilon}dr \leq 2\int_a^b \frac{1}{w'(r)}dr$$

for all $R < b$ and $\varepsilon > 0$ which in the limit $R \to b$, $\varepsilon \to 0$ yields the claim.

For the second claim, just note that

$$\int_r^R \frac{dt}{w'(t)} \geq \frac{(R-r)^2}{w(R)-w(r)}$$

since

$$(R-r)^2 \leq \left(\int_r^R w'(t)dt\right)\cdot\left(\int_r^R \frac{dt}{w'(t)}\right).$$

□

Now let us look for estimates in terms of the volume growth (instead of the energy growth). Given an $[a,b)$-exhaustion function, recall that $v(r) = m(\{a < \psi < r\})$ denotes the volume growth and recall that, in general, v is not absolutely continuous (not even continuous)! Let $v'(r)$ denote the density of the absolutely continuous part $v^0(dr)$ of $v(dr)$.

2.7 COROLLARY. *Let $A \subset X$ be given and assume that ψ is an $[a,b)$-exhaustion function with $\psi \leq a$ q.e. on A and $\mu_{\langle\psi\rangle} \leq m$ on $\{a < \psi < b\}$. Then*

$$Cap_0(A) \leq \frac{1}{2}\left(\int_a^b \frac{dr}{v'(r)}\right)^{-1} \leq \left(\int_a^b \frac{r-a}{v^0(r)}dr\right)^{-1} \leq \left(\int_a^b \frac{r-a}{v(r)}dr\right)^{-1}.$$

Proof. Since by assumption $\mu_{\langle\psi\rangle} \leq m$ on $\{a < \psi < b\}$, we obtain

$$w(dr) = (\psi_* \mu_{\langle\psi\rangle})(dr) \leq (\psi_* m)(dr) = v(dr)$$

and thus

$$w(dr) = w_0(dr) \leq v_0(dr) \leq v(dr).$$

This implies $w'(r) \leq v'(r)$. Hence, the claim follows from the previous Theorem 2.5. □

REMARKS. (i) *Operators on $I\!R^N$ and on Riemannian Manifolds.* For operators on $I\!R^n$, compact $A \subset I\!R^N$ and continuous ψ the capacity estimate of Theorem 2.5 appears already in Maz'ya [M], and for Laplace-Beltrami operators on Riemannian manifolds a similar estimate in Grigor'yan [G]. However, part (ii) of Theorem 2.5, namely, the fact that one always can achieve equality, seems to be entirely new even in this case.

(ii) *Hellinger Integral.* In the framework of strongly local Dirichlet spaces, capacity estimates similar to those of Corollary 2.7 have been derived by Sturm [S4], Okura [Ok] and Kaneko [Kn]. In these works, the capacity was always estimated in terms of the volume growth v, however, not using the density v' of the absolutely continuous part but the increase of the function v itself. This was achieved by means of the so-called Hellinger integral. These estimates are weaker than the estimates in terms of v'.

5.3.3 Criteria for Polarity

Recall that a set $A \subset X$ is *m-polar* (or *exceptional*) for $(\mathcal{E}, \mathcal{F})$ if and only if $\mathrm{Cap}_1(A) = 0$. Moreover, recall that $\mathrm{Cap}_1(A) = 0$ implies $m(A) = 0$. The converse is true if A is quasi-open but of course not in general.

2.8 THEOREM. *A set $A \subset X$ is m-polar if and only if there exists an $[a,b)$-exhaustion function ψ with $\psi \leq a$ q.e. on A and $\psi > a$ m-a.e. on X such that*

$$\int_0^\varepsilon \frac{dr}{w'_\psi(a+r)} = \infty \qquad \textit{for all (suff. small) } \varepsilon > 0. \qquad (*)$$

Condition $(*)$ *is satisfied whenever*

$$\int_0^\varepsilon \frac{r}{w_\psi(a+r)} dr = \infty \qquad \textit{for all (suff. small) } \varepsilon > 0$$

or

$$w_\psi(a+\varepsilon) \le C\varepsilon^2 \qquad \textit{for } \varepsilon \to 0.$$

Proof. If A is m-polar we can choose $\psi = \infty$. This proves the "only if" part for all $\varepsilon \in (0, b-a]$.

Now assume $(*)$. According to Theorem 2.5, it implies that A has zero 0-order capacity w.r.t. all the Dirichlet spaces $(\mathcal{E}^\varepsilon, \mathcal{F}^\varepsilon)$ obtained from $(\mathcal{E}, \mathcal{F})$ by restricting its domain to

$$\mathcal{F}^\varepsilon = \{u \in \mathcal{F} : \tilde{u} = 0 \text{ on } \{\psi \ge a+\varepsilon\}\}.$$

In other words, it implies that for every $\varepsilon > 0$ and every $\delta > 0$ there exists $u \in \mathcal{F}$ with $1_A \le \tilde{u} \le 1_{\{\psi < a+\varepsilon\}}$ q.e. and $\mathcal{E}(u) \le \delta$. Hence,

$$\mathrm{Cap}_1(A) \le \mathcal{E}_1(u) = \mathcal{E}(u) + \|u\|_2^2 \le \delta + m(\{\psi < a+\varepsilon\}).$$

Since $\psi > a$ m-a.e. on X one can choose $\varepsilon > 0$ such that $m(\{\psi < a+\varepsilon\}) \le \delta$ which implies $\mathrm{Cap}_1(A) \le 2\delta$. Since δ is arbitrary, this yields $\mathrm{Cap}_1(A) = 0$. □

REMARKS.

(i) *Exceptional Sets.* Again the condition "$\psi \le a$ q.e. on A" can be replaced by "$\psi \le a$ on A".

(ii) *Volume Growth Conditions.* If $\mu_{\langle\psi\rangle} \le m$ then in Theorem 2.8 the function $w_\psi(r)$ can be replaced by $v_\psi(r) = m(\{\psi < r\})$ (or its absolutely continuous part $v_\psi^0(r)$) and its derivative $w'_\psi(r)$ can be replaced by $v'_\psi(r)$, the derivative of $v_\psi^0(r)$.

(iii) *Irreducibility.* Assume that $\mathcal{E}$ is recurrent or transient (which is always the case if it is irreducible). Then the integral in $(*)$ is infinite for *all* $\varepsilon \in (0, b-a]$ provided the integral is infinite for *some* $\varepsilon \in (0, b-a]$.

Namely, if the integral in $(*)$ is infinite for some $\varepsilon_0 \in (0, b-a]$ then $\mathrm{Cap}_0(\{\psi < a+\varepsilon_0\}) = 0$ according to Theorem 2.5. If $\mathcal{E}$ is transient this implies $\mathrm{Cap}_1(\{\psi < a+\varepsilon_0\}) = 0$, i.e. A is m-polar. By the first part of the above proof, however, this implies that the integral in $(*)$ is infinite for *all* $\varepsilon \in (0, b-a]$.

On the other hand, if $\mathcal{E}$ is recurrent then $\mathrm{Cap}_0(A) = 0$ for any set $A \subset X$. Hence, by Theorem 2.5, the integral in $(*)$ is infinite for *all* $\varepsilon \in (0, b-a]$.

EXAMPLE: POLARITY OF POINTS. Let U be a domain in $\mathbb{R}^N$ and let $\varphi \in L^2_{loc}(U, dx)$. We will consider the Markovian form $(\mathcal{E}, C_0^\infty(U))$ with

$$\mathcal{E}(u,v) = \frac{1}{2}\int_U \nabla u \cdot \nabla v \cdot \varphi^2 \, dx$$

on the Hilbert space $L^2(U, \varphi^2 dx)$ with inner product

$$(u, v) = \int_U u \cdot v \cdot \varphi^2 \, dx.$$

We assume that this form is closable and denote its closure by $(\mathcal{E}, \mathcal{F})$. This is for instance the case if either $\varphi^{-1} \in L^2_{loc}(U, dx)$ or $\varphi \in H^1_{loc}(U)$ with $\varphi > 0$. See also Section 4.B.

The generator of this form is formally denoted by $-\frac{1}{2}\Delta - \nabla \log \varphi \nabla$. Note, however, that we do not require the existence of the gradient $\nabla \log \varphi$.

We want to discuss conditions which imply that the diffusion process associated with this generator will hit a given single point $z \in U$. Roughly speaken, the idea is that if $N \geq 2$ then the function φ must increase sufficiently fast at z in order to produce a large drift $-\nabla \log \varphi$ towards the point z.

PROPOSITION. *The point $z \in U$ is m-polar provided*

$$\int_0^\varepsilon \left(\int_{\partial B_r(z)} \varphi^2(x) \, \sigma_r(dx) \right)^{-1} dr = \infty \tag{1}$$

for all $\varepsilon > 0$ where $\sigma_r(dx)$ denotes the surface measure on the Euclidean sphere $\partial B_r(z)$.

REMARKS. (i) Condition (1) is satisfied whenever

$$\int_{B_r(z)} \varphi^2(x) \, dx \leq C \cdot r^2 \tag{2}$$

for $r \to 0$ and this in turn is satisfied whenever

$$\varphi \in L^p_{loc}(V, dx) \tag{3}$$

for some open set $V \ni z$ and with $p \geq \frac{2N}{N-2}$ if $N \geq 3$ and with $p = \infty$ if $N \leq 2$.

(ii) Finally, condition (3) is satisfied whenever

$$\varphi \in H^1_{loc}(V) \tag{4}$$

for some open set $V \ni z$.

(iii) Condition (1) is necessary and sufficient for m-polarity of z if φ is rotational invariant around the point z.

(iv) A necessary condition for m-polarity is that

$$\int_0^1 \varphi_r^{-2} \cdot r^{1-N} \, dr = \infty$$

where $\varphi_r = \sup\{\varphi(x) : x \in \partial B_r(z)\}$.

Proof. Choosing $\psi(x) = |x - z|$, most of the assertions follow immediately from Theorem 2.8 since

$$w_\psi(r) = \int_{B_r(z)} \varphi^2(x)dx$$

and thus

$$w'_\psi(r) = \int_{\partial B_r(z)} \varphi^2(x)\sigma_r(dx)$$

for a.e. $r > 0$ where $B_r(z)$ denotes the Euclidean ball of radius r around z.

For the remaining assertions, one just has to apply Hölder inequality and Sobolev embedding theorem. □

5.3.4 Criteria for Recurrence and Transience

DEFINITION. The Dirichlet form $(\mathcal{E}, \mathcal{F})$ is called *recurrent* if $Gf \in \{0, \infty\}$ m-a.e. on X for any nonnegative $f \in L^1(X, m)$. It is called *transient* if $Gf < \infty$ m-a.e. on X for any nonnegative $f \in L^1(X, m)$.

Here G denotes the potential kernel associated with $(\mathcal{E}, \mathcal{F})$, i.e. $Gf(x) = \int_0^\infty T_t f(x)\, dt$.

The connection to our previous results is established by the following characterization of recurrence/transience in terms of the 0-order capacity. It was proved in the framework of regular Dirichlet forms in Okura [Ok] but immediately extends to quasi-regular Dirichlet forms by the transfer principle and/or by regular representation.

2.9 PROPOSITION.

(i) *$(\mathcal{E}, \mathcal{F})$ is transient if and only if $Cap_0(F) > 0$ for any $F \subset X$ with $Cap_1(F) > 0$.*

(ii) *$(\mathcal{E}, \mathcal{F})$ is recurrent if and only if $Cap_0(F) = 0$ for any $F \subset X$.*

2.10 THEOREM. *The following assertions are equivalent:*

(i) *$\mathcal{E}$ is transient*

(ii) *For every $[a, b)$-exhaustion function ψ with $\widetilde{\inf}\, \psi < b$*

$$\int_R^b \frac{dr}{w'_\psi(r)} < \infty \qquad \textit{for some } R < b.$$

(ii) *For every $[a,b)$-exhaustion function ψ with* $\widetilde{\inf}\,\psi \leq a$

$$\int_a^b \frac{dr}{w'_\psi(r)} < \infty.$$

Proof. "(ii) $\Longrightarrow$ (i)" as well as "(iii) $\Longrightarrow$ (i)": Assume that $\mathcal{E}$ is not transient. Choose a version of X_d which is simultaneously quasi-open and quasi-closed and put $\psi = b1_{X_d} + a1_{X\setminus X_d}$. Then $(r-\psi)_+ = (r-a)_+ 1_{X\setminus X_d} \in \mathcal{F}_e$ and $w_\psi(r) = (r-a)_+^2 \mathcal{E}(1_{X\setminus X_d}) = 0$ for all $r \in [a,b]$. Hence, $w'_\psi \equiv 0$ on (a,b) and thus

$$\int_R^b \frac{dr}{w'_\psi(r)} = \infty \qquad \text{for all } R \in [a,b).$$

Moreover, if $\mathcal{E}$ is not transient then $m(X \setminus X_d) > 0$ ([FOT], Lemma 1.6.4), that is, $\widetilde{\inf}\,\psi = a < b$.

"(i) $\Longrightarrow$ (ii)": Assume that there exists an $[a,b)$-exhaustion function ψ with $\widetilde{\inf}\,\psi < b$ and

$$\int_R^b \frac{dr}{w'_\psi(r)} = \infty \qquad \text{for all } R < b.$$

According to Theorem 2.5 the latter implies $\mathrm{Cap}_0(\{\psi < R\}) = 0$ for all $R < b$. On the other hand, $\widetilde{\inf}\,\psi < b$ implies $m(\{\psi < R\}) > 0$ and thus $\mathrm{Cap}_1(\{\psi < R\}) = 0$ for some $R < b$. Hence, $\mathcal{E}$ is not transient.

"(i) and (ii) $\Longrightarrow$ (iii)": Let us assume that

$$\int_a^b \frac{dr}{w'_\psi(r)} = \infty$$

as well as

$$\int_R^b \frac{dr}{w'_\psi(r)} < \infty$$

for some $R \in (a,b)$.

Then Theorem 2.5 implies $\mathrm{Cap}_0(\{\psi < R\}) = 0$. If $\mathcal{E}$ would be transient this would imply $\mathrm{Cap}_1(\{\psi < R\}) = 0$ and thus $\widetilde{\inf}\,\psi \geq R$. This is a contradiction to $\widetilde{\inf}\,\psi \leq a$. Hence, $\mathcal{E}$ is not transient. □

REMARK. Let us restate the previous transience condition in order to see the analogy as well as the difference to the recurrence condition in Theorem 2.11 below. Note that recurrence implies non-transience. The converse is true if $\mathcal{E}$ is irreducible.

$\mathcal{E}$ is not transient if and only if there exists an $[a,b)$-exhaustion function ψ with $\tilde{\inf}\,\psi < b$ such that

$$\int_R^b \frac{dr}{w'_\psi(r)} = \infty \qquad \textit{for all } R < b. \tag{$*$}$$

2.11 THEOREM. *$\mathcal{E}$ is recurrent if and only if there exists an $[a,b)$-exhaustion function ψ with $\psi < b$ m-a.e. and*

$$\int_R^b \frac{dr}{w'_\psi(r)} = \infty \qquad \textit{for all } R < b. \tag{$*$}$$

Proof. Let us first prove the "if"-assertion. According to Theorem 2.5, the integral condition implies that $\text{Cap}_0(\{\psi < R\}) = 0$ for all $R < b$. Hence, $\text{Cap}_0(\{\psi < b\}) = 0$ and also $\text{Cap}_0(X') = 0$ for X' being the quasi-closure of $\{\psi < b\}$. But $\psi < b$ m-a.e. implies $m(X \setminus X') = 0$, hence, $\text{Cap}_1(X \setminus X') = 0$ (since $X \setminus X'$ is quasi-open). This finally yields $\text{Cap}_0(X) = 0$. That is, $\mathcal{E}$ is recurrent.

For the "only if"-assertion we proceed as in the proof of the "if"-assertion of Theorem 2.9. Defining as before $\psi = b1_{X_d} + a1_{X \setminus X_d}$ we conclude again that

$$\int_R^b \frac{dr}{w'_\psi(r)} = \infty \qquad \text{for all } R < b.$$

Now $\mathcal{E}$ is recurrent if and only if $m(X_d) = 0$ ([FOT], Lemma 1.6.4), that is, if and only if $\psi < b$ m-a.e. □

REMARKS.

(i) *Exceptional Sets.* The condition "$\psi < b$ m-a.e. on X" in Theorem 2.11, can always be replaced by "$\psi < b$ q.e. on X". (One only has to check the "only if"-part of the previous proof. If $\mathcal{E}$ is recurrent, then actually X_d is m-polar since it always can be chosen to be quasi-open.)

(ii) *Sufficient Conditions.* Condition $(*)$ is satisfied whenever

$$\int_R^b \frac{r}{w_\psi(r)} dr = \infty \qquad \text{for all } R < b \tag{$**$}$$

or if $b = \infty$ and

$$w_\psi(r) \leq Cr^2 \qquad \text{for } r \to \infty. \tag{$***$}$$

(iii) *Volume Growth Conditions.* If $\mu_{\langle\psi\rangle} \leq m$ then in Theorem 2.10 and in the previous Remark (ii), the function $w_\psi(r)$ can be replaced by $v_\psi(r) = m(\{\psi < r\})$ (or its absolutely continuous part $v_\psi^0(r)$) and its derivative $w'_\psi(r)$ can be replaced by $v'_\psi(r)$, the derivative of $v_\psi^0(r)$.

(iv) *Operators on $\mathbb{R}^N$ and on Riemannian Manifolds.* Up to now, necessary *and* sufficient conditons for recurrence were known only for Laplace-Beltrami operators on model spaces (i.e. rotational invariant spaces). There are many contributions which yield sufficient conditions for recurrence similar to our conditions. In all these conditions, the function ψ was chosen to be a (local) Lipschitz function, mostly, the distance function $d(.,x_0)$ from some fixed point $x_0 \in X$. See for instance, Ahlfors [Ah], Nevanlinna [Nc], Cheng and Yau [CY], Grigor'yan [G], Lyons and Sullivan [LS], Karp [K1] and Varopoulos [V].

For uniformly elliptic operators in divergence form on $\mathbb{R}^N$, the sufficiency of condition $(*)$ was proved by Ichihara [I].

(v) *Dirichlet Operators.* In the framework of strongly local Dirichlet spaces, the sufficiency of a condition (similar to) $(**)$ was proved by [S1]. The sufficiency of a condition (similar to) $(*)$ was proved by Sturm [S4], Okura [Ok], Kaneko [Kn]. Here again the Hellinger integral was used in order to handle the fact that the volume growth v is in general not absolutely continuous. These conditions are weaker than our new condition in terms of v'.

5.4 The Intrinsic Metric

We summarize some of the basic properties of the intrinsic metric and some results on the longtime behaviour of the semigroups (or stochastic processes) in terms of the intrinsic metric. For the proofs of these results and for other details we refer to [S1] and [S4].

$\longrightarrow$ Throughout this chapter we assume that $(\mathcal{E},\mathcal{F})$ is a strongly local, regular Dirichlet form on a locally compact separable state space X.

5.4.1 Basic Properties of the Intrinsic Metric

Recall that a Dirichlet form $(\mathcal{E},\mathcal{F})$ is called *regular* iff $\mathcal{F} \cap \mathsf{C}_0(X)$ is dense in $\mathsf{C}_0(X)$ (w.r.t. the uniform norm) and dense in $\mathcal{F}$ (w.r.t. the Hilbert norm $\sqrt{\mathcal{E}(.,.)+\|.\|^2}$). Here and in the sequel, for any open subset $U \subset X$ the set of continuous functions on U is denoted by $\mathsf{C}(U)$ and the subset of functions in $\mathsf{C}(U)$ with compact support in U is denoted by $\mathsf{C}_0(U)$.

For any open set $U \subset X$ let $\mathcal{F}_{loc}(U)$ denote the set of all m-measurable functions u on U for which for every relatively compact open set $V \subset \overline{V} \subset U$ there exists a function $u' \in \mathcal{F}$ with $u = u'$ m-a.e. on V.

The locality of $(\mathcal{E}, \mathcal{F})$ allows to extend the definition of the energy measure to $\mathcal{F}_{loc}(U)$. For $u, v \in \mathcal{F}_{loc}(U)$ the signed Radon measure $\mu_{\langle u,v\rangle}$ on U will be defined via its restriction to relatively compact open sets $V \subset \overline{V} \subset U$ by $1_V \, d\mu_{\langle u,v\rangle} = 1_V \, d\mu_{\langle u',v'\rangle}$ where u' and v' are suitably chosen functions in $\mathcal{F}$ which coincide on V with u and v, respectively.

The energy measure μ defines in an intrinsic way a pseudo metric ρ on X by

$$\rho(x,y) := \sup\left\{u(x) - u(y) : \; u \in \mathcal{F}_{loc}(X) \cap \mathsf{C}(X), \; \mu_{\langle u\rangle} \leq m \text{ on } X\right\},$$

called *intrinsic metric* or *Carathéodory metric*. Here the condition $\mu_{\langle u\rangle} \leq m$ of course means that the energy measure $\mu_{\langle u\rangle}$ is absolutely continuous w.r.t. the reference measure m with Radon-Nikodym derivative $\frac{d}{dm}\mu_{\langle u\rangle} \leq 1$ m-a.e. on X. The density $\frac{d}{dm}\mu_{\langle u\rangle}$ is the "carré du champ" of u and should be interpreted as the square of the (length of the) gradient of u.

ρ is always symmetric, positive and satisfies the triangle inequality. In general, however, ρ may be degenerate (i.e. $\rho(x,y) = \infty$ or $\rho(x,y) = 0$ for some $x \neq y$).

ASSUMPTION (A). The topology induced by ρ coincides with the original one.

For any $r > 0$ and $x \in X$ we denote by $B_r(x) = \{y \in X : \rho(x,y) < r\}$ the open ball of radius r with respect to the pseudo metric ρ. Assumption (A) means that the system $\{B_r(x) : r > 0, x \in X\}$ is a base of the (original) topology of X. Recall that X is assumed to be a locally compact separable Hausdorff space. Hence, for any $y \in X$ there exists $R > 0$ such that for all $r \leq R$ the balls $B_r(y)$ are relatively compact.

Moreover, Assumption (A) implies that $B_r(x)$ is connected and $\overline{B_r}(x) = \{y \in X : \rho(x,y) \leq r\}$. In other words, the latter states that the boundary of the open ball $B_r(x) = \{y : \rho(x,y) < r\}$ is given by the sphere $S_r(x) = \{y : \rho(x,y) = r\}$ (which in general metric spaces might be a much bigger set than the boundary of $B_r(x)$).

3.1 THEOREM. *Under Assumption (A) the following assertions hold true*

(i) *X is connected if and only if $\rho(x,y) < \infty$ for all $x, y \in X$.*

(ii) *(X, ρ) is complete if and only if all balls $B_r(x)$ are relatively compact ($r > 0, x \in X$).*

(iii) *(X, ρ) is a length space.*

Recall that a (pseudo) metric space (E, d) is called *length space* if for any $x, y \in E$

$$d(x,y) = \inf\{L(\gamma) : \gamma \in \mathsf{C}([0,1], E), \gamma(0) = x, \gamma(1) = y\}$$

with

$$L(\gamma) = \sup\left\{\sum_{i=1}^{n} d(\gamma(t_i), \gamma(t_{i-1})) : n \in I\!N, a \le t_0 < t_1 < \cdots < t_n \le b\right\}$$

being the length of the arc $\gamma[0,1] \to E$.

3.2 Proposition. *Let X be connected and Assumption (A) be fulfilled. Then for every $y \in X$ the distance function $\rho_y : x \mapsto \rho(x,y)$ satisfies $\rho_y \in \mathcal{F}_{\mathrm{loc}}(X) \cap \mathsf{C}(X)$ and*

$$\mu_{\langle \rho_y \rangle} \le m. \tag{$*$}$$

Examples.

(i) Let $(\mathcal{E}, \mathcal{F})$ be the canonical Dirichlet form on a smooth Riemannian manifold (X, g). Then $\mu_{\langle \rho_y \rangle} = m$ for any $x \in X$.

(ii) Let $(\mathbf{D}, H^1(I\!R^N))$ be the classical Dirichlet form on $L^2(I\!R^N, dx)$. Now replace the reference measure dx by $m(dx) = dx + \sigma(dy)$ where σ denotes the surface measure on the unit sphere $\partial B_1(0)$ (which is a smooth measure w.r.t. the classical Dirichlet form). This is still a regular, strongly local Dirichlet form with intrinsic metric $\rho(x,y) = |x - y|$. In this case, the inequality $(*)$ is no equality.

Another intrinsically defined pseudo metric on X is given by

$$\rho^0(x,y) = \sup\left\{u(x) - u(y) : u \in \mathcal{F} \cap \mathsf{C}_0(X), \ \mu_{\langle u \rangle} \le m \text{ on } X\right\}.$$

Of course $\rho \ge \rho^0$.

Let us briefly assume that the topology induced by ρ^0 is equivalent to the original topology on X. If X is connected then the pseudo metric ρ^0 shares all the properties of ρ. However, in general, the finiteness of ρ^0 on X does not imply that X is connected. Moreover, if X is not connected it may happen that $S_r(x) \neq \partial B_r(x)$ and that $B_r(x)$ is not connected.

3.3 Proposition. *(i) ρ induces the original topology on X if and only if ρ^0 induces the original topology on X.*

(ii) Now let Assumption (A) be fulfilled. Then for every $x \in X$ and $r > 0$ the ball $B_r(x)$ is relatively compact if and only if the ball $B_r^0(x)$ is relatively compact and in this case $B_r(x) = B_r^0(x)$.

For all $x, y \in X$

$$\rho^0(x,y) = \rho(x,y) \wedge (\rho(x,\infty) + \rho(y,\infty))$$

with

$$\rho(x,\infty) := \sup\{r > 0 : \ B_r(x) \text{ is relatively compact } \subset X\}.$$

Finally, ρ *is complete and finite on* X *if and only if* ρ^0 *is complete and finite on* X *and in this case* $\rho \equiv \rho^0$.

5.4.2 Global Form Properties and Geometry at Infinity

We investigate several global properties of the Dirichlet form $(\mathcal{E}, \mathcal{F})$ which have to do with its "behaviour at ∞". For each of these properties we give a sufficient condition in terms of the metric ρ and the volume growth $v : r \mapsto m(B_r(x_0))$. Here $v(r) = m(B_r(x_0))$ denotes the volume of balls centered at some point $x_0 \in X$ (arbitrary but fixed) and v' denotes the (a.e. defined) derivative of the absolutely continuous part $v^{(0)}$ of v.

$\longrightarrow$ Throughout this section, let $(\mathcal{E}, \mathcal{F})$ be a strongly local, regular Dirichlet form which satisfies Assumption (A). Moreover, we assume that (X, ρ) is complete and connected.

The latter is no serious restriction since all of these global questions can be treated separately on each component. However, it helps to simplify notations.

RECURRENCE. Recall from Section 2.D the definition of recurrence and the following basic result in terms of the volume growth.

3.4 THEOREM. $(\mathcal{E}, \mathcal{F})$ *is recurrent whenever*

$$\int_1^\infty \frac{1}{v'(r)} dr = \infty. \qquad (*)$$

Of course, $(*)$ is satisfied if $\int_1^\infty \frac{r}{v(r)} dr = \infty$ or if $v(r) \le C \cdot r^2$ for $r \to \infty$.

POSITIVITY OF THE SPECTRUM. Recall that λ denotes the infimum of the spectrum of the positive self-adjoint operator A on $L^2(X, m)$. By spectral theory, $||T_t||_{2,2} = \exp(-\lambda \cdot t)$ for all $t > 0$. We say that the Dirichlet form $(\mathcal{E}, \mathcal{F})$ has *spectral bound 0* if $\lambda = 0$.

It is worthwhile to mention that the spectral bound being 0 is preserved under quasi-isometric changes. This is immediate from the definition of λ as a Rayleigh-Ritz quotient.

If $(\mathcal{E}, \mathcal{F})$ is recurrent then it has spectral bound 0. The converse is not true. For instance, let $(\mathcal{E}, \mathcal{F})$ be the classical Dirichlet form on $\mathbb{R}^N$ with $N \geq 3$.

Note that $\lambda = 0$ implies $||T_t||_{2,2} = 1$ for all $t > 0$ and hence by interpolation and symmetry $||T_t||_{p,p} = 1$ for all $p \in [1, \infty]$ and $t > 0$. Recall that the semigroup $(T_t)_{t>0}$ is called *exponentially stable* on $L^p(X, m)$ if $||T_t||_{p,p} \to 0$ for $t \to \infty$ or, equivalently, if $||T_t||_{p,p} < 1$ for some $t > 0$.

3.5 THEOREM.

$$\liminf_{r\to\infty} \frac{1}{r} \log v(r) \leq k \qquad \Longrightarrow \qquad \lambda \leq \frac{k^2}{8}.$$

COROLLARY. *If v grows subexponentially then $\lambda = 0$ and thus the semigroup $(T_t)_{t>0}$ is exponentially instable on each of the spaces $L^p(X, m)$, $p \in [1, \infty]$.*

REMARK. The assumptions in Theorems 3.5 are sharp in the following sense. Let v be as above and assume in addition that it is smooth. If $\liminf_{r\to\infty} \frac{1}{r} \log v(r) > k$ then there exists a complete Riemannian manifold (X, g) with $v(r) \geq m(B_r(x))$ for some $x \in X$ such that the canonical Dirichlet form $(\mathcal{E}, \mathcal{F})$ on X satisfies $\lambda > \frac{k^2}{8}$; see [Br].

CONSERVATIVENESS. The Dirichlet form $(\mathcal{E}, \mathcal{F})$ is called *conservative* if $T_t 1 = 1$ for all $t > 0$.

REMARKS.

(i) If $(\mathcal{E}, \mathcal{F})$ is recurrent then it is conservative. The converse is obviously not true. Think of the classical Dirichlet form on $\mathbb{R}^N$ with $N \geq 3$.

(ii) If $(\mathcal{E}, \mathcal{F})$ has spectral bound 0 then $||T_t||_{\infty,\infty} = 1$ for all $t > 0$ but the converse is not true. Even conservativeness of $(\mathcal{E}, \mathcal{F})$ does not imply that it has spectral bound 0.

A counterexample is given by the canonical Dirichlet form on the hyperbolic space which is conservative but for which $\lambda > 0$.

(iii) If $(\mathcal{E}, \mathcal{F})$ is conservative then $||T_t||_{\infty,\infty} = 1$ for all $t > 0$ but the converse is not true. Even spectral bound 0 does not imply conservativeness.

A counterexample is given by the Dirichlet form $(\mathbf{D}, H_0^1(U))$ where $U = \{x \in \mathbb{R}^N : |x| > 1\}$. This is obviously not conservative but it has spectral bound 0.

3.6 THEOREM. *$(\mathcal{E},\mathcal{F})$ is conservative whenever*

$$\int_1^\infty \frac{r}{\log v(r)} dr = \infty. \tag{$**$}$$

REMARKS. (i) The conservativeness condition $(**)$ is satisfied whenever

$$m\left(B_{r_n}(x)\right) \le e^{C\cdot r_n^2}$$

(or $m\left(B_{r_n}(x)\right) \le r_n^{C\cdot r_n^2}$) for a point $x \in X$, a sequence $r_n \to \infty$ and a constant C.

(ii) The assumptions in Theorem 3.6 are sharp in the following sense: given a smooth function v (with $v, v', \log(v')' \ge 0$) which does not satisfy $(**)$ then there exists a complete Riemannian manifold (X,g) with $v(r) \ge m(B_r(x))$ for some $x \in X$ such that the canonical Dirichlet form on X is *not conservative.* See [G2].

(iii) Note that the volume growth assumption in Theorem 3.6 is invariant under quasi-isometric changes but that conservativeness is *not* preserved under quasi-isometric changes, see Lyons [L2].

5.5 Examples

The most easiest example of a Dirichlet form is $(\mathcal{E},\mathcal{H})$ with $\mathcal{E}(u,v) \equiv 0$. This Dirichlet form is regular and strongly local. It does not satisfy Assumption (A) (namely, $\rho(x,y) = \infty$ for all $x \ne y$). In the sequel, we will present some typical non-trivial examples.

5.5.1 The Laplacian on $\mathbb{R}^N$

THE FREE LAPLACIAN. The *classical Dirichlet form* on $L^2(\mathbb{R}^N, dx)$ is given by $(\mathbf{D}, H^1(\mathbb{R}^N))$ where

$$\mathbf{D}(u,v) = \frac{1}{2}\sum_{i=1}^N \int_{R^N} \frac{\partial u}{\partial x_i}\frac{\partial v}{\partial x_i} dx$$

is the classical Dirichlet integral and

$$H^1(\mathbb{R}^N) = \{u \in L^2(\mathbb{R}^N, dx) : \frac{\partial u}{\partial x_i} \in L^2(\mathbb{R}^N, dx) \text{ for } i = 1, .., N\}$$

is the Sobolev space of order 1. Here dx denotes the Lebesgue measure on $\mathbb{R}^N$, $N \geq 1$, and $\frac{\partial u}{\partial x_i}$ (for $i = 1, .., N$) is considered in the sense of Schwartz distributions. For instance, $(\mathbf{D}, H^1(\mathbb{R}^N))$ can be obtained as the closure of the closable Markovian form $(\mathbf{D}, C_0^\infty(\mathbb{R}^N))$. Note that there is only one closed extension of this closable form.

The classical Dirichlet form is regular and strongly local and satisfies Assumption (A). Its energy measure is given by

$$\mu_{\langle u,v\rangle}(dx) = \nabla u(x) \cdot \nabla v(x)\, dx$$

for any $u, v \in H^1(\mathbb{R}^N)$ and its intrinsic metric is the Euclidean metric, i.e.

$$\rho(x, y) = |x - y|$$

for any $x, y \in \mathbb{R}^N$.

The generator $(A, \mathcal{D}(A))$ of the above classical Dirichlet form is $-\frac{1}{2}$ times the Friedrichs extension of the *Laplacian* $(\Delta, C_0^\infty(U))$ on $L^2(\mathbb{R}^N, dx)$. The associated semigroup is exactly the *Gaussian semigroup.*

THE LAPLACIAN WITH DIRICHLET CONDITIONS. Now let U be an *arbitrary Borel subset* of $\mathbb{R}^N$, $N \geq 1$, and consider the form $(\mathbf{D}, H_0^1(U))$ with

$$H_0^1(U) := \{u \in H^1(\mathbb{R}^N) : \tilde{u} = 0 \text{ q.e. on } \mathbb{R}^N \setminus U\},$$

regarded as a subspace of $L^2(U, dx)$. This is always a quasi-regular, strongly local Dirichlet form.

Assume that U is open. Then the form $(\mathbf{D}, H_0^1(U))$ is regular and satisfies Assumption (A) . The associated intrinsic metric locally coincides with the Euclidean metric. Globally, $\rho(x, y)$ can be characterized as the infimum of the Euclidean lengths of arcs in U *connecting x and y.*

¿From this, it is clear that $\rho < \infty$ on $U \times U$ if and only if U is connected (otherwise $\rho(x, y) = \infty$ for all x, y in different components of U). Moreover, $\rho(x, y) = |x - y|$ for all $x, y \in U$ if and only if U is convex.

If U is open then the Dirichlet form $(\mathbf{D}, H_0^1(U))$ is the closure of the closable Markovian form $(\mathbf{D}, C_0^\infty(U))$. The generator $(A, \mathcal{D}(A))$ is the Friedrichs extension of $(-\frac{1}{2}\Delta,\ C_0^\infty(U))$, i.e. $-\frac{1}{2}$ times the Laplacian on $L^2(U, dx)$ with Dirichlet boundary conditions on ∂U.

5.5.2 Laplacians with Weights

GENERAL CASE. Let U be a domain in $\mathbb{R}^N$ and let $\varphi \in L^2_{loc}(U, dx)$ and $\psi \in L^\infty_{loc}(U, dx)$ with $\psi^{-1} \in L^\infty_{loc}(U, dx)$. In this section we will consider the Markovian form $(\mathcal{E}, \mathsf{C}_0^\infty(U))$ with

$$\mathcal{E}(u,v) = \frac{1}{2}\int_U \nabla u \cdot \nabla v \cdot \varphi^2 \, dx$$

on the Hilbert space $L^2(U, \varphi^2\psi^2 dx)$ with inner product

$$(u,v) = \int_U u \cdot v \cdot \varphi^2 \cdot \psi^2 \, dx.$$

We have written the weight in the inner product $(.,.)$ in the form $\varphi^2 \cdot \psi^2$ in order to have nice expressions for the metric and for the process associated with this form.

4.1 THEOREM. *(i) The form $(\mathcal{E}, \mathsf{C}_0^\infty(U))$ is closable if there exists an open set $V \subset U$ with $\int_{U\setminus V} dx = 0$ such that either*

$$\varphi^{-1} \in L^2_{loc}(V, dx) \tag{1}$$

or

$$\varphi \in H^1_{loc}(V). \tag{2}$$

In both cases the closure $(\mathcal{E}, \mathcal{F})$ is a regular and strongly local Dirichlet form.

(ii) If in addition to (1) $\varphi^{-1} \in L^2_{loc}(U, dx)$ or in addition to (2) $\varphi > 0$ on U then $(\mathcal{E}, \mathcal{F})$ satisfies Assumption (A).

If moreover $\psi \in \mathsf{C}(U)$ with $\psi > 0$ then the intrinsic metric is given by

$$\rho(x,y) = \inf_\gamma \int_0^1 |\dot\gamma|^2(s)\psi(\gamma(s))ds$$

where the infimum is taken over all arcs $\gamma \in \mathsf{C}^1([0,1] \to U)$ with $\gamma(0) = x$ and $\gamma(1) = y$.

Proof. (i) In the case $\varphi^{-1} \in L^2_{loc}(V, dx)$ the closability as well as the fact that its closure is a Dirichlet form follow from [MR]: sect. II.2(a). Obviously, this Dirichlet form is regular and strongly local. In the case $\varphi \in H^1_{loc}(V)$ the closability as well as the fact that its closure is a strongly local, regular Dirichlet form was proven in [FOT], Thms. 3.1.1 - 3.1.3.

(ii) In order to check Assumption (A), first of all note that this is a local property. Thus without restriction we may assume that there is a constant C such that $C^{-1} \le \psi \le C$ on U. But this implies that

$$C^{-1/2} \cdot \rho_1 \le \rho \le C^{1/2} \cdot \rho_1$$

on U where ρ_1 is the intrinsic metric which one obtains from the above Dirichlet form replacing ψ by 1. Hence, without restriction we may assume that already $\psi \equiv 1$ on U. In this case, we have to prove that

$$\rho(x, y) = |x - y|$$

provided x and y are close together.

The lower estimate is obvious from the fact that for any $z \in \mathbb{R}^N$ the map

$$u : x \to \langle x, z \rangle$$

(with $\langle .,. \rangle$ being the Euclidean scalar product) lies in $\mathcal{F}_{loc}(U) \cap \mathrm{C}(U)$. (The fact that $u \in \mathcal{F}_{loc}(U)$ follows by approximation of $u \in \mathrm{C}^\infty(U)$ by means of $u_n \in \mathrm{C}_0^\infty(U)$, $n \in \mathbb{N}$, in $\sqrt{\mathcal{E}_1}$-norm.) For given $y, y' \in U$ $(y \neq y')$ one then just has to choose $z = (y - y')/|y - y'|$ in order to obtain $\mu_{\langle u \rangle}(dx) = dx$ and $u(y) - u(y') = |y - y'|$.

Now assume conversely that $u \in \mathcal{F}_{loc}(U) \cap \mathrm{C}(U)$ with

$$\mu_{\langle u \rangle}(dx) \leq \varphi^2 dx. \tag{3}$$

We have to prove that $u \in H^1_{loc}(U)$ and

$$|\nabla u| \leq 1 \quad \text{on } U.$$

Let $\{u_n\}_n$ be a sequence in $\mathrm{C}_0^\infty(U)$ which converges (for $n \to \infty$) to u locally in $\sqrt{\mathcal{E}_1}$-norm and locally uniformly. Since $\{\frac{\partial}{\partial x_i} u_n\}_n$ is a Cauchy sequence in $L^2_{loc}(U, \varphi^2 dx)$ there exists $v_i \in L^2_{loc}(U, \varphi^2 dx)$ such that $\frac{\partial}{\partial x_i} u_n \to v_i$ in $L^2_{loc}(U, \varphi^2 dx)$. Then for any bounded Borel set F with $\overline{F} \subset U$

$$\int_F \mu_{\langle u \rangle}(dx) = \lim_{n \to \infty} \int_F \mu_{\langle u_n \rangle}(dx) = \lim_{n \to \infty} \int_F |\nabla u_n|^2 \varphi^2(dx) = \int_F v^2 \varphi^2(dx)$$

where $v^2 = \sum_{i=1}^N v_i^2$. Together with (3) this implies

$$\int_F v^2 \varphi^2(dx) \leq \int_F \varphi^2 dx$$

and thus

$$|v| \leq 1 \tag{4}$$

a.e. on U. In particular, v_i lies also in $L^2_{loc}(U, dx)$.

Now assume (1) with $V = U$. Then $\frac{\partial}{\partial x_i} u_n \to v_i$ in in $L^1_{loc}(U, dx)$ $(i = 1, \dots, N)$ since

$$\int_F |\frac{\partial}{\partial x_i} u_n - v_i| dx \leq \left(\int_F |\frac{\partial}{\partial x_i} u_n - v_i|^2 \varphi^2 dx\right)^{1/2} \cdot \left(\int_F \varphi^{-2} dx\right)^{1/2}$$

for any bounded Borel set F with $\overline{F} \subset U$.

Fix arbitrary points $x, y \in U$. Without restriction, let $x = (0, 0, \dots, 0)$ and $y = (R, 0, \dots, 0)$. Let B'_ε be the Euclidean ball in $I\!R^{N-1}$ with radius ε around the origin and put $S_\varepsilon = [0, R] \times B_\varepsilon(0) = \{z = (r, z') \in I\!R^N : 0 \leq r \leq R, |z'| < \varepsilon\}$. Moreover, let $d'z$ be the $(N-1)$-dimensional Lebesgue measure. Then

$$\begin{aligned}
u(x) - u(y) &= \lim_{\varepsilon \to 0} \frac{c_N}{\varepsilon^{N-1}} \int_{B'_\varepsilon} [u(0, z') - u(R, z')]\, d'z' \\
&\quad \text{(since } u \text{ is continuous)} \\
&= \lim_{\varepsilon \to 0} \frac{c_N}{\varepsilon^{N-1}} \lim_{n \to \infty} \int_{B'_\varepsilon} [u_n(0, z') - u_n(R, z')]\, d'z' \\
&\quad \text{(since } u_n \to u \text{ locally uniformly)} \\
&= \lim_{\varepsilon \to 0} \frac{c_N}{\varepsilon^{N-1}} \lim_{n \to \infty} \int_{S_\varepsilon} \frac{\partial}{\partial x_1} u_n(z) dz \\
&\quad \text{(since } u_n \text{ is smooth)} \\
&= \lim_{\varepsilon \to 0} \frac{c_N}{\varepsilon^{N-1}} \int_{S_\varepsilon} v_1(z) dz \\
&\quad \text{(since } \tfrac{\partial}{\partial x_1} u_n \to v_1 \text{ in } L^1_{loc}(U, dx)) \\
&\leq \lim_{\varepsilon \to 0} \frac{c_N}{\varepsilon^{N-1}} \int_{S_\varepsilon} dz \\
&\quad \text{(according to (4))} \\
&= R.
\end{aligned}$$

In other words, $|u(x) - (y)| \leq |x - y|$ for all $x, y \in U$ which can be connected by a straight line in U. This is the claim.

Now assume $\varphi \in H^1_{loc}(U)$ with $\varphi > 0$ on U. Since we only have to deal with local questions we may assume without restriction $U = I\!R^N$. In this case, Markov uniqueness for $(\mathcal{E}, \mathcal{F})$ was proven in [RZ2] which implies that $(\mathcal{E}, \mathcal{F})$ corresponds to the drift transformation $(X_t, I\!P_x)$ of standard Brownian motion considered in [Fi], see [T2]. According to [Fi], Cor. 4.12, we then see

$$\mathcal{F} = \{v \in \dot{H}^1_{loc}(I\!R^N) \cap L^2(I\!R^N, \varphi^2 dx) : \int |\nabla v|^2 \varphi^2\, dx < \infty\}$$

and for $u \in \mathcal{F}$

$$\mu_{\langle u \rangle}(dx) = |\nabla u|^2 \varphi^2 \, dx$$

(where in both places $|\nabla u|^2 dx$ has to be interpreted as the energy measure of the classical Dirichlet form $(\mathbf{D}, H^1(\mathbb{R}^N))$.) Together with (3) this yields $|\nabla u| \leq 1$ on $\mathbb{R}^N$. The latter of course also implies $u \in H^1_{loc}(\mathbb{R}^N)$.

(iii) This follows easily from the fact that (U, ρ) is a length space and that locally ψ can be replaced by constants. But for constant ψ the claim is obvious. □

COROLLARY (COMPLETENESS). *The space (U, ρ) is complete if and only if*

$$\int_0^1 |\dot{\gamma}|^2(s)\psi(\gamma(s))ds = \infty \tag{5}$$

for any $\gamma \in \mathsf{C}^1([0,1] \to \overline{U})$ with $\gamma(0) \in U$ and $\gamma(1) \in \partial U$.

Due to the triangle inequality, it suffices to check (5) for all arcs as above starting in some fixed point $x_0 \in U$, i.e. with $\gamma(0) = x_0$.

Note that the geometry is independent of φ! In particular, in order to study the geometry one can choose φ arbitrarily. We first treat the case $\varphi = \psi^{\frac{N-2}{2}}$ which leads to a Riemannian geometry.

ROTATIONALLY INVARIANT CASE. Let $U = \{x \in \mathbb{R}^N : |x| < R\}$ be the Euclidean ball around the origin with radius $R \in]0, \infty]$ and let $\psi(x) = \Psi(|x|)$ and $\varphi(x) = \Phi(|x|)$ with continuous functions $\Psi, \Phi : [0, R[\to]0, \infty[$. Then

$$\rho(x, 0) = \int_0^{|x|} \Psi(s)ds$$

and $\rho(x, \partial U) = \int_{|x|}^R \Psi(s)ds$. Moreover, if one chooses $u(x) = \|x\|$ then

$$w_u(r) = \int_{\{u<r\}} |\nabla u|^2 \varphi^2 dm = c \int_0^r \Phi^2(s) s^{N-1} ds$$

and

$$v_u(r) = \int_{\{u<r\}} \varphi^2 \psi^2 dm = c \int_0^r \Phi^2(t)\Psi^2(t) t^{N-1} dt.$$

The investigation of the 1-dimensional diffusion equation yields the following results:

(i) The metric space (U, ρ) is complete if and only if

$$\int^R \Psi(t)dt = \infty.$$

(ii) The Dirichlet form is conservative if and only if

$$\int^R \int^R 1_{\{t<s\}} \frac{\Phi^2(t)\Psi^2(t)t^{N-1}}{\Phi^2(s)s^{N-1}} ds\, dt = \infty.$$

This reduces to

$$\int^R \Psi^2(t)t\, dt = \infty$$

if $\Phi(s) \equiv s^k$ with $k > 1 - N/2$ or $\Phi(s)\Psi(s) \equiv s^k$ with $k > -N/2$, in particular, if $\Phi \equiv 1$ or $\Phi\Psi \equiv 1$.

(iii) The Dirichlet form is recurrent if and only if

$$\int^R \frac{1}{\Phi^2(s)s^{N-1}} ds = \infty.$$

(iv) The origin $0 \in U$ is m-polar if and only if

$$\int_0 \frac{1}{\Phi^2(s)s^{N-1}} ds = \infty.$$

RIEMANNIAN CASE.

Now assume $\varphi = \psi^{\frac{N-2}{2}}$. Then the Dirichlet form

$$\mathcal{E}[u] = \frac{1}{2} \int_U |\nabla u|^2 \cdot \psi^{-2} \cdot \psi^N\, dx$$

together with the inner product

$$||u||^2 = \int_U |u|^2 \cdot \psi^N\, dx$$

defines a Riemannian geometry on U with a global chart $\{x^1, \dots, x^N\}$ given by the Euclidean coordinates and in this chart the Riemannian metric is given by

$$g_{ij} = \psi^2 \cdot \delta_{ij}.$$

If $\psi \in H^1_{loc}(U) \cap \mathsf{C}(U)$ then the generator on $L^2(U, \psi^N dx)$ of this Dirichlet form is the Friedrichs extension of $(-\frac{1}{2\psi^N}\nabla\left(\psi^{N-2}\nabla\right), \mathsf{C}_0^\infty(U))$. It is $-\frac{1}{2}$ times the Laplace-Beltrami of the (not necessarily smooth) Riemannian manifold (U, g).

EXAMPLE (HYPERBOLIC SPACE). If we choose $\psi(x) = \frac{2}{[1-(\frac{x}{R})^2]}$ on $U := \{x \in \mathbb{R}^N : |x| < R\}$ then we obtain the Poincaré model of the hyperbolic space.

Note that this Dirichlet form is always *conservative* but it is *recurrent* if and only if $N = 1$. If $N \neq 1$ then the associated diffusion process $(X_t, I\!P_x)$ on U never leaves U but with positive probability it leaves each compact set $K \subset U$ several times and finally never returns to K.

Moreover, note that $||T_t||_{\infty,\infty} = 1$ for all $t > 0$ but that $||T_t||_{2,2} = e^{-\lambda t}$ with $\lambda = \frac{(N-1)^2}{8R^2}$ which is > 0 whenever $N \neq 1$.

5.5.3 Elliptic and Subelliptic Operators on $I\!R^N$

Throughout this section, let U be an open set in $I\!R^N$ and let $(a_{ij})_{i,j}$ be a $(N \times N)$-matrix of functions $a_{ij} \in L^\infty(U, dx)$ which is assumed to be symmetric and positive semidefinite, i.e. $a_{ij} = a_{ji}$ $(\forall i, j = 1, \ldots N)$ on U and

$$\sum_{i,j=1}^{N} a_{ij}(x)\xi_i\xi_j \geq 0$$

for all $\xi \in I\!R^N$ and dx-a.e. $x \in U$. The latter property is also called *degenerate ellipticily*.

4.2 THEOREM.

$$\mathcal{E}(u, v) = \frac{1}{2} \sum_{i,j} \int_U a_{ij}(x) \frac{\partial u}{\partial x_i}(x) \frac{\partial v}{\partial x_j}(x)\, dx \tag{$*$}$$

with domain $\mathcal{D}(\mathcal{E}) = \mathsf{C}_0^\infty(U)$ defines a symmetric form on $L^2(U, dx)$. If it is closable then its closure is a Dirichlet form which is regular and strongly local. The intrinsic metric satisfies

$$\rho(x, y) \geq c_0 \cdot |x - y|$$

for some constant c_0 and all $x, y \in I\!R^N$.

Proof. [FOT]: sect. 3.1. □

The form $(\mathcal{E}, \mathsf{C}_0^\infty(U))$ is called *uniformly subelliptic* iff there exist constants $\varepsilon > 0$ and $C \in I\!R$ such that

$$\mathcal{E}(u, u) + ||u||^2 \geq \frac{1}{C} \cdot ||u||^2_{H^\varepsilon} \qquad \forall u \in \mathsf{C}_0^\infty(U).$$

Here $||u||^2_{H^\varepsilon} = \int_{I\!R^N} |\hat{u}(\xi)|^2 \cdot (1 + |\xi|^2)^\varepsilon d\xi$ for any $\varepsilon > 0$ (with $\hat{u}$ being the Fourier transform of u) and $H^\varepsilon(I\!R^N) = \{u \in L^2(I\!R^N, dx) : ||u||_\varepsilon < \infty\}$ is the fractional Sobolev space of order ε.

The form $(\mathcal{E}, C_0^\infty(U))$ is called *subelliptic* iff $(\mathcal{E}, C_0^\infty(V))$ is uniformly subelliptic for every bounded open set V with $\overline{V} \subset U$. It is called *(uniformly) elliptic* if it is (uniformly) subelliptic with $\varepsilon = 1$.

4.3 THEOREM. *(i) If the symmetric form $(\mathcal{E}, C_0^\infty(U))$ is subelliptic then it is closable. Its closure $(\mathcal{E}, \mathcal{F})$ is a regular and strongly local Dirichlet form which satisfies Assumption (A). The intrinsic metric coincides with the metric defined by subunit curves on U.*

(ii) Let $U = \mathbb{R}^N$. The uniform subellipticity condition with some $\varepsilon > 0$ holds true if and only if there are constants $r_0 > 0$ and C_0 such that

$$\rho(x,y) \le C_0 \cdot |x-y|^\varepsilon$$

for all $x, y \in \mathbb{R}^N$ with $|x-y| < r_0$.

Proof. [FeP]; [JS], Prop. 3.1 and Thm. 2.3. □

If the matrix (a_{ij}) is elliptic then its inverse (g_{ij}) is well-defined. It determines a Riemannian geometry on U which corresponds to the Dirichlet form

$$\mathcal{E}^*(u,v) = \frac{1}{2}\sum_{i,j} a_{ij}(x)\frac{\partial u}{\partial x_i}(x)\frac{\partial v}{\partial x_j}(x) \cdot \mathbf{a}^{-1/2}(x)\, dx$$

on the Hilbert space

$$\mathcal{H}^* = L^2(U, \mathbf{a}^{-1/2}dx)$$

where $\mathbf{a}$ denotes the determinant of the matrix $(a_{ij})_{i,j}$, see also the next section. The intrinsic metric ρ^* defined by this form obviously coincides with the intrinsic metric ρ previously defined by the form $(*)$ on $L^2(U, dx)$ (since plugging in the same weigth $\mathbf{a}^{-1/2}$ simultaneously into the inner product and into the form leaves the intrinsic metric unchanged, at least if $\mathbf{a} \in L^\infty_{loc}(U, dx)$ and $\mathbf{a}^{-1} \in L^\infty_{loc}(U, dx)$.) Note that if the matrix (a_{ij}) is continuous then ρ^* is just the Riemannian distance of the manifold (U, g).

5.5.4 Laplace-Beltrami Operators on Riemannian Manifolds

SMOOTH MANIFOLDS. Let (X, g) be a smooth N-dimensional Riemannian manifold. Denote the gradient of a smooth function u on X by ∇u and the inner product in the tangent space $T_x(X)$ at $x \in X$ by $\langle .,. \rangle(x)$. Let $\mathcal{H}$ be the Hilbert space $L^2(X, m)$ where m denotes the Riemannian volume measure on X.

4.4 PROPOSITION. *$(\mathcal{E}, C_0^\infty(X))$ with*

$$\mathcal{E}(u,v) = \frac{1}{2}\int_X \langle \nabla u, \nabla v\rangle(x)\, m(dx)$$

defines a closable Markovian form on $\mathcal{H}$. Its closure $(\mathcal{E},\mathcal{F})$ is called canonical Dirichlet form on (X,g)*. The generator $(A,\mathcal{D}(A))$ of this Dirichlet form $(\mathcal{E},\mathcal{F})$ is $-\frac{1}{2}$ times the* Laplace-Beltrami operator *on (X,g). The canonical Dirichlet form on (X,g) is strongly local and regular and satisfies Assumption (A). Its intrinsic metric is the Riemannian distance.*

Proof. All of these results are well-known. In particular, the characterization of the Riemannian distance as the intrinsic metric of the canonical Dirichlet form is "classical". The proof runs as in the Euclidean case. The details are left to the reader. □

Note that in local coordinates $\{x^i\}_{i=1,\dots,N}$ one has

$$m(dx) = \sqrt{\mathbf{g}}(x)dx^1\cdots dx^N$$

and

$$\mu_{\langle u,v\rangle}(dx) = \langle \nabla u, \nabla v\rangle(x)\, vol(dx) = \sum_{i,j} g^{ij}(x)\frac{\partial u}{\partial x^i}(x)\frac{\partial v}{\partial x^j}(x)\cdot\sqrt{\mathbf{g}}(x)\, dx^1\cdots dx^N$$

where $\mathbf{g}$ denotes the determinant of the matrix $(g_{ij})_{i,j} := \left(\langle\frac{\partial}{\partial x^i},\frac{\partial}{\partial x^j}\rangle\right)_{i,j}$ and $(g^{ij})_{i,j}$ its inverse. Hence,

$$\Gamma(u,v) = \langle\nabla u, \nabla v\rangle = \sum_{i,j} g^{ij}\frac{\partial u}{\partial x^i}\frac{\partial v}{\partial x^j}$$

and

$$A = \frac{-1}{2\sqrt{\mathbf{g}}}\sum_{i,j}\frac{\partial}{\partial x^i}\left(\sqrt{\mathbf{g}}g^{ij}\frac{\partial}{\partial x^j}\right).$$

Moreover,

$$\rho(x,y) = \inf\left\{\int_0^1 |\dot\gamma|^2(t)dt:\ \gamma\in C^1([0,1]\to U), \gamma(0)=x, \gamma(1)=y\right\}$$

where

$$|\dot\gamma|^2(t) = \sum_{i,j} g^{ij}(\gamma(t))\frac{d\gamma_i(t)}{dt}\frac{d\gamma_j(t)}{dt}.$$

GLUEING TOGETHER MANIFOLDS – AN EXAMPLE. Let X_1, X_2 and X_3 be three two-dimensional half planes embedded in $I\!R^3$ in such a way that $X_i \cap X_j = \{0\} \times \{0\} \times I\!R =: Z$ for any $i \neq j \in \{1, 2, 3\}$. Let $X = X_1 \cup X_2 \cup X_3$ and let m be the two-dimensional Lebesgue measure on X (with $m(Z) = 0$). Choose real numbers $\alpha_1, \alpha_2, \alpha_3 \geq 0$ (not all $= 0$) and define

$$\mathcal{E}(u,u) := \frac{1}{2} \sum_{i=1}^{3} \int_{X_i^0} |\nabla u|^2 \alpha_i dm$$

and

$$\|u\|^2 := \sum_{i=1}^{3} \int_{X_i^0} |u|^2 \alpha_i dm.$$

Again we choose as form domain the set $\mathsf{C}_0^{Lip}(X)$ of compactly supported functions on X which are Lipschitz continuous on X (in particular, Lipschitz continuous on the common boundary Z of the X_i and restricted to any of the open sets X_i^0 lying in the space $H^1(X_i^0)$). Then $(\mathcal{E}, \mathsf{C}_0^{Lip}(X))$ is closable. Its closure is a regular and strongly local Dirichlet form $(\mathcal{E}, \mathcal{F})$. If all the α_i are > 0 then the Dirichlet form satisfies Assumption (A) and the intrinsic distance $\rho(x,y)$ between two points x and y is the Euclidean length of the shortest arc in the surface $X \subset I\!R^3$ which connects x and y.

The stochastic process $(X_t, I\!P_x)$ associated with $(\mathcal{E}, \mathcal{F})$ can be described as follows: as long as the "particle" moves in the interior of one of the parts X_i, it behaves as a Brownian motion. We emphasize that the process $(X_t, I\!P_x)$ is *entirely independent* of the choice of the α_i as long as $X_t \notin Z$. Moreover, it moves on any of the open sets X_i^0 with the same speed.

If it is in the set Z then it "chooses" each of the 3 half planes X_i with probability

$$p_i = \frac{\alpha_i}{\alpha_1 + \alpha_2 + \alpha_3}.$$

Of course, "choosing" one of the half planes can not be understood literally since a typical path passes the set Z infinitely often as soon as it passes it once.

In order to make this vague formulation precise, let us introduce "coordinates" in X. Let

$$X = \{(i, y, z) : \ i \in \{1, 2, 3\}, y \in [0, \infty[, z \in]-\infty, \infty[\}$$

such that $X_j = \{(i, y, z) \in X : i = j\}$ for any $j = 1, 2, 3$ and $Z = \{(i, 0, z) \in X$ independently of i. The stochastic process $(X_t)_t$ can be written as (I_t, Y_t, Z_t). Here $(Z_t)_t$ is a 1-dimensional Brownian motion with state space $I\!R$ and (Y_t) is a reflected 1-dimensional Brownian motion with state space $[0, \infty[$. $(I_t)_t$ is right continuous stochastic process with state space $\{1,2,3\}$ which only jumps at times t at which $Y_t = 0$. The process $(Z_t)_t$ plays no role in the sequel since it is independent of $(I_t, Y_t)_t$, or in other words, since we can entirely ignore the z-coordinate.

For $r > 0$ let ϑ_r be the stopping time

$$\vartheta_r = \inf\{Y_t \geq r\}.$$

With this notations the rigorous formulation of the above intuitive description reads as follows.

4.5 PROPOSITION. *For every $x \in Z$, any $r > 0$ and any $i \in \{1,2,3\}$*

$$\mathbb{P}_x\left(I_{\vartheta_r} = i\right) = \frac{\alpha_i}{\alpha_1 + \alpha_2 + \alpha_3}.$$

Recall that the process $(X_t)_t$ moves with the same speed on each of the sets X_i^0. Hence, at time ϑ_r it will be in a fixed half plane X_i with probability p_i if and only if each time when the process is in Z then it "chooses" this half plane X_i with probability p_i. This justifies the above formulation. Note that the above formula includes the possibility that one or two of these half planes are polar (not attainable), namely, if one or two of the α_i vanish.

Proof. We sketch the main arguments. (i) For any $r > 0$ and any $j \in \{1,2,3\}$ the function

$$u : x \mapsto \mathbb{P}_x\left(I_{\vartheta_r} = j\right)$$

is A-harmonic in the set $U_r = \{(i,y,z) \in X : y < r\}$.

(ii) The latter means that u is a local weak solution of the equation $Au = 0$ in U_r in the sense that

$$\sum_i \int_{X_i^0} \left(\frac{\partial}{\partial y} u(i,y,z) \cdot \frac{\partial}{\partial y}\varphi(i,y,z) + \frac{\partial}{\partial z} u(i,y,z) \cdot \frac{\partial}{\partial z}\varphi(i,y,z)\right) \cdot \alpha_i \, dx$$

for every $\varphi \in \mathsf{C}_0^{Lip}(X)$ with $\operatorname{supp}(\varphi) \subset U_r$.

(iii) Choosing suitable φ_i one can deduce that u does not depend on z and that $y \to u((i,y,z))$ is linear in $y \in [0, r[$ (for any fixed i).

(iv) It follows that

$$\sum_i \frac{\partial u(i,y,z)}{\partial y} \cdot \alpha_i = 0 \qquad (*)$$

for all y, z.

(v) u is continuous on $\overline{U_r}$ with

$$u(i,r,z) = \begin{cases} 0, & \text{if } i \neq j \\ 1, & \text{if } i = j. \end{cases} \tag{$**$}$$

(independently of z).

(vi) Combining $(*)$ and $(**)$ one deduces that for $x \in Z$

$$u(x) = \frac{\alpha_j}{\alpha_1 + \alpha_2 + \alpha_3}.$$

□

Bibliography

[Ah] L. V. Ahlfors: *Sur le type d'une surface de Riemann.* C.R.Acad.Sci.Paris **201** (1935), 30-32

[AKR] S. Albeverio, Yu. Kondratiev, M. Röckner: *Analysis and geometry on configuration spaces.* Preprint Bielefeld 1997

[Be] A. Bendikov: *Potential theory on infinite dimensional Abelian groups.* De Gruyter 1995

[BD1] A. Beurling, J. Deny: *Espaces de Dirichlet I. Le cas elementaire.* Acta Math. **99** (1958), 203-224

[BD2] A. Beurling, J. Deny: *Dirichlet spaces.* Proc. Nat. Acad. Sci. **45** (1959), 208-215

[BM1] M. Biroli, U. Mosco: *Formes de Dirichlet et estimations structurelles dans les milieux discontinus.* C. R. Acad. Sci. Paris **313** (1991), 593-598

[BM2] M. Biroli, U. Mosco: *A Saint-Venant Principle for Dirichlet forms on discontinuous media.* Annali Mat. Pura Appl.**169** (1995), 125-181

[BH] N. Bouleau, F. Hirsch: *Dirichlet forms and analysis on Wiener space.* De Gruyter 1991

[Br] R. Brooks: *A relation between growth and the spectrum of the Laplacian.* Math. Z. **178** (1981), 501-508

[CKS] E. A. Carlen, S. Kusuoka, D. W. Stroock: *Upper bounds for symmetric Markov transition functions.* Ann. Inst. H. Poincaré **2** (1987), 245-287

[CY] S. Y. Cheng, S. T. Yau: *Differential Equations on Riemannian manifolds and their geometric applications.* Comm. Pure Appl. Math. **28** (1975), 333-354

[CG] F. Cipriani, G. Grillo: *L^p-Exponential decay for solutions to functional equations in local Dirichlet spaces.* Preprint 1997

[D] E. B. Davies: *Explicit constants for Gaussian upper bounds on heat kernels.* Amer. J. Math. **109** (1987),319-334

[D1] E. B. Davies: *Heat kernels and spectral theory.* Cambridge University Press 1989

[D2] E. B. Davies: *L^1 properties of second order elliptic operators.* Bull. London Math. Soc. **17** (1985), 417-436

[FeP] C. L. Fefferman, D. Phong: *Subelliptic eigenvalue problems.* p. 590-606 in: Conference on Harmonic Analysis, Chicago. (Ed. W. Beckner et al.) Wadsworth, 1981

[FeS] C. L. Fefferman, A. Sanchez-Calle: *Fundamental solutions for second order subelliptic operators.* Ann. of Math. **124** (1986), 247-272

[Fi] P. J. Fitzsimmons: *Absolute continuity of symmetric diffusions.* Preprint 1994

[F1] M. Fukushima: *Dirichlet spaces and strong Markov processes.* Trans. Amer. Math. Soc. **162** (1971), 185-224

[F2] M. Fukushima: *Dirichlet forms and Markov processes.* North Holland and Kodansha 1980

[F3] M. Fukushima: *On recurrence criteria in the Dirichlet space theory.* In: From local time to global property, control and physics. (ed. K.D.Elworthy) Research Notes in Math. **150** Longman 1987

[FOT] M. Fukushima, Y. Oshima, M. Takeda: *Dirichlet forms and symmetric Markov processes.* De Gruyter 1994

[Ga] M. P. Gaffney: *The conservation property of the heat equation on Riemannian manifolds.* Comm. Pure Appl. Math. **12** (1959), 1-11

[G] A. A. Grigor'yan: *On the existence of positive fundamental solution of the Laplace equation on Riemannian manifolds.* Mat. Sbornik *128* (1985), 354-363 and Math. USSR. Sb. **56** (1987), 349-358

[**G1**] A. A. Grigor'yan: *On stochastically complete manifolds.* Soviet Math. Dokl. **34** (1987), 310-313

[**G2**] A. A. Grigor'yan: *Stochastically complete manifolds and summable harmonic functions.* Math. USSR Izvestiya **33** (1989), 425-431

[**G3**] A. A. Grigor'yan: *Integral maximum principle and its applications.* Proc. Roy. Soc. Edinburgh **124A** (1994), 353-362

[**G4**] A. A. Grigor'yan: *Heat kernel upper bounds on a complete non-compact manifold.* Revista Mathematica Iberoamericana **10** (1994), 395-452

[**G5**] A. A. Grigor'yan: *Heat kernel on a manifold with a local Harnack inequality.* Comm. Anal. Geom. **2** (1994), 111-138

[**I**] K. Ichihara: *Explosion problem for symmetric diffusion processes.* Trans. Amer. Math. Soc. **298** (1986), 515-536

[**Je**] D. Jerison: *The Poincaré inequality for vector fields satisfying an Hörmander's condition.* Duke J. Math. **53** (1986), 503-523

[**JS**] D. Jerison, A. Sanchez-Calle: *Estimates for the heat kernel for a sum of squares of vector fields.* Indiana Univ. J. of Math. **35** (1986), 835-854

[**Jo**] J. Jost: *Generalized Dirichlet forms and harmonic maps.* Calc. Var. **5** (1997), 1-19

[**Kn**] H. Kaneko: *Liouville theorems based on symmetric diffusions.* Bull. Soc. Math. France **124** (1996), 545-557

[**K1**] L. Karp: *Subharmonic functions, harmonic mappings and isometric immersions.* In "Seminar on Differential Geometry" (ed. by S. T. Yau). Ann. Math. Studies **102**, Princeton 1982

[**K2**] L. Karp: *Subharmonic functions on real and complex manifolds.* Math. Z. **179** (1982), 535-554

[**KL**] L. Karp, P. Li: *The heat equation on complete Riemannian manifolds.* (Unpublished) Preprint, Stanford and Ann Arbor 1983

[**Le**] Y. Lejan: *Mesures associées a une forme de Dirichlet. Applications.* Bull. Soc. Math. France **106** (1978), 61-112

[**Li**] P. Li: *Uniqueness of L^1 solutions for the Laplace equation and the heat equation on Riemannnian manifolds.* J. Diff. Geom. **20** (1984), 447-457

[**LS**] P. Li, R. Schoen: *L^p and mean value properties of subharmonic functions on Riemannian manifolds.* Acta Math. **153** (1984), 279-301

[**LY**] P. Li, S. T. Yau: *On the parabolic kernel of the Schrödinger operator.* Acta Math. **156** (1986), 153-201

[**L1**] T. Lyons: *Instability of the Liouville property for quasi-isometric Riemannian manifolds and reversible Markov chains.* J. Diff. Geom. **19** (1987), 33-66

[**L2**] T. Lyons: *Instability of the conservative property under quasi-isometry.* J. Diff. Geom. **34** (1991), 483-489

[**LS**] T. Lyons, Sullivan: *Function theory, random paths and covering spaces.* J. Diff. Geom. **19** (1984), 299-323

[**M**] V. G. Maz'ya: *Sobolev spaces.* Springer 1985

[**MR**] Z. Ma, M. Röckner: *Introduction to the theory of (non-symmetric) Dirichlet forms.* Universitext, Springer-Verlag, Berlin/New York, 1992

[**M**] U. Mosco: *Composite Media and asymptotic Dirichlet forms.* J. Funct. Anal. **123** (1994), 368-421

[**Ne**] R. Nevanlinna: *Ein Satz über offene Riemannsche Flächen.* Ann. Acad. Sci. Fenn. Series A **54** (1940), 1-18

[**Ok**] H. Okura: *Capacitary inequalities and global properties of symmetric Dirichlet forms.* In: Dirichlet forms and stochastic processes. (Edts. Z. M. Ma et al.), pp. 291-304. De Gruyter 1995

[**O1**] Y. Oshima: *Lectures on Dirichlet forms.* Lecture Notes, Erlangen 1988

[**O2**] Y. Oshima: *On conservativeness and recurrence criteria for Markov processes.* Potential Analysis **1** (1992), 115-131

[**OR**] L. Overbeck, M. Röckner: *Geometric aspects of finite and infinite dimensional Fleming-Viot processes.* To appear in Random Operators and Stochastic Equations.

[**RZ1**] M. Röckner, T. S. Zhang: *Uniqueness of generalized Schrödinger operators and applications.* J. Funct. Anal. **105** (1992), 187-231

[**RZ2**] M. Röckner, T. S. Zhang: *Uniqueness of generalized Schrödinger operators and applications II.* J. Funct. Anal. **119** (1994), 445-467

[Sc] A. Schied: *On geometric aspects of Fleming-Viot and Dawson-Watanabe processes.* To appear in Ann. Prob.

[Si1] M. L. Silverstein: *Symmetric Markov processes.* Lecture Notes in Math. **426**. Springer 1974

[Si2] M. L. Silverstein: *Boundary theory for symmetric Markov processes.* Lecture Notes in Math. **516**. Springer 1976

[S1] K.-Th. Sturm: *Analysis on local Dirichlet spaces — I. Recurrence, conservativeness and L^p-Liouville properties.* J. Reine Angew. Math. **456** (1994), 173-196

[S2] K.-Th. Sturm: *Analysis on local Dirichlet spaces – II. Upper Gaussian estimates for the fundamental solutions of parabolic equations.* Osaka J. Math. **32** (1995), 275-312

[S3] K.-Th. Sturm: *Analysis on local Dirichlet spaces – III. The parabolic Harnack inequality.* J. Math. Pures Appl. **75** (1996), 273-297

[S4] K.-Th. Sturm: *On the geometry defined by Dirichlet forms.* In "Seminar on Stochastic Analysis, Random Fields and Applications (Ascona 1993)" (ed. by E. Bolthausen et al.), pp. 231-242. Birkhäuser 1995

[S5] K.-Th. Sturm: *Sharp estimates for capacities and applications to symmetric diffusions.* Prob. Theory Related Fields **102** (1995), 73-89

[S6] K.-Th. Sturm: *Monotone approximation of energy functionals for mappings into metric spaces. I.* J. reine angew. Math. **486** (1997), 129-151

[S7] K.-Th. Sturm: *Monotone approximation of energy functionals for mappings into metric spaces. II.* Preprint MPI Leipzig 1997, to appear in Potential Analysis

[S8] K.-Th. Sturm: *How to construct diffusion processes on metric spaces.* Potential Analysis (1998)

[S9] K.-Th. Sturm: *Diffusion processes and heat kernels on length spaces.* Annals of Probab. **26** (1998), 1-55

[T] M. Takeda: *On the conservativeness of the Brownian motion on a Riemannian manifold.* Bull. London Math. Soc. **23** (1991), 86-88

[T1] M. Takeda: *Two classes of extensions for generalized Schrödinger operators.* Potential Analysis **5** (1996), 1-13

[T2] M. Takeda: *Transformations of local Dirichlet forms by supermartingale multiplicative functionals.* In: Dirichlet forms and stochastic processes. (Edts. Z. M. Ma et al.), pp. 363-374. De Gruyter 1995

[V] N. Th. Varopoulos: *The Poisson kernel on positively curved manifolds.* J. Funct. Anal. **44** (1981), 359-380

[V1] N. Th. Varopoulos: *Potential theory and diffusions on Riemannian manifolds.* In: Conference on harmonic analysis in honor of Antoni Zygmund. pp. 821-837. Wadsworth 1983

[Y1] S. T. Yau: *Harmonic functions on complete Riemannian manifolds.* Comm. Pure Appl. Math. **28** (1975), 201-228

[Y2] S. T. Yau: *Some function-theoretic properties of complete Riemannian manifolds and their applications to geometry.* Indiana Univ. Math. J. **25** (1976), 659-670